Yeasts in Natural and Artificial Habitats

Springer
Berlin
Heidelberg
New York
Barcelona
Budapest
Hong Kong
London
Milan
Paris
Santa Clara
Singapore
Tokyo

J.F.T. Spencer   D.M. Spencer[†]   (Eds.)

Yeasts in Natural and Artificial Habitats

With 59 Figures and 28 Tables

 Springer

John F.T. Spencer
Dorothy M. Spencer[†]

PROIMI
Avenida Belgrano y Pasaje Caseros
4000 S.M. de Tucumán
Argentina

ISBN 3-540-56820-4 Springer-Verlag Berlin Heidelberg New York

Library of Congress Cataloging-in-Publication Data. Spencer, John F.T. Yeasts in natural and artificial habitats/J.F.T. Spencer, D.M. Spencer.   p. cm. Includes bibliographical references and index. ISBN 3-540-56820-4 (hardcover) 1. Yeast fungi. I. Spencer, Dorothy M. II. Title. QK617.5.S65 1997 579.5'63—dc21 96-50359

© Springer-Verlag Berlin Heidelberg 1997
Printed in Germany

Cover design: Design & Production, Heidelberg

Typesetting: Best-set Typesetter Ltd., Hong Kong

SPIN: 10047436   31/3137/SPS – 5 4 3 2 1 0 – Printed on acid-free paper

Dedication

This book is dedicated to our many friends throughout the world who share our enthusiasm for the yeasts. It is dedicated in particular to my coauthor and devoted wife, Dorothy Spencer, who, besides her part in the writing of the book, did a great deal of the work of correcting my numerous typographical and similar errors, and who died of cancer before she could have the satisfaction of seeing that her labors were not in vain.

J.F.T. Spencer

Contents

List of Contributors

Berardi, Enrico, Dipartimento di Biologie Agrarie ed Ambientale, Università degli Studi di Ancona, 60131 Ancona, Italia

de Figueroa, L.I.C., PROIMI, Avenida Belgrano y Pasaje Caseros, 4000 S.M. de Tucumán, Argentina

Höfer, Milan, Botanisches Institut, Rheinische Friedrich-Wilhelms-Universität, Kirschallee 1, 53115 Bonn, Germany

Krishnaiah, M.M., Central Food Technological Research Institute, Microbiology & Fermentations, Mysore 570013, India

Lonsane, B.K., Central Food Technological Research Institute, Microbiology & Fermentations, Mysore 570013, India

Palková, Z., Dept. of Genetics and Microbiology, Faculty of Science, Charles University, Vinična 5, 128 44 Prague, Czech Republic

Spencer[†], Dorothy M., PROIMI, Avenida Belgrano y Pasaje Caseros, 4000 S.M. de Tucumán, Argentina

Spencer, John F.T., PROIMI, Avenida Belgrano y Pasaje Caseros, 4000 S.M. de Tucumán, Argentina

Vijayalakshmi, G., Central Food Technological Research Institute, Microbiology & Fermentations, Mysore 570013, India

Vondrejs, V., Dept. of Genetics and Microbiology, Faculty of Science, Charles University, Vinična 5, 128 44 Prague, Czech Republic

A Guide to the World of the Yeasts

J.F.T. Spencer and D.M. Spencer†

As the well-known authority on yeasts, the late Professor Rose, frequently pointed out, it is impossible for one person to present, in a single volume, the details of the life, composiotion, habitats, relationships, and actual and potential uses to mankind of the 500 (at last count) known species of yeasts. This book confirms the truth of this statement. However, our aim is actually more modest than that, and this book is an attempt to introduce the general reader, and possibly some interested specialists, to the lives of the yeasts in their natural and more artificial habitats, their use by human beings, and to give some idea of the wonderfully complex activities within the yeast cell, the characteristics of the metabolism and molecular biology of yeasts, and the applications of these characteristics to life in the present-day world of human existence. The book proceeds from a brief chapter on what is and is not known of the origins and early history of the yeasts, through a description of their classification, relationships, habitats and general life style, their external morphology and internal structures and mechanisms within their cells, the regulatory mechanisms controlling processes such as signal transmission, mating, cell fusion, and many others.

To prove the earlier contention that it is impossible for one person to describe adequately everything about every yeasts, there are special sections dealing with yeast membranes, killer factor, and in the part of the book dealing with the impact of yeasts on the human condition, we have included sections on fundamental and applied aspects of the methylotrophic yeasts, yeasts in fermented foods, and use of yeasts in production of fuel alcohol. As the originators and authors of the book, we are extremely grateful to these specialists for their valuable contributions to the volume.

In conclusion, we would like to thank our friends, once again, who have made substantial contributions to this task, in the hope that they will enjoy seeing the book in print and reading it. We thank them for their patience and understanding of the task we have had as authors ourselves, and as editors of their work. We would like also to express our thanks to Dr. Faustino Siñeriz and all our friends at PROIMI for their advice and encouragement during our work on the book.

J.F.T. Spencer/D.M. Spencer (eds)
Yeasts in Natural and Artificial Habitats
© Springer-Verlag Berlin Heidelberg 1997

CHAPTER 1

Historical Introduction: Yeasts and Man in the Past

J.F.T. Spencer and D.M. Spencer

1 From Earliest Times to the Beginning of the Modern Era

Fossil yeasts are unknown. Fungal spores have been found in fossiliferous materials, and yeasts may have been associated with them, but they have not been recognized. The sight of a yeast ascus containing ascospores, or of the zigzag configuration of the arthrospores of a fossilized *Trichosporon* species, would enable a paleobotanist to place the yeasts more accurately in time. There may be a long wait for such a discovery, though ancient diatoms, algae, and other microorganisms have been seen, and eventually a yeast ascus may be found, resting peacefully in its amber tomb (Poinar et al. 1993).

Yeasts have been known, indirectly, "only" for a few thousand years. The first peoples to eat leavened bread and drink beverages such as wines were either the Egyptians or one of the other civiizations of the Fertile Crescent. They knew the uses of yeasts but not their identity. Beer may have been discovered at about the same time, possibly by some farmer who was soaking a pot of grain in water, to make it more palatable to his domestic animals. It is known that the Egyptians knew bread, beer, and wine. The knowledge was probably shared by the other peoples of the region, and by those of the land with whose inhabitants they traded – India.

The use of yeasts in breadmaking may have arisen through some of the early methods of cooking raw flour. For instance, in making salt-rising bread, the flour and a weak brine are mixed and kept warm, probably at about 30 °C, until the volume of the mixture approximately doubles, and this is then mixed with more flour and warm water and allowed to rise again. The normal practice in all breadmaking, when commercial yeast or spent yeast from breweries is not available, is to save a portion of the uncooked dough to start the next batch. The better batches of starter would probably be propagated as long as possible, and would have consisted predominantly of yeasts. In this way, the characteristic odor of yeast starters would have been recognized, very early, and would have been favored by the women who did the real work of baking; they would in all probability have traded the "good" starters, which would have been largely yeasts.

Unleavened bread is known from flat loaves dating from at least 6000 B.C. in Egypt. If a dough mix of this sort were left to stand for any reason, fermented spontaneously, and was then baked, the result would be naturally leavened bread. From there to deliberately leavened loaves is only a short step. The odor of rising

J.F.T. Spencer/D.M. Spencer (eds)
Yeasts in Natural and Artificial Habitats
© Springer-Verlag Berlin Heidelberg 1997

bread is similar to that of fermenting beers. The use of spent yeast from the brewery vessels in the bakery was probably also a short step. Models of a bakery and a brewery, found in an Egyptian tomb, show that both fermentation and breadmaking were well established by 4000 years B.P.

Certainly, brewing technology, without the knowledge of the actual agent of fermentation, was well developed in the earliest civilizations. Winemaking was naturally confined to those countries where the climate favored the production of grapes. For a time, even England was a producer of wine on a small scale. However, the dark, gloomy forests of Germany and the cleared fields of that country gave the barley growers of the region an advantage over the vineyard husbandmen, the climate was likewise against the latter, and the Germanic tribes who fought the Roman legions were drinkers of beer.

Brewing developed more rapidly than winemaking where the climate was unfavorable to viticulture, and beers of several varieties developed in England, the Netherlands, and the Scandinavian countries, as well as in Germany. One of these, no longer brewed, was an unhopped beer made with rye malt, whose flavor must be left to the imagination. Winemaking predominated in France, and a vigorous trade in wine grew up between England and the southwest of France. In Scotland, where the climate was even more inclement and chilling than in England, the natives preferred a more warming beverage, and produced the strong-flavored whisky, distilled from fermented barley in pot stills. In all of this production of fermented beverages, it had still not been noticed that all depended on a living organism – yeast.

Probably the first person actually to see a yeast cell, through his unusual microscopes, constructed of a single tiny sphere of polished glass, was the Dutchman, Antonie van Leeuwenhoek (1632–1723; in Phaff et al. 1979). He observed globular bodies, round or oval, in a drop of fermenting beer – his "little animalcules". He reported his observations to the Royal Society of London, where presumably they aroused great interest, but it is not known whether and how much anyone speculated on the nature of these organisms, or on what their role was in the fermentation of beers and wines.

Erxleben (1818; in Phaff et al. 1979) expressed the view that yeast was a living organism, responsible for fermentation. Cagniard-Latour (1835; in Phaff et al. 1979) in France, and Schwann and Kutzing (1837; in Phaff et al. 1979) in Germany, observed the presence of single-celled organisms in the sediment found in fermentation vessels, and though the organisms were not identified, stated that fermentation resulted from their activities during growth.

Finally, Pasteur (1822–1895; in Phaff et al. 1979), in his works *Etudes sur la Bière* and *Etudes sur le Vin*, showed that the presence of these organisms was essential to the fermentation process. Without them, fermentation did not occur, and if other organisms, morphologically different, were present, fermentation of the desired type did not occur and the wine was spoiled. Pasteur dealt the final death blow to the idea of spontaneous generation as a viable theory.

Pasteur's views were not immediately accepted, as was noted earlier. Liebig and Wohler and their school of chemists were violently opposed to Pasteur's conclusions, and held to the view that fermentation was the result of a series of purely chemical reactions, and ridiculed the idea of a living organism such as yeast being responsible. Who is to say that Liebig was entirely wrong? Certainly, it is now known that fermentation and formation of new cells proceed through a series of reactions, catalyzed by enzymes, leading either to the formation of ethanol from glucose, or to the biosynthesis of proteins, nucleic acids, and other cellular components. All of these reactions are controlled by the coordinated action of a series of enzymes, which in turn are formed by other enzymes, whose formation and action are entirely coded and controlled by a series of instructions on a long and intricate tape, composed of another series of chemical compounds. Perhaps Pasteur and Liebig were both right.

2 From Hansen to Winge

At about the same time, Hansen, in Denmark, was investigating the nature of brewing and baking yeasts. He made numerous isolations of pure cultures of yeasts of the genus *Saccharomyces*, though at that time the brewers preferred a mixture of strains as being less susceptible to chance variation. Hansen, using and perfecting Pasteur's methods for obtaining pure cultures, spent 30 years in a study of the morphology and some physiological characteristics of yeasts, and differentiated numerous species, some of which are still recognized today. He is recognized as the originator of the study of the morphological characteristics of yeasts, and established the first comprehensive system of yeast taxonomy, in 1896 (in Phaff et al. 1979).

After Hansen, Guilliermond (1928) expanded the study of the systematics and life cycles of yeasts, and contributed a great deal to the understanding of the phylogenetic relations among yeast species, devising dichotomous keys for use in the identification of yeasts.

Following the work of Guilliermond (1928), the monographs of the Delft School, by Stelling-Dekker (1930; in Phaff et al. 1979), by Lodder (1934, 1941; in Phaff et al. 1979) and Diddens and Lodder (1941; in Phaff et al. 1979), brought some order into the chaotic field of yeast taxonomy. However, the first comprehensive classification of the sporogenous and asporogenous yeasts was published by Lodder and Kreger-van Rij (1950). These authors classified 1317 strains of yeasts from the Yeast Division of the Centraalbureau voor Schimmelcultures in Delft into 165 species and 17 varieties. This system of identification was inadequate to handle the rapidly increasing number of new yeast species, and 20 years later, Kreger-van Rij (1970) edited a further volume on yeast taxonomy, with a larger set of assimilation tests as a criterion, using a list of carbon compounds proposed by Wickerham (1951), and including 341 species. By 1984, the third edition included 500 species, many new species had been added, and the volume had become massive. This edition was out of date almost before it appeared, and a fourth is planned for the near future. It is no longer possible for a single worker or a group of three or four to produce such a volume. Future taxonomic guides will be the work of a group of editors and writers, although Barnett et al. (1983) have produced a series of volumes for the identification and classification of yeasts, using only biochemical properties of the yeast strain to assign the isolate to a species. This, too, is a weighty tome, full of information which is frequently useful for the muscular taxonomist. In 1993, Barnett et al. produced a computer program for yeast identification, which gives reasonably satisfactory results. Used with the most recent edition of *Yeast Taxonomy*, when it appears, identification of new strains should be facilitated.

The work of organization of the known yeasts into schemes of classification could proceed only as fast as observation and experimental work permitted. Originally, the name *Saccharomyces*, suggested by Meyen (1837; in Phaff et al. 1979), was applied by Reess (1870; in Phaff et al. 1979) to the spore-forming yeasts, and included them in the Ascomycetes, De Bary having compared the spore-containing bodies of yeasts with the spore sac of spore-forming Ascomycetes. However, it became obvious that other genera and species of yeasts existed and must be recognized and named. Much of the early work, as mentioned, was done by the Dutch school, and resulted in the publication of the volumes by Stelling-Dekker, Lodder, Diddens and Lodder, and N.J.W. Kreger-Van Rij (in Phaff et al. 1979). In the beginning, identification and classification of yeasts were based almost entirely on the morphology of the cell, the ability of the culture to form spores, the morphology of the spores, if present, and the ability to utilize and ferment half a dozen sugars and to utilize nitrate as a nitrogen source. This led to confusion of investigators who wished to identify yeast strains isolated from sources other than industrial fermentations, and which did not fit into any of the categories to be found in the first edition of *The Yeasts: a Taxonomic Study*.

At this point, enter Lynferd J. Wickerham. Wickerham's interest was in the ecology and taxonomy of yeasts, particularly those isolated from natural sources such as plants, and, in particular, trees; water, insects, or almost any other natural habitat where yeasts may be found. Wickerham's pioneering work made the USDA laboratory in Peoria one of the major centers in the world for investigations on yeasts. He soon recognized that the then current criteria for identification of yeasts were far too narrow, and set up a much enlarged list of carbon compounds to be tested for assimilation and, to a lesser extent, for fermentation. The list of nitrogen compounds which were considered significant in this respect remained short. At the same time, various workers had observed that one of the morphological criteria for identification of imperfect yeasts, the ability to form pseudomycelium and its morphology, was unimportant, since different strains of the same species may or may not form pseudomycelium. The character is no longer used, and yeasts of the genus *Torulopsis* have been transferred to the genus *Candida*.

2.1 Early Investigations on the Biochemical Properties of Yeasts

These began early in this century. Buchner (in Phaff et al. 1979) noted that a "press juice" from yeast, free of intact cells, would carry out the fermentation of sugar in much the same way as in the presence of whole cells. At first, this discovery was thought to vindicate Liebig's view of fermentation as purely a simple chemical process, but eventually it was observed that the fermentation carried out by the Buchner brothers' "press juice" differed greatly from that carried out by intact yeast cells. Harden and Young (Harden 1932) also observed fermentation by a yeast "press-juice", which was a relatively cell-free extract of yeast. Neuberg, Meyerhof, Warburg, Wieland, Parnass, Embden, and numerous others eventually elucidated the pathway of fermentation of glucose through glycolysis in yeasts and other organisms (in Phaff et al. 1979).

2.2 Later Developments in Yeast Taxonomy and Ecology

The knowledge of yeast taxonomy has increased and improved greatly and rapidly in recent years. Emil Mrak, H.J. Phaff, and M.W. Miller of the University of California at Davis began their investigations of the taxonomy and ecology of yeasts at about the same time as Wickerham had done. Part of their interest was in associations between yeasts and plants, yeasts and insects, yeasts and aquatic animals, which involved them deeply in taxonomy as well. As a result, the laboratory in the University of California at Davis became another of the major world centers for the study of yeast taxonomy, and a steady stream of students, postdoctoral fellows, and established scientists on sabbatical leave passed through it. The Northern Regional Research Laboratory in Peoria, under L.J. Wickerham, the Dutch school at Delft, being part of the Centraalbureau voor Schimmelcultures, and the laboratories in Czechslovakia and Pretoria, South Africa, the latter under J.-P. van der Walt, also played a very significant role in this work.

All of these laboratories participated in the search for new taxonomic criteria. It had become obvious, as Barnett pointed out, that the criteria in use, depending as they did on fermentation and assimilation of various carbon compounds, gave clues to the nature of only a small part of the total yeast genome, and the search began for other taxonomic clues. At first, the genome was not directly accessible, and two methods, involving the serological characteristics of the yeast strain, and the chemical structure of the cell wall mannans, respectively, were suggested, by Tsuchiya et al. (1974) in Japan, and Gorin and Spencer (1970) at the Prairie Regional Laboratory in Canada. These have been criticized as being too subject to variation and to changes due to mutations, somewhat unfairly as it turned out when methods for the direct determination of similarities and differences in the genomic and mitochondrial DNA became available. Both critics and proponents of the two methods failed to take into consideration the existence of strains identified as belonging to the same species, but having different serological characteristics and mannans of different chemical structures, which occurs in species such as *Candida tropicalis*. Critics also failed to recognize that not only serological characteristics and chemical structure of cell wall components are subject to changes due to mutations, but that all possible taxonomic characteristics of yeasts are subject to the same sort of changes.

About this time, the chemical characterization of yeast DNA became possible, first by the determination of the base ratios, by determination of melting points, and then by DNA-RNA and DNA-DNA reassociation. It soon became obvious that base ratios in the genomic DNA gave only a rough idea of relationships, serving as a means of exclusion rather than of positive indication of relationships, much as could be determined using magnetic resonance spectra. However, DNA reassociation was a much finer tool for determination of such relationships, and was soon used for establishing the proper taxonomic place in nature of many yeasts, and settling such controversies as the reality of such species as *Saccharomyces pastorianus* and *Saccharomyces bayanus*, and their relationship to *Saccharomyces cerevisiae*. At the same time, the technique has been used to show that a number of yeasts isolated from cactus rot pockets, resembling *Pichia membranaefaciens*, were in fact quite distinct species. The power of the method is far greater than that of any other so far available to the yeast taxonomist, though other methods such as serological characterization and comparison of NMR spectra of yeast cell wall mannans may often serve as a "fingerprint" which will reduce the time and effort required to place an unknown isolate in the correct category for positive identification by means of DNA reassociation.

Recently, Kurtzman (1994), at the US Department of Agriculture laboratory in Peoria, USA, and Lachance (1990) at the University of Western Ontario, Canada, have compared the ribosomal RNAs and ribosomal DNAs of a number of yeast species, to determine the phylogenetic distances between the species. The method permits detailed determination of the similarities of yeast cultures. Fell (1993) has developed a method for pairwise comparison of yeast strains using PCR (polymerase chain reaction) and three primers, which should prove of great value. Use of random amplified polymorphic DNA (RAPD) methodology (Hadrys et al. 1992) is another promising method for identification on unknown yeast strains.

Separation of the chromosomes by pulsed-field gel electrophoresis, and further identification by use of labeled probes, assignable to particular chromosomes, has recently become posible and may become another powerful tool for use in yeast taxonomy. From its first crude beginnings in the early 1980s, the technique has been developed to the point where in the most recently developed systems, individual industrial strains of wine yeasts (*Saccharomyces cerevisiae*) can be positively identified and distinguished from other similar strains. The individual chromosomes in any set of yeast strains can be compared by several methods (Chu et al. 1986).

3 From Genetics to Molecular Biology

Ojvind Winge began his career as a classical geneticist at the Veterinary and Agricultural University in Copenhagen, and was appointed to the directorship of the Physiology Department at the Carlsberg laboratory in 1933. In 1935, he made the first genetic analysis of yeast, in which he established that yeast (*Saccharomyces cerevisiae*) had a haplophase and diplophase and showed normal eukaryotic alternation of generations. Winge and Lautsten (1939) then did the first tetrad analysis in *Saccharomyces cerevisiae*, demonstrated the existence of Mendelian segregation in *S. cerevisiae*, and constructed the first interspecific hybrids, selected on the basis of fermentative properties and morphology of giant colonies. They later showed that the haploid spores isolated from asci of heterothallic strains of *S. cerevisiae*, and the haploid vegetative cells arising from them would fuse and restore the diploid state. They also dissected tetrads from another yeast species, *Saccharomycodes ludwigii*, concluded that the species was heterothallic, and demonstrated segregation of lethality and cell morphology. This key observation formed the foundation on which the structure of the science of yeast genetics was raised.

Winge and Roberts (1954) and the Lindegrens in the United States isolated haploid heterothallic strains of *Saccharomyces cerevisiae*, made auxotrophic strains of these cultures, and essentially began the study of recombination in yeast as one of the simplest eukaryotes accessible to genetic analysis (Lindegren 1949; Lindegren and Lindegren 1951). Many of the yeast strains used in genetic studies today are the descendants of a half dozen strains first isolated in the Lindegrens' laboratory. Lindegren was a controversial figure, and his real contributions were sometimes obscured by his unorthodox views on other subjects. For instance, he isolated haploid strains of *S. cerevisiae*, some of which are still in use in genetic investigations, and maintained, in the face of contradictory views by Winge, that the phenomenon of gene conversion was real and not ascribable to experimental error. These contributions were often overshadowed by his views on cytogenes and "vacuolar chromosomes". However, Lindegren and his wife can be listed alongside Winge and his collaborators as originators of yeast genetics. Since then, studies of the genetics of yeasts have been done, in particular, in the Genetics Department at the University of Washington in Seattle, in the University of California laboratories at Berkeley, Davis, and Riverside, at the University at Corvallis in Oregon, at the University of Rochester in New York State, at University College, London, in the laboratory of Professor David Wilkie, mostly concerned with the genetics of mitochondria; at Monash University and the Australian National University in Australia, in the laboratories of A.W. Linnane and G.D. Clark-Walker, also concerned with mitochondria, and in the Universities of Edinburgh (I.W. Dawes) and Strathclyde (J.R. Johnston). There have been significant contributions from other laboratories, too numerous to mention, but the contributions to the genetics of industrial yeasts by the investigators at the Carlsberg Laboratories in Denmark have been especially worthy of note.

Molecular biology came slightly later to the yeast world than to that of the bacteria. Transformation in bacteria, as a mysterious phenomenon involving passage of different strains of pneumococcus through mice, was early demonstrated by Griffiths in London in 1938 (see Phaff et al. 1979), and the transforming principle was isolated and identified as nucleic acid by Avery and coworkers in 1944. The determination of the structure of DNA by Watson and Crick (1953) followed, together with a host of related phenomena, but it was not until about 1960 that Oppenoorth (1960), in Holland, reported transformation in yeast. Unfortunately, the report was premature, because the data were erroneous. It was not until 1978, after the discovery of restriction enzymes, which permitted the construction of the necessary plasmids, that transformation of yeast protoplasts, followed by regeneration of the intact but transformed yeast cells with walls, was accomplished by Hinnen et al. (1978), in Fink's laboratory in the United States, and independently by Jean Beggs (1978) at Imperial College in London.

Since that time, the use of yeasts in biotechnology has grown at an ever-increasing rate. Yeasts have certain advantages over bacteria for the production of heterologous proteins, and almost any protein, synthesized by a gene small enough to be included in the genome of a yeast cell, can be formed in yeast and excreted if desired. Most recently, two important developments have occurred; the inclusion of larger and larger sequences of DNA in yeast artificial chromosomes and their transformation into yeast, and the use of yeasts other than *Saccharomyces cerevisiae* for production of heterologous proteins. In particular, the methylotrophic yeasts, *Pichia pastoris*, *Hansenula polymorpha*, and some others show great promise; the desired proteins being synthesized by a gene being substituted for that for alcohol oxidase in the yeast genome, and then being induced by switching the carbon source to methanol, so that much higher yields of the reasonably pure protein can be synthesized. The use of yeasts in biotechnology, for this and other purposes, is in its infancy.

References

Barnett JA, Payne RW, Yarrow D (1983) Yeasts: characteristics and identification. Cambridge University Press, Cambridge

Barnett JA, Payne RW, Yarrow D (1993) Computer program for identifying yeasts

Beggs J (1978) Transformation of yeast by a replicating hybrid plasmid. Nature 275:104

Chu G, Vollrath D, Davis RW (1986) Separation of large DNA molecules by contour-clamped homogeneous electric fields. Science 234:1582–1585

Fell JW (1993) Rapid identification of yeast species using three primers in a polymerase chain reaction. Mol Mar Biol Biotechnol 2:174–180

Gorin PAJ, Spencer JFT (1970) Proton magnetic resonance spectroscopy – an aid in identification and chemotaxonomy of yeasts. Adv Appl Microbiol 13:85–98

Guilliermond A (1928) Clef dichotomique pour la détermination des levures. Librarie le Francoise, Paris

Hadrys H, Balick M, Schierwater B (1992) Applications of random amplified polymorphic DNA (RAPD) in molecular ecology. Mol Ecol 1:55–63

Harden A (1932) Alcoholic fermentation, 4th edn. Longmans, Green, London

Hinnen A, Hicks JB, Fink G (1978) Transformation of yeast. Proc Natl Acad Sci USA 75:1929

Kreger-van Rij NJW (1970) The yeasts, a taxonomic study. North Holland, Amsterdam

Kreger-van Rij NJW (1984) The yeasts, a taxonomic study. Elsevier, Amsterdam

Kurtzman CP (1994) Molecular taxonomy of the yeasts. Yeast 10:1727–1749

Kurtzman CP, Phaff HJ (1987) Molecular taxonomy. In: Rose AH, Harrison JH (eds) The yeasts, vol 1. Biology of yeasts. Academic Press, New York, pp 63–94

Lachance M-A (1990) Ribosomal spacer variation in the cactophilic yeast *Clavispora opuntiae*. Mol Biol Evol 7:178–198

Lindegren CC (1949) The yeast cell, its genetics and cytology. Educational Publishers, St Louis

Lindegren CC, Lindegren G (1951) Linkage relationships in *Saccharomyces* of genes controlling the fermentation of carbohydrates and the synthesis of vitamins, amino acids and nucleic acid components. Indian Phytopathol 4:11–20

Lodder J, Kreger-van Rij NJW (1950) The yeasts, a taxonomic study. North Holland, Amsterdam

Oppenoorth WFF (1960) Modification of the hereditary character of yeast by ingestion of cell-free extracts. Antonie Leeuwenhoek J Microbiol Serol 26:129–168

Phaff HJ, Miller MW, Mrak EM (1979) The life of yeasts. Harvard University Press, Cambridge

Poinar GO, Waggoner BM, Bauer U-C (1993) Terrestrial soft-bodied protists and other microorganisms in Triassic amber. Science 259:222–224

Tsuchiya T, Fukazawa Y, Taguchi M, Nakase T, Shinoda T (1974) Serological aspects of yeast classification. Mycopathol Mycol Appl 53:77–91

Vaughan Martini A, Kurtzman CP (1985) Deoxyribonucleic acid relatedness among species of the genus *Saccharomyces sensu stricto*. Int J Syst Bacteriol 35:508–511

Watson JD, Crick FHC (1953) The structure of DNA. Cold Spring Harbor Symp Quant Biol 18:123–131

Wickerham LJ (1951) Taxonomy of yeasts. US Dept Agric Tech Bull 1029:1–50

Winge O, Lautsten O (1939) On fourteen new yeast types produced by hybridisation. CR Trav Lab Carlsberg Ser Physiol 22:337–332

Winge O, Roberts C (1954) Causes of deviation from 2:2 segregations in the tetrads of monohybrid yeasts. CR Trav Lab Carlsberg Ser Physiol 25:285–329

CHAPTER 2

Taxonomy: The Names of the Yeasts

J.F.T. Spencer and D.M. Spencer

1 Introduction

Approximately 500 species of "true" yeasts, in 69 genera, are described in the most recent edition of *The Yeasts: a Taxonomic Study* (Kreger-van Rij 1984). There are a number of other relatively significant organisms which have a yeast-like phase under some conditions of growth. Indeed, one of the better definitions of the yeasts in general is "those fungi, basidiomycetes or ascomycetes, whose vegetative stage is unicellular, which multiply by budding or fission, which may or may not form spores during a sexual stage, and which have not been named as some other type of fungus". A definition of the yeasts says as much about what they are not, as about what they are. Nevertheless, the different species of yeasts must be given names.

The place of yeasts in the Eumycotina and the genera of the ascomycetous, basidiomycetous, and imperfect yeasts (Deuteromycotina) are given in Tables 2.1–2.4. Two factors which often increase the difficulty of identification and classification of the yeasts are, first, the inclusion of so many of the yeasts in three huge genera; 196 species of *Candida*, 56 species of *Pichia*, and 30 of *Hansenula*. If the genera *Hansenula* and *Pichia* are lumped together, there are 86 species of *Pichia* and 196 of *Candida* (many *Candida* species are anamorphs of perfect species belonging to a variety of genera), for a total of 282 species, or well over half of the known yeast species. The next largest genus, *Cryptococcus*, has 19 species, and *Trichosporon*, 15. On the other hand, there are 25 genera consisting of a single species each.

The other difficulty occurs because of the few distinctive features of the yeasts. *Cryptococcus* species form starch, and the colonies stain blue with iodine. *Rhodotorula* species and similar basidiomycetous yeasts form red carotenoid pigments. Some of these form ballistospores. Colonies of all basidiomycetous yeasts stain violet with diazonium blue (DBB). *Metschnikowia* species form club-shaped asci containing needle-shaped spores, if the strain can be induced to sporulate. *Pachysolen tannophilus* forms an ascus at the end of a very long (30 μm) tube. The apiculate yeasts form lemon-shaped cells, but nevertheless are often not easy to identify to the species level. The fission yeasts (*Schizosaccharomyces*) are easy to identify at the genus level, and the peculiar triangular shape of the cells of *Trigonopsis* is unique. Having progressed through this list of yeasts having distinctive and unequivocal characteristics, how do we identify the remaining 90% (or

J.F.T. Spencer/D.M. Spencer (eds)
Yeasts in Natural and Artificial Habitats
© Springer-Verlag Berlin Heidelberg 1997

Table 2.1. Place of yeasts in the Eumycotina

Ascomycotina	
Hemiascomycetes	
Endomycetales	Spermophthoraceae
	(with needle-shaped spores)
	Coccidiascus
	Metschnikowia
	Nematospora
	Saccharomycetaceae
	(all other ascomycetous yeasts)
Basidiomycotina	Filobasidiiaceae
Ustilaginales	Teliospore-forming
	yeasts
Tremellales	Sirobasidiaceae
	Tremellaceae
Deuteromycotina	
Blastomycetes	Cryptococcaceae
	Sporobolomycetaceae

Table 2.2. Major groupings in the ascosporogenous yeasts (Ascomycotina)

Spermophthoraceae (needle-shaped spores)
    Coccidiascus
    Metschnikowia
    Nematospora
Saccharomycetaceae
  Schizosaccharomycetoideae (Multiplication by fission)
    Schizosaccharomyces
  Nadsonioideae (Vegetative multiplication by bipolar budding. Lemon-shaped cells)
    Hanseniaspora
    Nadsonia
    Saccharomycodes
    Wickerhamia
  Lipomycetoideae (Habitat soil. Asci usually multispored)
    Lipomyces
  Saccharomycetoideae (All of the others)

Ambrosiozyma	*Pachysolen*
Arthroascus	*Pachytichospora*
Citeromyces	*Pichia*
Clavispora	*Saccharomyces*
Cyniclomyces	*Saccharomycopsis*
Debaryomyces	*Schwanniomyces*
Dekkera	*Sporopachydermia*
Guilliermondella	*Stephanoascus*
Hansenula	*Torulaspora*
Issatchenkia	*Wickerhamiella*
Kluyveromyces	*Wingea*
Lodderomyces	*Zygosaccharomyces*

Table 2.3. Numbers of species per genus in ascomycetous, basidiomycetous, and imperfect genera of the yeasts

1 Ascomycetes

1 Species/genus	2–5 Species/genus	6–20 Species/genus	>20 Species genus
Arthroascus	Ambrosiozyma (4)	Debaryomyces (9)	Hansenula (30)
Citeromyces	Dekkera (2)	Hanseniaspora (6)	Pichia (56)
Clavispora	Issatchenkia (4)	Kluyveromyces (11)	
Coccidiascus	Lipomyces (5)	Metschnikowia (6)	
Cyniclomyces	Nadsonia (3)	Saccharomyces (7)	
Guilliermondella	Schizosaccharomyces (4)	Zygosaccharomyces (8)	
Lodderomyces	Sporopachydermia (2)		
Nematospora	Torulaspora (3)		
Pachysolen			
Pachytichospora			
Schwanniomyces			
Stephanoascus			
Wickerhamia			
Wickerhamiella			
Wingea			

2 Basidiomycetes

  A. Filobasidiaceae

    Chionosphaera Filobasidium (3)
    Filobasidiella
    (*Cryptococcus neoformans*)

  B. Teliospore-forming yeasts
    Leucosporidium (5)

                                            Rhodosporidium (8)

    Sporidiobolus (4)

  C. Sirobasidiaceae and Tremellaceae
    Fibulobasidium   *Tremella* (9)
    Sirobasidium
    Holtermannia

3 Fungi Imperfecti

1 Species/genus	2–5 Species/genus	6–20 Species/genus	>20 Species genus
Aciculoconidium	Malassezia (2)	Brettanomyces (9)	Candida (196)
Oosporidium		Bullera (6)	
Phaffia		Cryptococcus (19)	
Sarcinosporon		Kloeckera (6)	
Schizoblastosporion		Rhodotorula (10)	
		Sporobolomyces (7)	
Sympodiomyces		Sterigmatomyces (6)	
Trigonopsis		Trichosporon (15)	

thereabouts) of white or cream-colored yeasts, having round or oval cells, forming either round or relatively hat-shaped spores, or none at all, and in general, looking like all the rest of the yeasts? It is not an easy task.

The known Ascomycetous yeast species are placed in the families Saccharomycetaceae and Spermophthoraceae. The former comprises the subfamilies Schizosaccharomycetoideae (one genus, *Schizosaccharomyces*),

Table 2.4. Unmistakably distinctive features of yeasts

1 Spores
Needle-shaped	*Coccidiascus, Nematospora, Metschnikowia*
Clavate	*Clavispora*
Bean-shaped	*Kluyveromyces* (some species)
Ballistospores	*Bullera, Sporobolomyces*
Round, oval, hat- or helmet-shaped, etc.	All the rest

2 Asci
Club-shaped	*Metschnikowia*
Banana- or crescent-shaped	*Coccidiascus*
Thick-walled with a long tube	*Pachysolen*
Elongate	*Nematospora.* Eight spores/ascus
Nothing unusual	All the rest

3 Vegetative cells
Triangular	*Trigonopsis*
Fission	*Schizosaccharomyces*
Apiculate	*Hanseniaspora, Kloeckera, Nadsonia, Saccharomycodes, Schizoblastosporion, Wickerhamia*
Arthrospores	*Trichsporon* (some)
Multilateral budding, with or without mycelium or pseudomycelium	All the rest
True mycelium with clamp connections	*Chionosphaera, Filobasidiella, Filobasidium, Leucosporidium, Rhodosporidium, Sporidiobolus*
Cells on sterigmata	*Sterigmatomyces*
Pigments (red or yellow carinoid)	*Cryptococcus* (some), *Rhodotorula*
Pigments (red to purple)	*Metschnikowia pulcherrima,* some *Kluyveromyces* spp.

Saccharomycetoideae (the largest group, comprising 26 genera, including the largest genus, *Pichia*, especially now that the genus *Hansenula* has been transferred to it), Lipomycetoideae (one genus, *Lipomyces*), and Nadsonioideae (four genera; *Hanseniaspora, Nadsonia, Saccharomycodes*, and *Wickerhamia*, all multiplying vegetatively by bipolar budding on a broad base). The latter family, Spermophthoraceae, contains the genera *Coccidiascus, Metschnikowia*, and *Nematospora*, all of which form needle- or spindle-shaped spores, in a rather large ascus. All of the species are parasitic on arthropods and/or produce inhibitory compounds.

The group of basidiosporogenous yeasts includes the group forming teliospores (*Leucosporidium, Rhodosporidium*, and *Sporidiobolus*), the Filobasidiaceae (*Filobasidiella* and *Filobasidium*), which form basidiospores, and such unclassified yeasts as *Sterigmatosporidium*, which multiplies by formation of budding cells on sterigmata. Most of the species of basidiosporogenous yeasts, when in the dikaryotic state, form true mycelium with clamp connections.

The group of imperfect yeasts (not forming sexual spores) are classified in the Sporobolomycetaceae (genera *Bullera* and *Sporobolomyces*, which form ballistospores), and Cryptococcaceae, comprising 16 genera. The best known of these are *Brettanomyces* (which produce acid copiously), *Candida* (containing all imperfect species which are not classified as something else), *Cryptococcus* (basidiomycetous yeasts, which assimilate inositol, form starch-like compounds and mostly do not form red or orange pigments), *Kloeckera* (bipolar budding, imperfect forms of *Hanseniaspora*), *Malassezia* (previously known as *Pityrosporum*, a commensal residing on human skin), *Phaffia* (the only red-pigmented yeast which ferments a sugar), *Rhodotorula* (red yeasts, anamorphs of *Rhodosporidium*, nonfermentative, not using inositol, not producing starch-like compounds), *Sterigmatomyces* (anamorph of *Sterigmatosporidium*), *Trichosporon* (forming true mycelium and arthrospores, and containing ascomycetous and basidiomycetous genera), and *Trigonopsis variabilis*, an unusual yeast species forming triangular cells. The ascomycetous and basidiomycetous species can be distinguished by the diazonium blue test, and by both the internal structure, as determined by transmission electron microscopy, and the chemical composition and chemical structure of the cell wall and extracellular polysaccharides. The guanosine + cytosine (GC) content of the DNA usually distinguishes the ascomycetous yeasts from the basidiomycetous ones (Kurtzman and Phaff 1987; Kurtzman and Robnett 1991), since ascomycetous species have GC contents ranging from approximately 27–50%, and basidiomycetous yeasts 50–70%. There is some overlapping in the range 48–52% GC content, but the taxonomic affinities of yeasts having GC contents in this range can be determined by other methods.

As long as the only yeast used for commercial purposes or research was *Saccharomyces cerevisiae*, taxonomy was relatively unimportant. Even though other yeast species existed, they could often be considered as being of no significance; unless they were associated with some pathological condition or food spoilage, they were important only to some specialist.

Now, however, this view of yeast taxonomy is far from correct. It is now known that yeasts other than *Saccharomyces cerevisiae* can produce a number of polyhydroxy alcohols, glycolipids, vitamins, and other extracellular metabolites. Some can transform steroids and other lipids into more valuable products; others can convert alkanes into biomass or other materials, some produce enzymes such as proteases and inulinases, and still others are useful for production of heterologous proteins (hormones, for instance) of considerable pharmaceutical value. Thus, first, it is important scientifically to know the identity of the yeast used because of its peculiar characteristics, and second, it is important financially and legally to know its name for patent purposes. A patent is of little value if it refers to an unknown object.

Finally, even *Saccharomyces cerevisiae* has been split into at least three species, not necessarily interfertile, and having DNAs which are distinctive enough that they do not reassociate. Also, when the chromosomes are separated on pulsed field gels, *Saccharomyces cerevisiae* proves to be composed of distinct and identifiable strains according to the nature of its chromosomes.

The genera of the yeasts, as accepted at present, with their outstanding characteristics, are given in Tables 2.1–2.3.

2 Methods in Yeast Taxonomy

Originally, yeasts were classified according to the morphological characteristics of the vegetative cells and spores. Hansen (see Phaff et al. 1979), who produced the first comprehensive system of taxonomy of the then known yeasts, based it on morphology, as did Guilliermond (see Phaff et al. 1979), who included physiological characteristics as well. By the time the volume on classification of the sporogenous and asporogenous yeasts by Lodder and Kreger-van Rij (see Phaff et al. 1979) appeared in 1952, the criteria for identification of the yeasts included morphology of the vegetative cell, including size as well as shape, morphology, and mode of formation of the spores, if any, characteristics of the colony, surface growth on liquid medium, ability to grow on nitrite or nitrate as sole source of nitrogen, and ability to ferment and/or assimilate six sugars; glucose, galactose, maltose, sucrose, lactose, raffinose, and, implicitly, melibiose. These criteria were more or less adequate for identification of an unknown yeast, if it happened to belong to one of the known species studied by these authors.

The inadequacy of these rather limited criteria became more and more obvious as the number of new species of yeast increased, and by 1951, Wickerham (1951) had introduced the use of more than 30 carbon compounds, whose use as sole C sources allowed the separation of many more yeast species. Wickerham himself was partly responsible for the isolation of many of these species, as he was involved in the isolation of yeasts from coniferous and deciduous trees, and, in particular, in the phylogeny of the genus Hansenula associated with plants and similar habitats. (For a description of the media used by Wickerham, and the carbon and nitrogen compounds tested, see Kreger-Van Rij 1984.)

However, investigators still found these criteria inadequate, and the search for new methods of identification and classification of yeasts continued. The macromolecules of the yeast cell suggested themselves as logical contenders for criteria for this purpose, and included the nucleic acids (RNA and DNA), proteins, and the polysaccharides of the cell wall. Investigation of all three classes of compounds went on nearly simultaneously, as it turned out.

2.1 Molecular Taxonomy

Chemical structures and characteristics of reassociation of nucleic acids and other macromolecules are used as taxonomic criteria in yeast.

The principles and methods of molecular biology are increasingly important in the identification, classification, and relationships of organisms from yeasts to apes and men.

The first methods investigated were reassociation of RNA and DNA, and the determination of the GC content of both genomic and mitochondrial DNA. Stenderup (see Kurtzman et al. 1983), in Aarhus, determined the degree of reassociation of RNA of one species with DNA from another, for numerous species of yeast. Nakase and Komagata (see Kurtzman et al. 1983), and Phaff and his coworkers (Kurtzman et al. 1983) determined the GC content of a large number of yeast species and grouped them according to their results. GC content was generally determined from the "melting point" of the genomic DNA. This method was most useful as a sort of "exclusion" taxonomy, whereby differences in GC content were an almost certain indication that the species were not identical, but the same GC content gave no indication whatever of possible relationships or similarity.

The reassociation method is now used by most yeast taxonomists who have equipment to determine relationships between different yeast species (Kurtzman et al. 1983; Kurtzman and Phaff 1987; Kurtzman 1990). The method is to a great extent an "all or nothing" one, as a high percentage of reassociation of the genomic DNAs usually means that the strains are conspecific, and low values mean that they are probably unrelated, though there have been intermediate values reported between groups of strains isolated, especially from cactus in different regions, and having the physiological characteristics of *Pichia membranaefaciens*.

This state of affairs being somewhat unsatisfactory, the sequences of the ribosomal RNAs (rRNA) and ribosomal DNAs (rDNA) have been investigated as taxonomic criteria (Guého et al. 1990; Kurtzman and Robnett 1991; Kurtzman 1992; Molina et al. 1992, 1993; Mendonça-Hagler et al. 1993). These highly conserved sequences allow the determination of the evolutionary distance between yeast species. Both methods are based on fragmentation of the rRNA or rDNA with restriction enzymes and separation of the fragments by gel electrophoresis for comparison. Repeated sequences such as those encoding ribosomal RNA are enhanced and appear in the gels as discrete bands, while random, single sequences are not visible. The patterns of repeated sequences are characteristic and can serve as a fingerprint for initial identification, and the DNA can be isolated for further investigation. When libraries of eletrophoretic patterns of restriction digests of genomic DNA of known yeast species are available, tentative identifications of unknown isolates of yeasts may be possible directly within a day or two. Coupled with the method described below using three primers in a PCR system (Fell 1993), the method should be even more powerful.

This system permits the determination of the phyletic distances between different yeast species and groups. It has been applied to the genera *Kluyveromyces* and *Clavispora* by Lachance and coworkers, (1990), and to other species (Kurtzman and Robnett 1991; Kurtzman 1992), and is satisfactory.

Methods for identifying yeast strains and for determining their relationships and phyletic distances between them are still undergoing development. Fell and Kurtzman (1990) have determined sequence variation in the large subunit (LSU) ribosomal RNA of a number of basidiomycetous yeasts inhabiting marine environments, and have shown that these sequences can be used for actual identification of different species.

Recently, Fell (1993) has described a method of particular interest for rapid identification of unknown yeast species, using three primers in a polymerase chain reaction. Two of the three primers delimit a segment of DNA, known to be universally present, at least in the group of yeasts which is believed to include the unknown one. The third primer is species-specific and found only in the species to which the unknown strain is believed to belong. Total genomic DNA is extracted from the yeast cells and subjected to PCR (polymerase chain reaction) in the presence of the three primers. If there is not a matching (target) sequence to the internal primer sequence, only the large universal sequence is amplified. However, if there is a matching internal target sequence, the large universal sequence is not amplified, and a shorter one, delimited by the internal sequence and one of the universal primers, is amplified and detected by gel electrophoresis.

Thus, the unknown strain can be matched with a high degree of accuracy to a type strain on the basis of its genomic DNA sequences. The method is rapid, and if DNA from the appropriate type strains is available, the unknown strain can be identified within 1 day after the cells have been grown to the desired stage. The power of these techniques is very great, and allows the determination of taxonomic relationships in yeasts with great accuracy.

Recently, a method has been described for fingerprinting genomes, using PCR with arbitrary primers (Welsh and McClelland 1990) and applied by Bostock et al. (1993) to typing of strains of *Candida albicans*. In comparison with typing with restriction fragment length polymorphisms (RLFPs) and electrophoretic karyo-typing by pulsed field gel electrophoresis (PFGE), RAPD (random amplification of polymorphic DNA) was convenient, quick, and economical, and had as good resolving power as PFGE.

Other methods used in taxonomic investigations of yeasts which are based on determination of the structure of their macromolecules include determination of the NMR (PMR; proton magnetic resonance) spectra of the cell wall mannans, and study of the soluble proteins of yeasts, particularly the spectra of isoenzymes. Gorin and Spencer (1970) took the PMR spectra of the mannans of most of the yeast species known then, and grouped the species according to similarities in the spectra. However, differences in the spectra were more significant than similarities, and showed that the isolates being compared were different strains or species. Similarities in the spectra did not indicate relationships. In addition, single-gene changes in some yeasts, at least, led to significant changes in the PMR spectrum, which they felt made the technique no more definitive than the determination of the utilization of a list of carbon compounds. However, Fukazawa and coworkers (in Kurtzman and Phaff 1987) showed that serologically distinct strains of *Candida tropicalis*, separated according to serotype, corresponded to those obtained using the proton magnetic resonance (PMR) spectra of the cell wall mannans, and that the character was stable. Therefore, this method of identification of yeasts is valid and valuable, and yields a characteristic fingerprint which allows comparison with the spectra of the mannans from any type strains of yeasts which are available.

The soluble proteins of the yeast cytoplasm can also be used for identification and classification. In another sense, these proteins are already used as criteria for yeast identification, as the proteins comprise the enzymes which are used by the organisms to metabolize the standard compounds for identification. The cellular proteins can be used in another way, to resolve problems in the classification and relationships among (yeasts such as the *Kluyveromyces*, by separation of the different isozymes by gel electrophoresis. Lachance and Phaff (in Kurtzman et al. 1983) determined the exo-β-glucanases formed by this group, and found that the results were not conclusive, since the enzyme was poorly conserved and the immunological distances between most species were too great for reliable determination of evolutionary relationships in this group of yeasts. However, if more than one alloenzyme was used, species of *Cryptococcus* could be separated taxonomically, and comparison of seven enzymes in a group of isolates of *Rhodotorula* and *Rhodosporidium* species permitted the taxonomic relationships among them to be established. The type strain of *Candida parapsilosis* was shown to be unrelated to the proposed perfect stage, *Lodderomyces elongisporus*. Gorin and Spencer (1970) had shown that the cell wall mannans of these species were dissimilar, and Kurtzman and Phaff (1987) had confirmed the absence of relationship by determination of DNA homologies.

The banding patterns of 14 enzymes were used by Holzschu (in Kurtzman et al. 1983) to determine the evolutionary relationships of a number of cactophilic strains of *Pichia*, and separated a number of apparently similar species.

Finally, by use of PAAG electrophoresis of 11 isofunctional enzymes, 13 species were established in the genus *Kluyveromyces*, as was determined earlier by DNA homology (Martini and Vaughan-Martini 1990). Thus, the investigation of alloenzymes is an extremely useful tool, especially for use in conjunction with DNA homology, in determining the taxonomic and hence evolutionary position of yeast species.

All of these methods, however, depend on determination of the genotype of the yeast indirectly, so are probably less powerful than those which determine the similarities and differences in the nucleic acids directly.

The use of DNA reassociation between the genomic DNAs of the unknown yeast and any type strains which appear as possibilities is extensively used to determine relationships among yeast species. Phaff and coworkers at Davis, Kurtzman at Peoria, and the Martinis at the University of Perugia in Italy, have shown, for instance, that the relationships among the *Saccharomyces* species, *S. cerevisiae*, *S. bayanus*, and *S. pastorianus* (*carlbergensis*) are complex, that the first two are not related at all as far as DNA reassociation is concerned, and that the third is apparently related to both of the other two, though not closely, and may be a natural hybrid between them. However, this method cannot be used to identify a single unknown yeast strain in isolation. For this, some other method, such as the comparison of PMR spectra with a file of spectra from type strains, must be used. Another method which may be used to yield a fingerprint is the separation of yeast chromosome by pulsed field gel electrophoresis. The CHEF (Contour-clamped Electrode Field) and TAFE (Transverse Alternating Field Electrophoresis) systems appear to give the best separations. Strains of *Saccharomyces cerevisiae* used in winemaking in France have proven to have distinctive and repeatable chromosome patterns, and the method is applicable to other yeast species. Other yeasts often have fewer and larger chromosomes, but separation by the TAFE system should reduce the number of possible species, for an unknown isolate, to a manageable number.

Thus, molecular taxonomy, in the form of DNA-DNA and DNA-rRNA homology, comparison of rRNAs and rDNAs, and enzyme electrophoresis of isofunctional enzymes, perhaps assisted by NMR spectroscopy of cell wall mannans and other major macromolecular components of the cell, has become the major tool in yeast taxonomy, even as the value of the traditional criteria of cell and spore morphology and utilization and fermentation of different carbon compounds has decreased. NMR spectra can also provide an independent fingerprint for placing a given yeast isolate in a small group of species, thus reducing the time and labor required for final identification.

Characteristics such as the presence or absence of pseudohyphae have been discarded in the latest editions of the taxonomic guides, and ascospore topography has recently increased, rather than decreased, the number of taxonomic puzzles to be solved. Although wartiness of the spores is consistent within the genus *Schwanniomyces*, in *Torulaspora* and *Debaryomyces* it could not be used as a distinguishing character between the two genera, and in *Pichia ohmeri* and *Yarrowia lipolytica* the spore shape depends upon mating type. Therefore, spore morphology cannot always be used even to prove differences between yeast species. It appears that only molecular taxonomy can give definite answers. Or so we think.

Other related methods which have been developed for use in determining taxonomic relationships include differentiation by staining with mixed dyes, and restriction analysis of mitochondrial DNA, which has been used with considerable success by Meyer and her coworkers (Su and Meyer 1991). The traditional methods of species determination according to fermentation and assimilation patterns are valuable for the initial identification of unknown yeasts. These have been used recently by Lachance and coworkers (1990), for instance, for identifying the new species, *Metschnikowia hawaiiensis*, and by Phaff and his colleagues (1992) for identification of *Pichia carabaea*, isolated from rotting cacti in the Caribbean region.

3 Yeast Species of Particular Interest

3.1 The *Saccharomyces cerevisiae* Group

This group is the one which has been most closely associated with humankind. *S. cerevisiae* and its near relatives have long been used by humans for breadmaking, brewing, winemaking, and similar purposes. So far, it is the best-understood and most-investigated of the yeast species. It is of great industrial value, having been used for many centuries in the traditional yeast-based industries, and it is now used in the production of heterologous proteins of many sorts. It has been claimed almost without exaggeration that the gene for any desired protein of pharmaceutical or industrial interest can be cloned and expressed in yeast. Yeasts, not only *S. cerevisiae*, have certain advantages over bacteria as vehicles for the production of heterologous proteins, especially hormones and other proteins of pharmaceutical interest.

This group includes three very closely related species, *Saccharomyces cerevisiae*, *Saccharomyces carlsbergensis*, and *Saccharomyces bayanus*. *S. carlsbergensis* is said by taxonomists to be properly named *S. pastorianus*, this specific name having priority according to taxonomic rules. Brewers and other industrial users of these yeasts will probably continue to use the name, *S. carlsbergensis*.

S. bayanus has been mostly associated with the wine industry, though at least some strains of this species ferment grape must rather slowly, and the yield of ethanol is low.

There has been some doubt whether these yeasts constitute separate species or whether they should all be considered as strains of *S. cerevisiae*. The name *S. carlbergensis* has been maintained to distinguish "top" and "bottom", or ale and lager yeasts. Strains labeled *S. cerevisiae* utilize melibiose, while the lager yeasts, *S. carlsbergensis*, normally do not. Investigations using DNA reassociation have confirmed that the name "*S. cerevisiae*" includes at least three closely related but separate species, *S. pastorianus* possibly being a natural hybrid between the other two (Kurtzman and Phaff 1987). *Saccharomyces paradoxus* is another recently added member of the group. The DNA relatedness between *S. cerevisiae*, *S. bayanus*, and *S. carlsbergensis* is shown in Fig. 2.1.

Saccharomyces diastaticus is completely interfertile with *S. cerevisiae*, its genomic DNA reassociates completely with that from *S. cerevisiae*, and the species is considered a glucoamylase-positive variant of the latter species. The three STA genes and the SSG (sporulation-specific glucoamylase) gene are homologous, and the STA genes may have originated by replication of the SSG gene and transposition of it to the other sites.

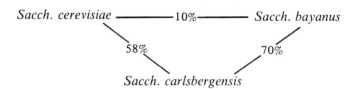

Relative genome size
Sacch. cerevisiae = 1.00
Sacch. bayanus = 1.15
Sacch. carlsbergenis = 1.46

Fig. 2.1. DNA relatedness between type strains of *Sacch. cerevisiae*, *Sacch. baynus*, and *Sacch. carlbergensis*. Comparison of genome sizes suggests *Sacch. carlsbergensis* to be a partial amphidiploid which may have arisen as a natural hybrid between *Sacch. cerevisiae* and *Sacch. bayanus*. (After Vaughan Martini and Kurtzman 1985)

Saccharomyces kluyveri is phenotypically similar to *S. cerevisiae* but does not form hybrids with it. Its chromosomes differ in number and size from those of *S. cerevisiae*, as shown by pulsed field gel electrophoresis. *S. kluyveri* has a very active α-galactosidase and metabolizes melibiose very strongly. The ascus is always tetrahedral, and the ascus wall is not dissolved by the enzymes used to release the spores of *S. cerevisiae*.

Saccharomyces exiguus differs from *S. cerevisiae* in being unable to metabolize maltose. A recent addition to the *S. exiguus* group is *Saccharomyces spencerorum* (A. Martini and A. Vaughan Martini, pers. comm.). The remaining species of *Saccharomyces*, namely *S. dairensis*, *S. servazzii*, *S. telluris*, and *S. unisporus*, differ from the previously mentioned species by their inability to utilize more than three or four carbon sources. None is of commercial importance.

3.2 The Osmotolerant Yeasts

These include the *Zygosaccharomyces* species, *Z. bailii*, *Z. bisporus*, *Z. cidri*, *Z. fermentati*, *Z. florentinus*, *Z. microellipsoides*, *Z. mrakii*, and *Z. rouxii*, and other osmotolerant species such as *Pichia farinosa*, *Hansenula anomala*, *Torulaspora delbrueckii*, *Debaryomyces hansenii*, a well-known spoilage agent on smoked sausages and other salt meats, and *Citeromyces matritensis*, which has been isolated from fruit in syrup, condensed milk, and other products of high sugar content. Of the four species and one variety of *Schizosaccharomyces* (*S. japonicus* var. *japonicus*, *S. japonicus* var. *versatilis*, *S. malidevorans*, *S. octosporus*, and *S. pombe*), only *S. japonicus* var. *japonicus* did not grow on 50% glucose medium. *Schizosaccharomyces octosporus* may be more osmotolerant, and is readily isolated from dried fruits such as currants, figs, raisins, and dates. *Candida mogii*, *Candida* (*Torulopsis*) *bombi*, *Candida* (*Torulopsis*) *bombicola*, *Candida* (*Torulopsis*) *lactis-condensi*, and *Candida* (*Torulopsis*) *magnoliae* were normally isolated from nectar, bumblebee nests, or in other ways were associated with bees of various types. *Candida* (*Torulopsis*) *lactis-condensi*, as the name indicates, was originally isolated from condensed milk.

3.2.1 The Genus *Zygosaccharomyces*

Members of the genus *Zygosaccharomyces* sporulate after conjugation of two haploid strains of opposite mating types. Two of the spores are found in one of the conjugating parents, and two in the other, giving the ascus a dumb-bell shape. *Z. rouxii*, *Z. bailii*, and *Z. bisporus* are the most highly osmotolerant of the species in this group, growing on 60% glucose-yeast extract agar. They are also spoilage yeasts, *Z. rouxii*, causing fermentation of "green" or unripe honey, in the comb or after extraction, and also causing explosive bursting of fondant-centered chocolates. Yeast growing in the fondant produced CO_2 until the chocolate burst. *Z. bailii* is resistant to preservatives, and grows readily in fruit juices and fruit drinks, for instance, which have been treated with benzoates, sorbates, and 1% acetic acid. *Z. bisporus* was isolated from spoiled, fermenting pickled cucumbers. All species of the genus *Zygosaccharomyces* are osmotolerant, though the remaining species are less so, and do not grow on medium containing 60% glucose.

The name *Z. rouxii* includes a number of strains previously known under other names. *Z. bailii* was originally named *Z. acidifaciens*, and was the first of the osmotolerant yeasts which was shown to produce polyhydroxy alcohols under appropriate conditions (Phaff et al. 1979).

3.2.2 The Genera *Pichia* and *Hansenula*

Pichia farinosa (originally named *Pichia miso*) was isolated from soy sauce and miso pastes in Japan by Onishi (1990). It is tolerant of the high concentrations of NaCl used in the manufacture of these products, and also produces reasonably high yields of xylitol (from xylose) and heptitols.

3.2.3 The Genus *Torulaspora*

The genus *Torulaspora* is characterized by small, round cells and the production of round ascospores. Some of the species are osmotolerant.

Torulaspora delbrueckii has been used in Japan as a baker's yeast, since its osmotolerance makes it useful for raising sweet breads and pastries. Its main disadvantage as a baker's yeast is the small size of its cells, which makes recovery of the biomass during production more difficult. The Japanese investigators therefore used protoplast fusion to construct diploid or polyploid strains which have larger cells. *T. pretoriensis* is also used as a baker's yeast.

T. delbrueckii includes the former species *Saccharomyces inconspicua*, *Saccharomyces rosei*, and *Saccharomyces vafer*. These species have been reduced to synonymy on the basis of reassociation of their DNAs.

3.2.4 The Genus *Debaryomyces*

This genus is now represented by nine species, all of which are osmotolerant. *Debaryomyces hansenii* is readily isolated from spoiled salt meats and fish. It produces polyhydroxy alcohols which act as compatible solutes and permit the cell to survive and grow under conditions of high osmotic tension.

3.2.5 Osmotolerant Yeasts of the Genus *Candida*

Most of these yeasts were originally classified in the genus *Torulopsis*. *Torulopsis bombi*, *Torulopsis bombicola*, *Torulopsis lactiscondensi*, and *Torulopsis magnoliae* do not form spores or pseudomycelium. They have now been transferred to the genus *Candida*.

All of these species produce polyhydroxy alcohols when grown in media having elevated osmotic tensions. *Candida* (*Torulopsis*) *magnoliae* produces erythritol, and the others varying amounts of glycerol.

Candida (*Torulopsis*) *bombicola* produces glycolipids composed of sophorose and a fatty acid hydroxylated on either the terminal or penultimate carbon atom. Lipids and vegetable oils, long-chain alkanes, and fatty acid esters added to the medium are converted directly to the corresponding hydroxy fatty acid and linked to the sophorose moiety to form the glycolipid molecule (Spencer et al. 1979).

The black "yeast", *Trichosporonoides oedocephalis*, is osmotolerant, and produces erythritol in submerged culture (see Sect. 3.10). Under some culture conditions, this species produces true mycelium, so it is not, strictly speaking, a yeast.

3.3 The Genus *Kluyveromyces*

The best-known members of the group are possibly the species *Kluyveromyces fragilis*, *Kluyveromyces marxianus*, and *Kluyveromyces lactis*. They form bean- or kidney-shaped spores, exhibit a high degree of thermotolerance, and usually ferment the disaccharide, lactose. They are often suggested for reducing the BOD of cheese whey and other dairy plant wastes. Some strains (or species) can metabolize inulin and can be used as a source of inulinase, and those strains which ferment lactose can be used for production of β-galactosidase. These and six other former species are now classified as varieties of *Kluyveromyces marxianus*.

Strains of *S. cerevisiae* which utilize lactose have been constructed both by transformation with plasmids carrying the genes coding for β-galactosidase and lactose permease and by protoplast fusion. These strains have not come into general use as yet, probably because of problems of instability. The strains of *Kl. marxianus* mentioned above are quite thermotolerant, which is useful in fermentations at elevated temperatures.

Other species recognized in the genus *Kluyveromyces* include *Kluyveromyces drosophilarum*, *Kluyveromyces thermotolerans*, *Kluyveromyces phaseolosporus*, *Kluyveromyces vanudenii*, *Kluyveromyces waltii*, *Kluyveromyces africanus*, and *Kluyveromyces polysporus*. Spore morphology is quite variable, ranging from bean- or kidney-shaped to crescentiform (lunate), round, or oval, and the spores are released early from the ascus, unlike those of *Saccharomyces* or *Zygosaccharomyces*. Some species of *Kluyveromyces* produce a diffusible red pigment related to pulcherrimin, which is produced by *Metschnikowia pulcherrima*.

Two species of *Kluyveromyces*, *Kl. africanus* and *Kl. polysporus*, form 16 or more spores per ascus, unlike the remaining species which form 1–4 spores/ascus, and which show other points of difference from the main group. Like other species of *Kluyveromyces*, these two species show early release of the spores.

3.4 The Methylotrophic Yeasts: *Hansenula polymorpha* (now *Pichia angusta*), *Pichia pastoris, Pichia pinus, Candida boidinii, Candida sonorensis,* and *Candida methanolica*

All of these yeasts can utilize methanol as sole carbon source. When the carbon source is switched from glucose to methanol (or alkanes), the cells become packed with microbodies (peroxisomes) which contain the enzyme, alcohol oxidase. This characteristic makes the methylotrophic yeasts ideal vehicles for production of heterologous proteins such as hormones – somatostatin, tumor necrosis factor, and many others. The gene coding for alcohol oxidase could be disrupted and the desired gene inserted at this point, and after switching the carbon source to methanol, the cells again were packed with microbodies stuffed with protein, only in this case the protein was the valuable one encoded by the heterologous gene which had been introduced. The major disadvantage was the slow growth of the yeast on methanol as substrate.

Hansenula polymorpha (*Pichia angusta*) and *Pichia pastoris* produce hat-shaped spores, usually two to four/ascus, and are amenable to genetic investigation. The species are homothallic, which causes some additional difficulty in genetic analysis. *Pichia pastoris* is a common component of the yeast microflora of slime fluxes of tree species. *Hansenula polymorpha* occurs in similar habitats, and has also been isolated from rotting tissue of cactus species.

Candida boidinii and *Candida sonorensis* do not form spores, and their genetics can only be investigated by use of mitotic recombination and protoplast fusion, or by recombinant DNA techniques.

All of the methylotrophic yeast species which have been isolated from rotting cactus tissue probably utilize the methanol liberated from pectins. Other methylotrophic species will undoubtedly be discovered.

3.5 *Yarrowia lipolytica, Candida bombicola, Candida bogoriensis,* and *Candida maltosa*

Yarrowia lipolytica was known for some time as the imperfect form, *Candida lipolytica*, until Wickerham et al. (1970) discovered mating reactions among the various strains. Finally, it was described by van der Walt and von Arx (1980) as *Yarrowia lipolytica*, the only species in this genus. *Candida maltosa* also metabolizes n-alkanes and utilizes very few sugars, but is unrelated to *Y. lipolytica*, the GC contents of their respective DNAs being greatly different.

Y. lipolytica metabolizes aliphatic, unbranched hydrocarbons, converting them to biomass. *Candida bombicola* also metabolized hydrocarbons, but converts them into sophorosides. This species was further described under osmotolerant yeasts (Sect. 3.2). *Candida bogoriensis* also forms sophorosides, but is not osmotolerant. The smut fungus, *Ustilago maydis*, forms sophorosides of similar structure, when grown in submerged culture in the yeast-like phase.

Y. lipolytica has been used by the British Petroleum Corporation for dewaxing of crude petroleum and simultaneous production of biomass, but the process is no longer in use.

3.6 *Candida utilis-Hansenula (Pichia) jadinii*

This species, under the name of Torula yeast, then *Torulopsis utilis*, has been used for many years as a fodder yeast supplement for animal and poultry feeds. It has been used to reduce the BOD of spent sulfite liquors from pulp mills using this process. The soluble sugars found in the waste liquors are converted to biomass (yeast cells), washed and dried, and added to feed mixes. The process was first used in Germany in 1914–1918, to alleviate the protein shortage then, and later in Wisconsin for disposal of pulp mill wastes. It is less used at present.

H. jadinii is the perfect form of *C. utilis*, and is a typical representative of the genus. It forms hat-shaped spores and utilizes nitrate as sole N source. It is thought to be weakly pathogenic, undesirable in a yeast used for food.

3.7 Lipolytic Yeasts: *Y. lipolytica, Candida ingens, Pichia mexicana,* and Others

The first of these has been discussed earlier (Sect. 3.5). *Candida ingens* and *Pichia mexicana* are so far the only lipolytic species isolated from cacti. *C. ingens* has a very strong lipolytic activity against monoesters of short-chain fatty acids (C_8, C_{10}, C_{12}) with triterpenes and sterols, which occur in relatively high concentrations in some species of giant cacti. The fatty acids thus released are readily metabolized by this yeast, and possibly by *P. mexicana* also.

Some species of *Rhodotorula* and similar yeast species will hydrolyze tributyrin and possibly other triglycerides as well.

3.8 Pentose-Fermenting Yeasts: *Pachysolen tannophilus, Pichia stipitis, Candida shehatae, Candida tropicalis,* and *Pichia farinosa*

Until recently, no yeasts which fermented xylose to ethanol were known. However, three species, *Pachysolen tannophilus, Pichia stipitis,* and *Candida shehatae,* are now known which ferment xylose. The fermentation of xylose is slow and the concentration of xylose tolerated is low.

Candida tropicalis and *Pichia farinosa* will convert xylose to biomass and xylitol, respectively. The latter is a pentitol which may be used as a sweetener in toothpastes, chewing gum, and similar products.

The best solution might be to construct a strain of *Saccharomyces cerevisiae* which was able to ferment xylose to ethanol. Attempts to do this have had limited success, as so far only the transfer of one or two genes of the xylose fermentation pathway has been reported.

P. stipitis and *P. farinosa* are tyical members of the genus, and form spores characteristic of the group. *Candida shehatae* is not the anamorph of *Pichia stipitis*, though the assimilation and fermentation patterns are the same.

Pachysolen tannophilus, originally isolated from tanning liquors, forms its spores in an ascus at the end of a long tube (ascophore) extending from the mother (vegetative) cell. The spores are hemispheroidal with a narrow ledge, up to four per ascus. Genetic analysis of both species of *Pichia* is possible, and mating strains have been isolated.

3.9 Amylolytic Yeasts: *Saccharomyces diastaticus*, *Schwanniomyces occidentalis*, *Saccharomycopsis fibuligera*, *Saccharomycopsis capsularis*, and Others

The ability to hydrolyze starch is reasonably widely distributed among the different genera and species of yeasts. *Saccharomyces diastaticus* is indistinguishable from *Saccharomyces cerevisiae*, the DNA homology being essentially 100%. It produces glucoamylase, but not a debranching enzyme, so that highly branched starches are hydrolyzed to a lesser degree and leave larger residues of unfermentable dextrins. *S. diastaticus* possesses three genes encoding glucoamylase, which are homologous with each other and with the SSG (sporulation-specific glucoamylase) gene.

Schwanniomyces occidentalis produces an α-amylase and ferments most starches completely. It has been suggested as a superyeast, capable of fermenting all starches, or as a source of starch-hydrolyzing enzymes (Ingledew 1987).

There is only one species in the genus, according to the DNA homology. Genetic analysis of this yeast is not possible by classical genetic methods, since it has no mating types, and forms only one spore (usually) per ascus.

Saccharomycopsis fibuligera and *Saccharomycopsis capsularis* form true mycelium as well as pseudomycelium, yeast-like vegetative cells, and hat-shaped spores. Classical genetic analysis is possible. *S. fibuligera* is the causal agent of "chalky" bread, and produces active amylases. It was used with *Candida utilis*, in the "Symba" process for treating wastes from potato processing plants and other starchy effluents. It produces several amylases.

Saccharomycopsis capsularis, besides producing amylases, also produces polyhydroxy alcohols, mostly D-arabitol.

A number of yeasts of the genus *Pichia* (now including *Hansenula*), such as *Hansenula capsulata*, also utilize and sometimes ferment starch, but so far none of these has been used as a source of genes encoding amylases.

3.10 Pectinolytic "Yeasts": *Aureobasidium pullulans*. Other Black Yeasts: *Trichosporonoides oedocephalis*

The ability to hydrolyze pectins is rare in the "true" yeasts, but occurs in the black "yeast", *Aureobasidium pullulans*, a yeast-like organism, widely distributed in nature, which occurs on flowers and other plant parts, in soil, and many other habitats. When first isolated, colonies are white or light rose-pink in appearance. They then develop rhizoid outgrowths, after which the colonies darken and usually turn dark green or black. A culture of *A. pullulans* often has a faintly sour and unpleasant odor. Recently, Gainvors and Belarbi (1995) have described pectinolytic activity in a strain of *Saccharomyces cerevisiae.*

Many strains have the ability to hydrolyze pectins and related polysaccharides, and produce a gum of their own, pullulan.

Trichosporonoides oedocephalis resembles a species of *Trichosporon* in producing budding yeast-like cells and true mycelium with arthrospores, and forms conidiospores on the swollen tips of aseptate sporophores. Colonies are deep black. In submerged culture, the species ferments glucose, galactose, maltose, and sucrose, and produces extracellular erythritol. It was isolated from a honeycomb, in Davis, California, and is probably osmotolerant.

3.11 Yeasts Metabolizing Aromatic Compounds. *Trichosporon cutaneum*

Numerous yeasts, ascomycetes as well as basidiomycetes, are able to utilize a number of aromatic compounds, including several flavonoids. Yeast species which metabolize flavonoids release CO during the process. Species of *Hansenula*, *Pichia*, and *Candida* can break down numerous aromatic materials. Numerous yeasts of this type have been isolated from citrus waste dumps.

Species of *Rhodotorula* and *Cryptococcus* can metabolize a wide range of aromatic compounds. Some species are quite selective in their ability to metabolize different isomers of hydroxybenzoic acids, one set of strains growing on one isomer only, and another set being able to metabolize only another isomer (Spencer et al. unpubl. data).

Trichosporon cutaneum is one of the most versatile yeasts in its ability to metabolize aromatic compounds. It will metabolize such fungicidal compounds as cinnamic acid and related compounds, when these are present as a cosubstrate in a medium containing glucose and an organic N source. The species is frequently isolated from waste waters and sewage treatment plants, where ability to metabolize such substrates probably gives it a competitive advantage.

3.12 Yeasts Forming Needle-Shaped Spores. The Genera *Coccidiascus,* *Nematospora,* and *Metschnikowia*

Many of these species are pathogenic for some plants or invertebrate animals. *Coccidiascus legeri* is a parasite of *Drosophila* species, infesting the intestinal cells. It produces eight fusiform ascospores intertwined in a helix.

Nematospora coryli is a pathogen of cotton, tomatoes, hazelnuts, citrus fruits, pomegranates, pecans, beans, soybeans, rice, and numerous other crop plants. *Ashbya gossypii*, a "yeast-like" organism, is a pathogen of cotton. This species also forms needle-shaped spores with a whip, in multispored asci (up to eight spores/

ascus), and both are transmitted by cotton-stainer bugs (*Dysdercus* spp.) and other hemipterous species.

Metschnikowia bicuspidata forms one or two needle-shaped spores/ascus. The two spores lie close to each other in the ascus and are often mistaken for a single spore. The spores are pointed and do not have a whip. The species has been isolated from saline waters and from brine shrimp, and from the salivary glands of a parasite of a snail. It is also parasitic on *Daphnia magna*. Three other species, *Metschnikowia australis*, *Metschnikowia krissii*, and *Metschnikowia zobellii*, have been isolated from marine sources.

Metschnikowia lunata, *Metschnikowia pulcherrima*, and *Metschnikowia reukaufii* are inhabitants of the nectar of flowers and honey stored in bumblebee nests, where the sugar concentration is not as high as in ripened honey in honeycomb of domestic bees. Recently, *Metschnikowia reukaufii* has been shown to inhibit pollen germination in infected flowers. *Metschnikowia hawaiiensis* was isolated from flowers of Hawaiian morning glory (Lachance et al. 1990).

The phylogenetic relationships among *Metschnikowia* species according to divergence among rRNA sequences showed that the aquatic species and terrestrial species form two groups within the genus. *M. lunata* and *M. hawaiiensis* do not belong to either group. *M. hawaiiensis* may belong to a new genus. All *Metschnikowia* species have a deletion in the large-subunit rRNA sequence, including nucleotides 434–483 (Mendonça-Hagler et al. 1993).

3.13 The Apiculate Yeasts

Cells are often lemon-shaped because of the mode of reproduction by bipolar budding. They include the genera *Hanseniaspora*, (imperfect form *Kloeckera*), *Nadsonia*, *Saccharomycodes*, *Schizoblastosporion*, and *Wickerhamia*. All of these yeasts, except for *Schizoblastosporion* and one species of *Nadsonia*, are fermentative. They are mostly associated with plants and soils, and *Hanseniaspora* and *Kloeckera* are often found in spoiled citrus and other fruits. They may have a weak pectinolytic activity. They are often found in winery fermentations which rely on the natural microflora as inoculum. In this case, *Kloeckera apiculata* usually predominates in the early stages of the fermentation, being replaced quite soon by *Saccharomyces cerevisiae*. The nectar-inhabiting yeast species, *Metschnikowia* (*Candida*) *pulcherrima*, is also frequently present, on the ripe grapes and in the must in the very early stages of the fermentation.

3.14 Other Basidiomycetous Yeasts: *Cryptococcus, Phaffia, Rhodotorula, Sporobolomyces, Bullera, Aessosporon, Filobasidium, Leucosporidium*, and *Rhodosporidium*

This group of yeasts is characterized by a cell wall having a much different composition and structure than is found in the ascomycetous yeasts. The perfect species in this group form teliospores rather than ascospores, and *Sporobolomyces* and *Bullera* form ballistospores which are discharged forcibly.

There are three major groups of basidiosporogenous yeasts, besides a number of anamorphs (imperfect forms) of the perfect species. These include the yeast-like genera of the Filobasidiaceae, *Chionosphaera*, *Filobasidiella*, and *Filobasidium*; of the teliospore-forming genera, *Leucosporidium*, *Rhodosporidium*, and *Sporidiobolus*; and yeast-like forms of Tremellales, the Sirobasidiaceae and Tremellaceae. The Sirobasidiaceae include *Fibulobasidium inconspicuum* and *Sirobasidium magnum*; the Tremellaceae, *Holtermannia corniformis* and nine species of *Tremella*. Most of the species in this group resemble *Cryptococcus* species in utilizing many carbon compounds including inositol, and in producing extracellular starch, though two or three do not. The two species of Sirobasiciaceae form ballistospores, while *Holtermannia corniformis* and the species of *Tremella* do not.

Many of the members of the genus *Rhodotorula* are imperfect forms of *Rhodosporidium* Banno, and form red carotenoid pigments, usually β-carotenes, though at least one species forms τ-carotene (Simpson et al. 1965, in Phaff et al. 1979). Species of this genus may be useful for production of carotenes. Some species of *Rhodotorula* also produce extracellular lipids and/or glycolipids. (Spencer et al. 1979).

Phaffia rhodozyma, the sole species in this genus, forms astaxanthine, a red pigment closely related chemically to β-carotene. It is valued as a food additive in fish farming, as it gives a red color to the flesh of trout and salmon. Chemically synthesized astaxanthine costs approximately US$2400/kg, so that a process for production of the pigment using this species of yeast may be commercially feasible. The maximum temperature for growth of the organism is approximately 26 °C, which is a disadvantage. It may be possible to develop strains growing at higher temperatures by genetic engineering.

As far as is known, there are no industrial uses for the products of any of the *Cryptococcus* species, though some strains produce proteases having milk-clotting activity. These may be useful in cheesemaking.

Most species produce extracellular gums which have unusual structures containing pentoses and glucuronic acid. *Cryptococcus neoformans* is a pathogen of warm-blooded animals, including man. Fortunately, it is of low infectivity, as it occurs in large numbers in the feces of birds, notably pigeons. These birds deposit amounts of *Cr. neoformans* reaching many tons per year in large cities.

Sporobolomyces, the third genus of imperfect red yeasts, occurs on the surfaces of leaves and other plant parts, and is among the few airborne yeasts. For this reason, it may occur as contaminants on microbiological media. *Leucosporidium* occurs in many aquatic environments, and has been isolated from Antarctic waters.

Other species of imperfect yeasts belonging in the basidiomycetous group include *Bullera* (ballistospores), *Malassezia* (a pathogenic form), *Sarcinosporon*, and *Sterigmatomyces*, which forms new cells at the ends of sterigma (short stalks). The genera *Candida*, *Cryptococcus*, and *Trichosporon* contain both ascomycetous and basidomycetous species.

Bullera singularis is unusual in producing tri- and tetrasaccharicdes when grown on lactose as a carbon source. It does not grow on galactose as a sole carbon source, and hydrolyzes lactose to its component sugars. The glucose is metabolized, and the galactose moiety is transferred to lactose *or another acceptor* by a galactosyl transferase. If suitable acceptors are provided, a whole series of galactose-containing oligosaccharides can be formed in this way, by the use of whole cells or cell-free extracts as sources of the enzyme (Gorin et al. 1964).

The genus *Candida*, which contains both basidiomycetous and ascomycetous yeasts, has its taxonomic boundaries set so wide that the genus acts as a repository for all the nonsporulating yeasts except those specifically defined as *not* being *Candida* species. It is a garbage can.

This makes it the largest genus, containing the greatest number of species, in the heterogeneous group of fungi called yeasts.

References

Bostock A, Khattak MN, Matthews R, Burnie J (1993) Comparison of PCR fingerprinting, by random amplification of polymorphic DNA, with other molecular typing methods for *Candida albicans*. J Gen Microbiol 139:2179–2184

Fell JW (1993) Rapid identification of yeast species using three primers in a polymerase chain reaction. Mol Mar Biol Biotechnol 2(3):174–180

Fell JW, Kurtzman CP (1990) Nucleotide sequence analysis of a variable region of the large subunit rRNA for identification of marine-occurring yeasts. Curr Microbiol 21:295–300

Gainvors A, Belarbi (1995) Detection method for polygalacturonase-producing strains of *Saccharomyces cerevisiae*. Yeast 11(15):1493–1499

Gorin PAJ, Spencer JFT (1970) Proton magnetic resonance spectroscopy – an aid in identification and chemotaxonomy of yeasts. Adv Appl Microbiol 13:25–89

Gorin PAJ, Spencer JFT, Phaff HJ (1964) The synthesis of β-galacto- and β-glucopyranosyl disaccharides by *Sporobolomyces singularis*. Can J Chem 42:2307–2317

Guého E, Kurtzman CP, Peterson SW (1990) Phylogenetic relationships among species of *Sterigmatomyces* and *Fellomyces* as determined from partial rRNA sequences. Int J Syst Bacteriol 40(1):60–65

Ingledew WM (1987) *Schwanniomyces* – a potential superyeast? CRC Crit Rev Biotechnol 5:159–176

Kreger-van Rij NJW (1984) The yeasts, a taxonomic study. Elsevier, Amsterdam

Kreger-van Rij NJW (1987) Classification of yeasts. In: Rose AH, Harrison JS (eds) The Yeasts, vol 1, 2nd edn. Academic Press, New York, pp 5–61

Kurtzman CP (1990) DNA relatedness among species of *Sterigmatomyces* and *Fellomyces*. Int J Syst Bacteriol 40(1):56–59

Kurtzman CP (1992) rRNA sequence comparisons for assessing phylogenetic relationships among yeasts. Int J Syst Bacteriol 42(1):1–6

Kurtzman CP, Phaff HJ (1987) Molecular taxonomy. In: Rose AH, Harrison JS (eds) The yeasts, vol 1, 2nd edn. Academic Press, New York, pp 63–94

Kurtzman CP, Robnett CJ (1991) Phylogenetic relationships among species of *Saccharomyces*, *Schizosaccharomyces*, *Debaryomyces*, and *Schwanniomyces*, determined from partial ribosomal RNA sequences. Yeast 7:61–72

Kurtzman CP, Phaff HJ, Meyer SA (1983) Nucleic acid relatedness among yeasts. In: Spencer JFT, Spencer DM, Smith ARW (eds) Yeast genetics, fundamental and applied aspects. Springer, Berlin Heidelberg New York, pp 140–166

Lachance M-A, Starmer WT, Phaff HJ (1990) *Metschnikowia hawaiiensis* sp nov, a heterothallic haploid yeast from Hawaiian morning glory and associated drosophilids. Int J Syst Bacteriol 40(4):415–420

Martini A, Vaughan Martini A (1990) Wine yeasts. In: Spencer JFT, Spencer DM (eds) Yeast technology. Springer, Berlin Heidelberg New York, pp 105–123

Mendonça-Hagler LC, Hagler AN, Kurtzman CP (1993) Phylogeny of *Metschnikowia* species estimated from partial rRNA sequences. Int J Syst Bacteriol 43(2):368–373

Molina FI, Inoue T, Jong S-C (1992) Ribosomal DNA restriction analysis reveals genetic heterogeneity in *Saccharomyces cerevisiae* Meyen ex Hansen. Int J Syst Bacteriol 42(3):499–502

Molina FI, Shen P, Jong S-C (1993) Validation of the species concept in the genus *Dekkera* by restriction analysis of genes coding for rRNA. Int J Syst Bacteriol 43(1):32–35

Onishi H (1990) Yeasts in Fermented foods. In: Spencer JFT, Spencer DM (eds) Yeast technology. Springer, Berlin Heidelberg New York, pp 167–198

Phaff HJ, Miller MW, Mrak EM (1979) The life of yeasts. Harvard University Press, Cambridge

Phaff HJ, Starmer WT, Lachance MA, Aberdeen V, Tredick-Kline J (1992) *Pichia carabaea*, a new species of yeast occurring in necrotic tissue of cacti in the Caribbean area. Int J Syst Bacteriol 42(3):459–462

Spencer JFT, Spencer DM, Tulloch AP (1979) Extracellular glycolipids of yeasts. In: Rose AH (ed) Economic microbiology, vol 3, Secondary products of metabolism. Academic Press, New York, pp 523–540

Su CS, Meyer SA (1991) Characterization of mitochondrial DNA in various *Candida* species: isolation, restriction endonuclease analysis, size, and base composition. Int J Syst Bacteriol 41(1):6–14

Van der Walt J-P, von Arx JA (1980) J Microbiol Serol 46:517

Vaughan Martini A, Kurtzman CP (1985) Deoxyribonucleic acid relatedness among species of the genus *Saccharomyces sensu stricto*. Int J Syst Bacteriol 35:508–511

Welsh J, McClelland M (1990) Fingerprinting gemones using PCR with arbitrary primers. Nucleic Acids Res 18:7213–7218

Wickerham LJ (1951) Taxonomy of yeasts. Techn Bull 1029, US Dept Agr, Washington DC

Wickerham LJ, Kurtzman CP, Herman AI (1970) Science 167:1141

CHAPTER 3

Ecology: Where Yeasts Live

J.F.T. Spencer and D.M. Spencer

1 Introduction

The ecology of yeasts is important for its practical value and as the fundamental basis of the evolution of yeast species. New species of yeast are formed due to selection pressures exerted by the environment, and observation of the similarities and differences of the yeasts found in a particular environment can assist the investigator to see evolution as it happens.

The study of the ecology of yeasts involves not only the effects of the physical environment on the yeast cells, but also the interactions of the yeast species with the other microorganisms and with larger organisms, especially insects. The latter often feed on yeasts and act as transmission vectors for the different yeast species (Do Carmo-Sousa 1969).

Bacteria, fungi, and insects, including fly larvae and boring beetles, degrade dead and dying plant material to forms which can be metabolized by yeasts. Yeasts do not photosynthesize, and do not fix nitrogen, and depend on the other organisms to convert their substrates to available forms. Very few yeast species can hydrolyze hemicelluloses, cellulose, and pectins, and free the monomers for utilization by the yeasts. Much of this is done by bacteria, especially of the genus *Erwinia* (soft rots, etc.), and fungi. Likewise actinomycetes produce antibiotics, organic acids and other metabolic products which may inhibit yeast growth.

Yeasts are a major food source for both the larval and adult stages of numerous insects. *Drosophila* species are closely associated with yeasts in slime fluxes, bumblebees with certain yeast species (*Candida* spp., *Metschnikowia reukaufii*, *Metschnikowia pulcherrima*, and other species found in the nectar of flowers, and numerous *Hansenula* and *Pichia* species, with bark beetles (*Ips*, *Dendroctonus*). Recently, we have observed that some yeast species are introduced into citrus fruits (grapefruit, oranges, lemons, limes) by other fruit fly species (South American fruit fly and Mediterranean fruit fly; *Anastrepha fraterculus* and *Ceratitis capitata*). Most yeasts are rapidly digested by the flies after ingestion, but some cells probably survive long enough to pass through the gut. Others are transmitted during oviposition (Spencer et al. 1992).

J.F.T. Spencer/D.M. Spencer (eds)
Yeasts in Natural and Artificial Habitats
© Springer-Verlag Berlin Heidelberg 1997

2 Methodology

The study of the distribution of yeasts in nature requires the isolation of the yeast species present in a given microenvironment, and their identification and enumeration. The initial stages of collection are shown in Figs. 3.1–3.4. Samples of the materials containing the yeasts are then pretreated, if necessary, by rehydration of dry material, weighing out a small amount into a dilution medium, water, peptone solution, or other such solution, and plating out an aliquot on acidified yeast extract-malt extract agar, adjusted to pH 3.7 with 1N HCl. Peptone water, 0.1%, phosphate buffer, or physiological saline solution are often used as diluents for isolation and enumeration of yeasts in spoiled foods.

Growth of bacteria is suppressed by the acidification of the medium or with antibiotics. Isolation on potato-glucose agar, a weak medium, containing 0.003% Rose Bengal or 1% oxgall, is sometimes used to inhibit growth of fungi, which grow in small tufts rather than spreading over the surface of the plate. Enrichment cultures in shaken medium may be used, and the fungal pellet filtered off after a few days' growth. However, the use of enrichment techniques makes cell counts impossible.

Separation of the yeasts present into colony types can be done, and relative numbers of each determined. An estimate of the numbers of each yeast species present can be made if enough isolates are tested. The initial identification can still only be done for large numbers of yeasts by determination of their fermentative and assimilative abilities, since methods for determination of DNA base composition, DNA reassociation, NMR spectra of yeast mannans, and ribosomal nucleic acid (rRNA and rRDA) sequences require much more time and labor. Mitochondrial DNA can be extracted from yeast isolates, digested with restriction enzymes, and the patterns obtained by gel electrophoresis compared. Methods based on amplification of DNA by PCR are faster and are rapidly coming into use.

The use of traditional methods of identification of isolates of unknown species is limited by the fact that numerous yeast species have phenotypes so similar that they cannot be differentiated by these methods. For instance, some yeasts, phenotypically assigned to the species *Pichia membranaefaciens*, have DNAs showing no reassociation with the DNA of this species. Use of traditional methods only would not reveal the fact that these were new, and probably evolving, species.

Specialized groups of yeasts can usually be isolated by use of selective media, containing particular carbon or nitrogen sources. Methanol-utilizing, inositol-utilizing, melibiose-utilizing, sulfur-containing amino acid-requiring, glucosamine-utilizing, N-acetylglucosamine-utilizing, xylose-utilizing, and nitrate-, creatine-, creatinine-, nitrite-, lysine-, and ethylamine-utilizing species can be isolated by using these compounds in the medium, sometimes by replica plating the colonies on the original isolation plates to the desired selective medium. Media having a high sugar content can be used for isolation of osmotolerant yeasts: *Zygosaccharomyces rouxii*, *Zygosaccharomyces bailii*, other *Zygosaccharomyces* species, *Schizosaccharomyces octosporus*, and other species of *Schizosaccharomyces*. Species of *Brettanomyces* (perfect form, *Dekkera*) grow slowly even on complex media, but better if the medium is supplemented with thiamin, $10\,mg\,l^{-1}$. They produce relatively large quantities of acid, which can be detected by using Custer's medium, which contains $CaCO_3$. Thermotolerant and also thermosensitive yeasts can be isolated by taking advantage of these characteristics; thermotolerant species by incubating the isolation plates at temperatures high enough to prevent growth of other yeasts, and thermosensitive species by replica plating the original yeast colonies to plates incubated at low and high temperatures.

Osmotolerant yeasts may not necessarily require salt in the growth medium. A few, such as some strains of *Metschnikowia bicuspidata*, from infected brine shrimp, required the presence of NaCl absolutely. Some sugar-tolerant yeasts, such as *Zygosaccharomyces bisporus* and *Zygosaccharomyces rouxii*, will grow on "normal" media, but after adaptation. This is not true of all strains of these yeasts, since we have isolated these species, as well as *Schizosaccharomyces octosporus*, directly from their original habitats. We did not, however, use a synthetic medium.

A few yeasts are identified by their specialized habitats, as was found in investigations of yeasts associated with cactus rots.

In general, any yeast which is able to metabolize a particular compound not used by other yeast species can be isolated by using this compound as sole carbon, nitrogen, or other source in the isolation medium.

3 Evolution in Action in Yeasts: Darwin's Finches and the World of the Yeasts

"The thirteen species of Darwin's finches form a closely-knit genealogical group of widely divergent life styles – a classic case of adaptive radiation into a series of roles and niches that would be filled by members of several bird families in more conventional, and crowded, continental situations". S.J. Gould, *The Flamingo's Smile*, Essay 23 (1985).

The same principle probably holds for yeasts. The interaction between the yeasts, their substrates in nature, and the host plants and other organisms, especially insects, may lead to evolutionary changes in the yeasts and their insect vectors, and possibly in the host plants as well.

The yeast-*Drosophila*-cactus-*Cactoblastis cactorum* system (Barker and Starmer 1982) is well known. The first invaders of damaged cactus tissue are probably pectolytic bacteria. These attract *Drosophila* species, which are vectors for the different cactophilic yeasts. The yeasts invade the rotting tissue of numerous species of the giant cacti, where they metabolize sugars and related compounds released by bacterial action, the *Drosophila* feed on the yeasts and lay their eggs in the rotting tissue, the cactus eventually dies, and the *Drosophila* move on to other rotting cactus, carrying the yeasts with them, and the cycle is repeated.

Species of *Opuntia* cactus were introduced into Australia in the 19th century and became weeds. There began a search for insects which might become predators of these spiny weeds. *Cactoblastis cactorum*, a small moth native to northern Argentina was successfully employed. The adult lays its eggs in a case attached to the *Opuntia* plant, and the caterpillars eat their way into the tissues of the cactus, destroying it.

Unfortunately, the insect was introduced to Florida, Texas, the Bahamas, and adjacent regions, where it attacks commercial varieties of *Opuntia*.

Other imported insects included *Drosophila* species, especially *Drosophila buzzatii*, and their accompanying yeasts.

All of the members of the complex, cacti, *Drosophila* spp., yeasts, and *Cactoblastis cactorum*, appear to be evolving into separate species.

Numerous cactophilic yeasts have been isolated, including *Pichia opuntiae*, *Candida sonorensis*, and *Cryptococcus* (now *poropachydermia*) *cereanus*. One group consists of species which have the phenotype of, but which are not, *Pichia membranaefaciens*, according to the DNA reassociation values. Of these, *Pichia heedii* is a homogeneous species, while both *Pichia amethionina* and *Pichia opuntiae* are split into two varieties, *P. amethionina* var. *pachycereanus* and *P. amethionina* var. *amethionina*, and *P. opuntiae* var. *thermotolerans* and *P. opuntiae* var. *opuntiae*. These yeasts were isolated from cacti of the family Pachycereinae (senita, saguaro, and cardon) and Stenocereinae (organ pipe, agria, and cina; see Table 3.1).

These species are phenotypically almost indistiguishable, but the degrees of reassociation of the genomic DNAs are greatly different. Apparently, the changes in the genomic DNA came first, the changes in the phenotype last.

Another yeast species, *Pichia cactophila*, is not restricted in its habitat to any particular species of cactus. It contains strains which phenotypically resemble each other very closely. There are no physiological parameters known for separating and grouping the strains into varieties. However, one divergent group within this species has DNA with only 34% homology with that of the other group of *P. cactophila* strains. This group is probably on the way to speciation as well.

Saccharomyces cerevisiae is familiar to most people. These yeasts have long been associated with human activities, but it has only recently been shown that it comprises at least three closely related species, one of which, *Saccharomyces pastorianus* (*Saccharomyces carlbergensis*), probably originated as a natural hybrid between the other two, *Saccharomyces cerevisiae* and *Saccharomyces bayanus*

Table 3.1. Specific and common names of giant cacti associated with yeasts

Specific name	Common name
Lophocereus schottii	Senita
Carnegia gigantea	Saguaro
Pachycereus pringlei	Cardon
Stenocereus thurberi	Organ pipe
Stenocereus gummosus	Agria
Stenocereus alamocensis	Cina
Backebergia sp.	–
Opuntia sp.	Prickly pear

Fig. 3.1A,B. Sampling rot pockets in cactus. **A** Cardon (species not known). (Prof. H.J. Phaff, University of California, Davis). **B** *Opuntia* sp. (Luis Ducrey, PROIMI, Tucumán, and Dr. M.-A. Lachance, University of Western Ontario, Canada)

(Fig. 3.1). A fourth species, *Saccharomyces paradoxus*, has recently been added. A fifth, *Saccharomyces kluyveri*, is phenotypically very similar to the first four, but is unrelated.

DNA base composition and DNA homology are the first two methods developed using molecular biology for showing evolutionary relationships and distances. Other techniques in molecular taxonomy have recently been added: determination of similarities in ribosomal RNA (Kurtzman and Phaff 1987) and rDNA (Shen and Lachance 1993). Methods using PCR amplification of genomic DNA sequences have been developed (Fell 1993; Bostock et al. 1993) and permit very rapid determination of similarities between yeast strains. A method of determining similarities between mitochondrial, DNA sequences has also been determined (Su and Meyer 1991).

4 Yeast Habitats

4.1 Plants and Insects

The association between yeasts and plants, insects, soil, and waters is extensive. Yeasts grow on the surfaces of leaves, in slime fluxes, in nectar of flowers, in decaying fruits, and other plant parts (where the nutrients metabolized by the yeasts are often released by enzymes produced by bacteria or insects), and in many

Table 3.2. Yeasts found in the crop and gut contents of bees collecting nectar and pollen. (California 1981)

Yeast species	Collecting: (no. of isolates)	
	Nectar	Pollen
Hanseniaspora guilliermondii	21	5
Hanseniaspora uvarum	3	7
Kluyveromyces fragilis	5	7
Metschnikowia reukaufii	46	8
Metschnikowia pulcherrima	19	6
Pichia terricola	15	4
Candida guilliermondii var. *guilliermondii*	15	6
Candida krusei	27	1
Candida pulcherrima	16	9
Other *Candida* spp.	63	15
Cryptococcus albidus var. *albidus*	14	12
Cryptococcus macerans	7	6
Cryptococcus laurentii	4	8
Kloeckera apiculata	15	1
Kloeckera apis (perfect stage, *H'spora guilliermondii*)	23	7
Rhodotorula glutinis	23	15
Torulopsis apicola	9	17
Torulopsis magnoliae	18	17
Torulopsis stellata	7	6

other plant-associated habitats. Insects carry yeasts from plant to plant. *Drosophila* species carry the yeasts found in slime fluxes and decaying fruits and vegetables. Bees, especially bumblebees, carry yeasts from flowers to the honey reserves in the nests (Spencer et al. 1970; Table 3.2). Bark beetles carry many yeast species, especially *Hansenula* and *Pichia* species, into their tunnels. Lacewing flies (*Chrysopa carnea*) have a yeast symbiont which provides essential amino acids to the adult, and some species of *Drosophila* and several species of anobiid beetles also have yeasts as intracellular symbionts.

Grapefruit and other citrus fruits infested with larvae of the South American fruit fly and the Mediterranean fruit fly (*Ceratitis*) are normally infected with a somewhat restricted range of yeasts (*Kloeckera* sp., *Pichia fermentans*, and *Pichia kluyveri*) and *Geotrichum candidum* (Spencer et al. 1992).

4.1.1 Yeasts on Surfaces of Leaves of Trees, Shrubs, Herbaceous Plants, and Grasses

Yeasts are often found on the phylloplane, mostly basidiomycetes (*Cryptococcus, Rhodotorula*, and especially *Sporobolomyces*, and their anamorphs). These yeasts may obtain their nutrients from the leaf exudates. Many tropical plants harbor a large population of nitrogen-fixing bacteria on the leaf surface, which may produce nitrogenous compounds, available to the yeasts. Others produce sticky, sugary materials, metabolized by the yeasts. Decaying leaves may also provide nutrients for a population of yeasts.

The yeasts inhabiting leaf surfaces include both their normal inhabitants, and accidental inhabitants. The yeasts normally inhabiting leaf surfaces of both broad-leaved plants and monocotyledonous grasses are usually basidiomycetes; *Sporobolomyces* spp., *Rhodotorula, Cryptococcus, Aureobasidium, Candida*, and some species of *Candida* and *Torulopsis* (now *Candida* also). *Candida javanicis, C. foliarum, C. diffluens, C. bogoriensis* (which produces a sophoroside glycolipid), and, among the ascomycetous yeasts, *Hansenula* species.

In New Zealand, *Cryptococcus laurentii, T. ingeniosa* (a basidiomycetous species), *Rhodotorula marina, Rhodotorula graminis*, and *Sporobolomyces roseus* were normal inhabitants of grasses. Normal soil yeasts did not comprise a significant fraction of the species inhabiting the phylloplane on grasses. The phylloplane yeasts increased in relative numbers in the soil, especially in the upper layers, in the summer season when total yeast numbers were high, but decreased later in the year (Table 3.3).

The yeast population found on elm leaves in California was quite different. Elm leaves are covered with a sticky, sugary exudate in late summer and early fall, which influences the yeast population on the leaves. Fermentative yeast species [*Torulaspora delbrueckii, T. pretoriensis*, and *Kloeckera veronae* (*Kloeckera thermotolerans*)] were among the species isolated.

Yeasts occurring on surfaces, whether those of leaves, skins of fruit, stems, flower petals and other flower parts other than in the nectaries, or of nonliving

Table 3.3. Genera of yeasts, associated with plants, insects and soils

Genus	Habitat and no. of species					
	Exudates	Cacti	Bark beetles	Other borers	Soils	Water/ sediment
Aureobasidium						1 –
Ambrosiozyma			5			
Candida		6	7	10	5	12, 4
Clavispora		1				
Cryptococcus		. 1			1	3, 2
Citeromyces	1					
Debaryomyces					1	1, 1
Geotrichum		1				1 –
Hansenula		1	2	8	1	1, 1
Hanseniaspora						1 –
Kluyveromyces	1					
Metschnikowia						1 –
Pichia	10	10	1	8	5	
Prototheca		1				1 –
Rhodotorula						5, 5
Saccharomyces	2				1	2 –
Saccharomycopsis			1	1		
Schizosaccharomyces	1					
Sporobolomyces						1 –
Sporopachydermia	1					
Sterigmatomyces			1			
Torulopsis			5	3	4	
Trichosporon				3	1	1, 1

materials such as soil particles, are usually bound to the surface by some type of gummy material. To determine accurate numbers of yeasts, the gum must be dissolved and the cells released. This can be done by shaking the material in liquid, percolating to exhaustion, or by ultrasonic treatment. Shaking or sonication in liquid in the presence of microbeads may improve the completeness of release. Too severe ultrasonic treatment may reduce the number of viable cells.

4.1.2 Yeasts Occurring in Other Plant Exudates
(Slime Fluxes and Necrotic Tissues, and Giant Cacti)

Slime fluxes occur where the tree has been mechanically injured. Other exudates, caused by disease (e.g., black knot in red oak) mostly support a more restricted yeast population of a more definite composition. A slime flux may persist for several years without necessarily causing damage to the tree. The flowing sap may form a thick gum, probably containing microbial polysaccharides.

Slime fluxes support a large and varied population of bacteria, yeast, protozoa, fly larvae, and other insects, but few filamentous fungi. Some of the earlier species of yeasts isolated from slime fluxes include *Saccharomycodes ludwigii* and species of *Nadsonia*, from exudates of birch and oak. Other species included *Pichia pastoris*, *Trichosporon penicillatum* (*Geotrichum penicillatum*), and the colorless alga, *Prototheca zopfii*, from elm, and *Arthroascus javanensis* from oak. Most of these species are widely distributed on tree exudates in European Russia, Japan, and the west coast of North America, though *Nadsonia* species were not found in Alaska or the west coast. Numerous new yeast species have been isolated from similar habitats in South Africa, Australia, Indonesia, and a few from South America.

Numerous other species of *Pichia* and *Hansenula* and their imperfect forms have been isolated from slime fluxes of deciduous trees and from conifers.

Phaffia rhodozyma and *Trichosporon pullulans*, which have low maximum growth temperatures (26 °C), are found in fluxes (birch) in Alaska, Japan, and eastern Europe, especially the Moscow region. *T. pullulans* occurs in fluxes of deciduous trees in the same regions. The yeast-like organism, *Dipodascus aggregatus*, was found only in Japan.

The identification of yeasts isolated from nature requires the determination of DNA base composition and DNA reassociation and other tools of molecular biology. Two series of isolates from alder (*Alnus rubra*) and from cottonwood (*Populus trichocarpa*) in British Columbia, Canada, contained three strains at first identified as *Hansenula wingei*, and five (from *P. trichocarpa*) as *Hansenula nonfermentans*. However, neither group showed any significant degree of reassociation of the DNAs with the type strains of *H. wingei* or *H. nonfermentans*, and they were eventually named *Hansenula alni* and *Hansenula populi*. *Hansenula dryadoides* is phenotypically similar to *H. nonfermentans*, but the DNA base compositions differ by 13%, so the two were not conspecific. Several strains resembling *Torulopsis molischiana* were isolated from poplar and birch, in the Pacific Northwest, and were also shown by DNA analysis to be a new species.

The yeasts isolated from rotting tissue of giant cactus species by Phaff and his coworkers were similar. These yeasts were first identified as *Pichia membranaefaciens*, but their DNAs were not homologous with *P. membranaefaciens* DNA, or with DNAs from groups isolated from different regions, so they were placed in new, distantly related species.

The genomic DNAs of these and other yeast genera should probably be correlated with the chemical structures of the cell wall mannans, determined by NMR spectroscopy. The known species of *Schizosaccharomyces*, *S. pombe*, *S. octosporus*, and *S. japonicus* show almost no reassociation among the nuclear DNAs, but the NMR spectra of the H-1 region of the cell wall mannans are almost indistinguishable. The relationship between the genomic DNAs and the components of the cell wall, and of the mannans in particular, may be taxonomically significant.

The ability to metabolize triterpenes may be a guide to the identity of yeasts in these habitats.

4.1.3 Yeasts Isolated from Tanning Liquors

Tanning solutions are extracts of the wood, bark, and leaves of trees and other woody plants, having a high tannin content. Oak, acacia, sumac, chestnuts, and quebracho are used to make liquors for tanning hides and skins in the leather industries. The solutions also contain such wood extractives as sugars, nitrogenous compounds, and mineral salts. The pH is generally in the range 4.0–5.9, permitting yeast growth. Several yeast species tolerant of tannins and similar compounds have been isolated from these solutions. These include *Pichia chambardii, Pichia rhodanensis, Pichia strasburgensis, Pichia toletana, Pichia pseudopolymorpha* (*Debaryomyces pseudopolymorpha*), *Candida boidinii* (a methylotrophic yeast), *Torulopsis* (*Candida*) *molischiana*, and the new species of *Pichia; P. abadieae, P. adzetii,* and *P. pignaliae.* The xylose-fermenting yeast species, *Pachysolen tannophilus,* was isolated from tanning liquors.

Most of these species have not been found in other habitats. They probably occur in small numbers on the trees used as sources of tanning solutions, and multiply only in the tanning liquors (Phaff and Starmer 1987). Slime fluxes of trees seldom contain yeasts found in tanning solutions.

4.1.4 Yeasts Associated with Cacti

Yeasts are associated with *Drosophila* species in pockets of rotting tissue in cacti, in the southern United States, Mexico, Central America, and some countries of South America (see Fig. 3.2). Investigation of this important ecosystem began with the work of Professor Herman J. Phaff of the University of California at Davis and his students and collaborators (Barker and Starmer 1982; Phaff and Starmer 1987).

Numerous yeast strains identified as *Pichia membranaefaciens* were first obtained. Comparison of the genomic DNAs of these strains showed that the isolates consisted of several groups of similar strains, but apparently of distinct species. Most of these were *Pichia* species or imperfect forms of them. Some methylotrophic yeasts were obtained, and at least one new species, *Sporopachydermia cereana,* (originally *Cryptococcus cereanus*) was found. This species utilized *myo*-inositol, but was renamed when Rodriguez de Miranda discovered the ascosporogenous stage in 1978 (Kreger-van Rij 1984). Strains of *Clavispora* spp., *Candida ingens, Candida mucilaginosa,* and *Candida sonorensis,* (a methylotrophic, cactus-specific imperfect species), were isolated from rotting cactus tissue (Lachance et al. 1986).

The isolates originally identified as *Pichia membranaefaciens* proved to belong to at least six distinct, cactus-specific species. *Pichia cactophila* was the most widespread. The others included *Pichia pseudocactophila, Pichia opuntiae, Pichia amethionina, Pichia heedii, Pichia antillensis, Pichia deserticola,* and a new species, *Pichia mexicana,* superficially resembling *Candida tenuis.*

The distribution of these yeasts among different species of cacti is governed partly by the composition of the host plants. Cacti of the subtribe Stenocereinae

Fig. 3.2A,B. Typical habitats for plant associated yeasts; cactus, *Opuntia* sp., attacked by cactus moth (*Cactoblastis cactorum*)

contain triterpene glycosides (TTGs), but not alkaloids. Saguaro, senita, *Opuntia* spp., and *Backebergia* spp. contain alkaloids, but not triterpene glycosides. *P. heedii* is sensitive to TTGs and is mostly found in senita and saguaro cacti. *P. cactophila* can apparently live in rots of all species of cacti. Strains of *P. heedii* inhabiting senita cactus utilize citrate, while those occurring in saguaro do not. Nevertheless, the DNAs of the two yeast strains show very high homology, and the two sets of strains are completely interfertile.

The difference in habitat may be caused ultimately by the requirements and preferences of the *Drosophila* species utilizing the different cacti. Senita rots contain a sterol which is required for development and fertility of *Drosophila pachea*, and also contain alkaloids toxic to *Drosophila nigrospiracula* and other species (*D. mettleri*) which breed in saguaro cacti. This may limit the citrate-positive strains of *P. heedii* to the senita cactus.

Similarly, *P. amethionina* var. *pachycereinae* is resistant to triterpene glycosides, while *P. amethionina* var. *amethionina* is not, so that the two varieties are limited in their distribution in the same way. In this case, a single temperature-sensitive gene is responsible for the difference in sensitivity. These two varieties show a DNA homology of approximately 65% and are closely related, but differ enough to indicate that they may be at the beginning of the speciation process.

The two varieties of *Pichia opuntiae*, *P. opuntiae* var. *opuntiae* and *P. opuntiae* var. *theromtolerans*, have progressed even farther along the path to speciation. The DNA homologies are much lower, reassociation being approximately 28%, and there is no interfertility whatsoever, neither zygotes nor asci being formed. Species status has been proposed for the two varieties (Barker and Starmer 1982).

Some cactophilic yeasts can metabolize methanol and other unusual compounds. *C. sonorensis* metabolizes methanol, which is probably a product of bacterial hydrolysis of cactus pectin. *Candida boidinii* and *Hansenula polymorpha* can also metabolize methanol, but these species are found less often in cactus rots. *P. cactophila*, the most widely distributed yeast associated with cacti, metabolizes D-glucosamine but few other carbon compounds. The presence of this sugar has not been demonstrated in rotting cactus tissue. *Candida ingens* is strongly lipolytic (*P. mexicana* is less so), liberates short-chain fatty acids (C_8–C_{12}) from triterpenes and sterols, and grows well on them. Since fatty acids having these chain lengths are often toxic to other yeasts, this may give *C. ingens* a selective advantage. Finally, *Cr. cereanus* (*Sporopachydermia cereana*) grows readily on *myo*-inositol. Inositol is a constituent of phytic acid in plants, and can be released by the action of phytase. Few other yeasts, except *Cryptococcus* species, can metabolize inositol, and these seldom occur in cactus rot pockets.

We have described the yeasts and their association with rotting cactus tissue and the *Drosophila* species which also inhabit this environment in some detail, because, first, it is well known and illustrates the evolutionary events leading to the appearance of new yeast species, and second, because of the economic importance of the system yeast-*Drosophila*-cactus-*Cactoblastis cactorum* (cactus moth). A better understanding of this and similar systems is important from the point of view of both its ecological value and its practical applications (Barker and Starmer 1982).

4.1.5 Yeasts and the Flowers and Fruit of Higher Plants

4.1.5.1 Nonpathogenic Yeasts

Most flowers provide two distinct habitats for yeast. The outer surfaces and the inside of the flower near the opening harbor a yeast population similar to that of leaf and stem surfaces; *Sporobolomyces* (*Sp. roseus*), *Rhodotorula* (*Rh. rubra, Rh. graminis, Rh. pilimanea, Rh. glutinis, Rh. aurantiaca*) *Cryptococcus* (*Cr. laurentii, Cr. macerans, Cr. infirmo-miniata*), and similar species, and especially *Aureobasidium pullulans* (Last and Price 1969). The nectaries, containing nectar of a high sugar content but low in nitrogen, have a different yeast population. Both groups of yeasts are probably transported by bees and other insects. Nectar collected by bumblebees has a yeast flora peculiar to this habitat (Spencer et al. 1970).

Two yeast species, *Candida pulcherrima* and *Candida reukaufii*, are fermentative yeasts which are frequently isolated from nectar and bumblebee nests. They are imperfect, haploid strains of the genus *Metschnikowia*, a genus previously known only from isolates from marine organisms and saline lakes (Pitt and Miller 1968). Yeasts of this genus have one or two needle-shaped spores in a large, elongated ascus. Golubev isolated another species, *Metschnikowia lunata* (Kreger-van Rij 1984), and recently, Lachance isolated *Metschnikowia hawaiiensis* from morning glory flowers in Hawaii (Lachance 1990).

Yeasts in fruit are usually found in decaying material, though there is a complex symbiotic association between a species of wasp, a variety of fig, a yeast, and a bacterium. The fruit of the fig is a hollow bag-like structure (synconium), lined with the true flowers. The inedible variety, caprifigs, produces three crops of fruit a year. The first spring crop is the only one which produces pistillate and staminate flowers. The fig wasp, *Blastophaga psenes*, lives inside the fig and carries pollen from the first profichi crop to the next crop so that the cycle repeats itself from year to year. The wasp also carries with it, besides pollen, a microbial population consisting of a bacterium, *Serratia plymuthica* (now *Serratia ficaria*) and a yeast species, *Candida guilliermondii* var. *carpophila* (do Carmo-Sousa 1969; Phaff and Starmer 1987).

The association is important commercially because of the necessity of "artificial" pollination of the Smyrna-type commercial fig variety. The fruit of this variety has only pistillate flowers in the synconium, and pollination is required for proper setting of the fruit. The fruit is pollinated by hanging bags of profichi caprifigs, containing wasps and pollen, in the orchards. The wasps, with their associated yeast and bacteria, invade and pollinate the Smyrna figs. Fortunately, while these microorganisms grow to a certain degree within the fig, they do not cause spoilage. This is an example of a complex ecological interrelationship and its practical application. This ecosystem includes man as a beneficial rather than a malefic component (do Carmo-Sousa 1969; Phaff and Starmer 1987).

Spoilage at this stage in figs occurs when the presence of the two microorganisms makes the fruit more attractive to *Drosophila* species, which introduce spoilage yeasts such as *Hanseniaspora* and *Kloeckera* species, *Torulopsis stellata*, and *Candida krusei*. These species produce ethanol, which is converted to acid by acetic acid bacteria. Dried fruit beetles (*Carpophilus hemipterus*) may invade fallen figs or attack sound fruits, introducing other molds and yeasts. The young green figs may be colonized by a specialist yeast, *Torulopsis fructus*. The mature fig exocarp may become infected with other yeast species, carried to the fruit by a number of *Drosophila* species. *Pichia membranaefaciens*, *Hanseniaspora valbyensis*, *Hanseniaspora uvarum*, *Kloeckera apiculata*, and *Candida sorboxylosa* have been identified among isolates from souring figs.

The apiculate yeasts are frequently identified as associated with spoilage in various fruits. *Kloeckera apiculata* has been isolated from injured strawberries, where it causes a leaky type of rot, and from grape clusters. It formed a significant part of the population of fermenting grape juice in the early stages of winemaking. Other yeasts associated with spoiled fruits include *Torulopsis fragaria* from strawberries and black currants, *Torulopsis pustulata* from black currants, *Torulopsis multis-gemmis* from raspberries, and *Torulopsis bacarum* from black currants, strawberries, and raspberries. *Cryptococcus albidus* and *Crytpococcus laurentii* have been isolated from strawberries, but may have originated in the soil, or possibly from the flowers.

Hanseniaspora and *Kloeckera* species (the second genus comprises the imperfect forms of the first) also play a considerable part in the fermentation of decaying tomatoes and oranges, accounting for more than half of the isolates from *Drosophila* species feeding on fermenting tomatoes, and from the tomatoes themselves. The other two species found were *Pichia kluyveri* and *Candida krusei*. In decaying oranges, one-third of the isolates consisted of strains of *Kloeckera apiculata*, the other species found being *Pichia fermentans*, *Pichia kluyveri*, and *Candida krusei*.

We found that decaying grapefruit, oranges, and mandarins investigated in the citrus-growing region of Tucumán, Argentina, contained mostly species of apiculate yeasts, (probably a species of *Kloeckera*), *Pichia* species, and *Geotrichum candidum*. The decaying regions of the fruit were infested with larvae of one of the larger fruit flies rather than *Drosophila*. Two of these species, the South American fruit fly and the Mediterranean fruit fly, are endemic to the region. We collected a number of the larvae, allowed them to pupate, and identified the emerging adults as the Mediterranean fruit fly (Spencer et al. 1992).

Cauliflower can be considered either a flower or a fruit, by a stretch of the imagination. A yeast, *Candida fusiformata*, has been isolated from macerated cauliflower florets.

Cactus fruits, used where the large commercial varieties of *Opuntia* cactus are grown, are also invaded by yeasts if injury occurs. However, the yeasts are not the species occurring in rotting stem tissue, but are usually species of apiculate yeasts, associated with other types of decaying fruit (*Hanseniaspora* and *Kloeckera* species).

4.1.5.2 Yeasts as Plant Pathogens

Few yeasts can be classed as plant pathogens. *Nematospora coryli* and *Ashbya gossypii* attack a number of valuable crops. They both produce needle-shaped ascospores having a whip-like part, attached at one end of the spore. They cause boll disease of cotton, being carried to the plant by bugs and other sucking insects. The disease causes staining of the cotton fibers and infection of the seeds.

The yeast cells occur in the stylet pouches of the bugs, especially cotton-stainer bugs, *Dysdercus* spp. Tomatoes, beans, citrus fruits, pomegranates, soybeans, rice, pecans, and other crops are damaged by *Nematospora coryli*, introduced by leaf-footed bugs, *Leptoglossus zonatus*, pumpkin bugs, green and brown stink bugs (*Acrosternum hilare* and *Euschistus servus*), and the rice stink bug, *Oebalus pugnax*.

Apiculate yeasts occurred in rotting citrus fruits in Tucumán, especially *Pichia* species and *Geotrichum candiddum*. The vectors were probably the Mediterranean fruit fly and the South American fruit fly.

The actual pathogenicity of the apiculate yeasts and possibly of the *Pichia* species, to citrus fruits, has not been determined. The rind and flesh of the infected fruits become much softened and discolored.

4.1.6 Other Associations of Yeasts and Insects

Yeasts may be found is association with numerous insect species; bark beetles (*Ips*, *Dendroctonus*, *Scolytus*), ambrosia beetles (having as symbionts ambrosia fungi and various yeasts), lacewings (*Chrysopa* spp.), bees, *Drosophila*, anobiid beetles, and other insects. The yeasts may be commensals, food for the larvae or adults, or they may be symbionts, intracellular or extracellular, of the insect.

4.1.6.1 Yeasts and Bark Beetles

Yeasts themselves, associated with bark beetles, do not damage the tree as some of the other fungi carried by these insects do. Bark beetles make galleries in the cambial layer, between the bark and the wood of coniferous trees, and carry filamentous fungi fo the genera *Ceratocystis* and *Graphium* (the imperfect forms) as well as yeasts into the tunnels. These fungi include causal agents of Dutch elm disease, blue-staining fungi, and others, which are responsible for killing of the trees, discoloration of the wood, and similar damage.

The yeasts may be food for both beetles and larvae. *Pichia pinus*, *Hansenula capsulata*, and *Candida silvicola* (perfect stage *Hansenual holstii*) are the principal yeast species associated with beetles infesting pine. Other yeasts [*Pichia haplophila*, *Torulopsis* (now *Candida*) *nitratophila*, *T. melibiosum*, *Candida curvata*, and *Rhodotorula crocea*] are present in much smaller numbers. *P. pinus* and *C. curvata* are not found in the phloem of *Pinus jeffreyi*. This species of pine contains n-heptane and oleoterpenes, n-heptane probably being inhibitory to these two yeast species.

Beetles of the species *Scolytus* infest firs (*Abies*, *Pseudotsugae*) and infect them with other yeasts. The yeasts are found with the beetles at all stages of the life cycle, from the early larval stage to the winged, migratory adults. Phaff and coworkers (Phaff and Starmer 1987) first found the species *Pichia scolyti* associated with two species of fir-engraver beetles (*Scolytus* spp.). *Hansenula wingei* and *Candida tenuis* have also been found there. (*H. wingei* has since been placed in synonymy with *Hansenula canadensis*.) *Candida tenuis* has been isolated from numerous coniferous tree and many species of beetles, as well as from cactus rots. These isolates may be strains of the same species, or strains of different species having similar physiological reactions. Some strains originally identified as *C. tenuis* were haploid, heterothallic strains of *Pichia mexicana*. These may constitute another group of phenotypically similar but genetically different yeast species, which, like those isolated from rotting cacti, may be another example of evolution and speciation in yeasts.

Other species of yeast associated with bark beetles are *Hansenula americana*, *H. bimundalis*, *Pichia amylophila*, and *Pichia mississippiensis*, isolated from species of pines and fir.

4.1.6.2 Yeasts and Ambrosia Beetles

Ambrosia beetles (Lymexylidae, Platypodidae, Scolytidae) are similar to the bark beetles. They do not live in the cambium or phloem layer, but bore directly into the heartwood of both coniferous and deciduous trees. Many of the species have specialized structures (mycangia) in which spores of the ambrosia fungi and yeast cells are carried. The spores and yeast cells germinate and grow on the walls of the tunnels, and beetles and larvae feed on the ambrosia fungi and yeasts. Yeasts such as *Ambrosiozyma* spp., one species of *Saccharomyces*, two of *Hansenula* and one of *Pichia*, numerous species of *Candida* and *Torulopsis*, and one of *Sterigmatomyces* are known (Phaff and Starmer 1987). *Hansenula capsulata*, *Debaryomyces vanrijii*, *Candida oregonensis*, and *Candida shehatae*, and one strain of *Torulopsis nitratophila* were associated with beetle larvae (Buprestidae) in Quebec province, Canada. (Pignal, in Phaff and Starmer (1987). Phaff and Starmer list many more species of *Pichia*, *Hansenula*, *Saccharomycopsis*, *Candida*, *Torulopsis*, and *Trichosporon*, from wood-boring insects or their borings or phloem frass. Phaff and Starmer (1987) give four tables listing yeasts associated with plants, and one of yeasts from soils, most of which belong to the genera *Pichia*, *Hansenula* and *Candida* (see Table 3.3).

4.1.6.3 Yeasts Associated with Lacewings

Larvae of lacewing flies (*Chrysopa carnea*) feed on aphids (plant lice) and other soft-bodied insects. The adults feed mostly on "honeydew" produced by the aphids, and require several essential amino acids which are present only in low concentrations in the honeydews. The adults carry a yeast symbiont in the crop, which apparently provides any additional nutrients required for life and egg production. The larvae do not carry the symbionts, and the adults probably acquire the yeasts, possibly by the sharing of a drop of regurgitated liquid before mating. The yeast grows quite vigorously in the insect crop. They are sometimes visible as agglutinated spheres, regurgitated from the crop in areas where oviposition has taken place. The yeast resembles *Torulopsis multisgemmis*, but is probably a new species, as the DNA base composition is different.

4.1.6.4 Bees and Yeasts; Yeasts and Bees

The association of yeasts with many species of bees is to be expected, since bees collect nectar, an ideal fluid for yeast growth and metabolism. Yeasts are responsible for spoilage of "unripened" honey stored in the combs of domestic bees (*Apis mellifera*), since the sugar concentration in the fresh nectar is of a lower sugar concentration than in the "ripened" honey. Honey stored by other species (bumblebees, *Bombus* spp.) is less concentrated, and is a good source for isolation of yeasts collected from flowers or normally associated with the bees (Spencer et al. 1970). Osmophilic yeasts predominate in isolates from honeys of various sources.

Relatively large numbers of yeasts can be isolated from honey, nectar, and the crop and gut contents of domestic (honey) bees (Table 3.2). Some of these species, (*Cryptococcus* spp. and other basidiomycetous yeasts), may not be normal residents in this habitat, though *Rhodotorula* species occur in pollen-collecting and nectar-collecting workers. However, *Metschnikowia hawaiiensis*, *Metschnikowia pulcherrima*, and *Metschnikowia reukaufii*, *Torulopsis apicola*, and *Torulopsis magnoliae*, are normal nectar-inhabiting species and are found in the crops of honey bees. The higher concentration of yeasts in the crop contents of the nectar-collecting bees than in the original nectar may be the result of evaporation of the nectar to a higher yeast (and sugar) concentration, and multiplication of the yeasts within the bee's crop. In California, the yeasts varied in numbers and species, with the season of the year, because of the changing flora and the nature of decaying fruit visited by the bees. (Phaff and Starmer 1987).

The highly osmotolerant species, *Zygosaccharomyces rouxii*, *Zygosaccharomyces bailii*, and *Saccharomyces rosei* (*Torulaspora delbrueckii*) were not found.

Bumblebees (*Bombus* spp.) do not store their honey in a comb, as do domestic bees, but in small wax pots in the nest. The numbers of bees per nest is much smaller than in the case of domestic bees, and the bumblebees' habits differ as well.

The honey is more dilute than ripe honey produced by domestic bees, and permits the survival and growth of a wider range of yeast species.

Spencer et al. (1970), collected honey samples from artificial nests in southern Alberta, Canada, and determined the numbers and species of yeasts found in the honey.

We found *Candida* (*Metschnikowia*) *reukaufii, Candida* (*Metschnikowia*) *pulcherrima, Torulopsis apicola,* a few isolates which we identified tentatively as *Debaryomyces phaffii,* and numerous isolates of a new species, *Torulopsis bombicola.* This species produces high yields of an extracellular glycolipid (a hydroxy fatty acid sophoroside) when certain triglycerides, long-chain fatty acid esters, or long-chain liquid hydrocarbons are added to the culture medium as precursors. These substances are surface-active, and may have important industrial uses.

The total numbers of yeasts in honey produced by bumblebees varied greatly, from a few cells per milliliter to many thousands. It is not known whether they are normal inhabitants of nectar and are acquired by the bees from there, or whether they are normally associated with the bees, live in the nest with the overwintering queens, and are carried to the flowers the following spring. Our initial isolation of *T. bombicola* was from nectar, but we never obtained the species from samples of nectar in such numbers as we found in the samples of bumblebee honey.

4.1.6.5 *Yeasts Associated with* Drosophila, *Particularly in the Gut*

Yeasts found in the gut contents and crop of *Drosophila* species are seldom the same species as those isolated from plants where the insects feed. So far, the feeding sites and the yeast species eaten have not been determined.

The Californian investigators have studied this habitat for some time. *Drosophila* species are dependent on yeasts for their survival. Their associated yeasts may be dependent on *Drosophila* for their own survival, by distribution to fresh substrate material.

Domestic species of *Drosophila* (*Drosophila melanogaster, Drosophila simulans*) feed on decaying tomatoes and other fruits and vegetables, but wild species may feed principally on yeasts. The yeasts most frequently isolated in California by Phaff and Starmer (1987), associated with *Drosophila,* were species of *Kluyveromyces* [*Kl. veronae* (*Kl. thermotolerans*), *Kl. drosophilarum, Kl. fragilis,* and *Kl. apiculata*]. *Hansenula polymorpha* was also isolated. Three new species, *Kluyveromyces drosophilarum, Kluyveromyces dobzhanskii,* and *Kluyveromyces phaseolosporus,* were isolated during the investigation. These yeasts, isolated from the crops of the flies, were quite different from those found in the probable feeding sites.

The same results were obtained in a survey of the flies in the Yosemite region (altitude 1300–3000 m). The yeast species isolated from the crops of the flies were completely different from those found in the feeding sites; the larvae fed on entirely different yeasts than the adults did.

The yeast species isolated most frequently were *Saccharomyces montanus*, *Saccharomyces cerevisiae* var. *tetrasporus*, *Kluyveromyces veronae*, *Kluyveromyces Drosophilarum*, *Hansenula polymorpha*, *Kloeckera apiculata*, *Kloeckera magna*, *Torulopsis stellata*, and *Candida krusei*. The new species *Kluyveromyces wickerhamii*, *Saccharomyces kluyveri*, and *Trichosporon aculeatum* were found, but were rare. *Kluyveromyces fragilis* was the only species found in the south but not in central California. Later, Kodama also isolated strains of *Saccharomyces montqnus* and *Saccharomyces cerevisiae* var. *tetrasporus* from tree exudates in Japan. These yeasts were common in the isolates from the crops of *Drosophila* in California. Phaff postulated that the *Drosophila* species investigated by Kodama in Japan fed directly on the slime exudates.

Drosophila are selective concerning the yeast species they feed on. They will not feed on *Cryptococcus* species, but adults and larvae alike feed readily on *Saccharomyces cerevisiae*.

In Africa, a primitive species of drosophilid (*Lissocephalus* sp., is associated with a yeast species, (*Torulopsis fructus*), a bacterium, and perhaps a fig wasp, the latter being a pollinator. The flies lay their eggs near the opening of the fig (ostiole), and the first-instar larvae enter the immature fig, possibly carrying the yeast cells. The larvae feed on the yeast and bacteria until during the third instar, when they leave the fig, drop to the ground, and pupate. The events are similar to the sequence occurring in the caprifig, but the yeast species found in the fig, the fly, and its larvae, are completely different.

Rowan (*Sorbus* sp.) berries and the stinkhorn fungus (*Phallus impudicus*) are unusual yeast substrates occurring in England. The stinkhorn fruiting body has an unforgettable appearance and stench, though the rowan berries are inoffensive. Nevertheless, two species of *Drosophila*, *D. subobscura*, which feeds on rotting vegetation and sap flows, and *D. phallerata*, which is a specialist in fungi, require yeasts occurring on these two substrates for larval growth and development. The particular yeast species seems not to be important.

We have described the yeasts associated with cactus earlier; these yeasts are essential to several species of *Drosophila* that feed on them. Generally, the flies feed on the yeasts peculiar to the host plant where the larvae matured. The most common yeast species found in the senita cactus are *Pichia heedii*, *Candida ingens*, and *Cryptococcus cereanus* (*Sporopachydermia cereana*), and constitute 89% of the yeasts isolated from *Drosophila pachea*, which feeds and oviposits mainly in this cactus.

Similarly, the major yeasts found in rot pockets of the agria cactus are *Pichia cactophila*, *Pichia amethionina* var. *amethionina*, and *Candida sonorensis*, and these yeasts consitute 76% of the isolates from adults of *Drosophila mojavensis*, which uses the agria cactus as a host plant. The major yeasts found in decaying saguaro and cardon cacti were the major species isolated from *Drosophila nigrospiracula* and *Drosophila mettleri* infesting these cacti. *Drosophila* larvae are discriminating and somewhat fastidious in their tastes and feed preferentially on particular yeast species.

The metabolic activities of some of the yeasts may affect the survival of the larvae. The organpipe cactus produces fatty acids of medium chain length, which inhibit the growth of larvae of *Drosophila mojavensis*. The yeast *Candida ingens* metabolizes these fatty acids, removing them from the decaying cactus tissue, and permitting the development of the of *Drosophila* larvae.

It is interesting to note that the larvae of *Drosophila* species whose adults are host specialists, feeding on a single host plant, feed on most of the yeast species present, while the larvae of those species in which the adults are generalists (polyphagic feeders) are selective in the yeast species eaten.

4.1.6.6 Yeasts as Intracellular Symbionts

A number of species of insects, mostly adults and larvae of anobiid beetles, carry yeasts as true intracellular symbionts, in specialized cells or structures (mycetocytes and mycetomes) (Phaff and Starmer 1987). One yeast species, *Coccidiascus legeri*, has been observed in intestinal epithelial cells of two species of *Drosophila*, *Drosophila funebris*, and *Drosophila melanogaster*. This yeast produced two spindle-shaped, flattened ascospores. It could not be cultivated on artificial media in the absence of the insect cells, so its physiological characteristics are unknown.

Three yeast species (*Torulopsis ernobii*, *Torulopsis karawaiewi*, and *Torulopsis xestobii*), have been cultivated without the anobiid host. The yeasts appear to synthesize vitamins and (possibly) sterols essential to the host beetle, and the beetle may supply amino acids (asparagine or glutamic acid) and RNA to the yeast. The taxonomy and physiology of these yeasts have not been investigated.

4.2 Soil and Water

Soil and water receive all or most of the nutrients they contain from external sources such as plants and animals, including man. Man-made materials are now one of the largest sources of yeast nutrients as well as a major pollution problem. Soils, too, receive increasing amounts of man-made pollutants, and have been used deliberately as a repository for human wastes of all kinds. Sanitary landfills are a prime example of such an abuse of the soil, so much so that suitable sites for landfill have become scarce enough to show that the method is unsatisfactory at best. The changing nature of the yeast population of soils at landfill sites has not been investigated, but should be. Some studies have been made of yeasts from soils taken from dumps used for disposal of citrus-processing wastes.

4.2.1 Soils

Phaff and Starmer (1987) observed that the yeast microflora of soils depends on the type of nutrients reaching them. Orchard soils enriched by decaying windfalls may support populations of species of *Hanseniaspora* and *Kloeckera*. Many other yeast species are transients, residing temporarily in the soil. Yeasts in soil are frequently dependent on the action of other microorganisms to break down the recalcitrant polymers found there.

However, some yeasts are permanent residents in the soil. These include *Schwanniomyces occidentalis, Lipomyces starkeyi, Lipomyces lipofera, Lipomyces kononenkoae, Lipomyces tetrasporus, Lipomyces anomalus, Schizoblastosporion starkeyi-henricii, Cryptococcus*, and a few species of *Rhodotorula* (*Rh. minuta*) and *Sporobolomyces* (*Sp. salmonicolor*), all of which have been isolated only from soil.

Some soils of the antarctic continent have been investigated for the presence of yeasts, species of *Cryptococcus* often predominating. Eight species of *Cryptococcus*, one of *Rhodotorula*, one of *Sporobolomyces*, four of *Leucosporidium*, one of *Candida* (*C. scottii*), and numerous isolates of *Aureobasidium pullulans* were obtained from samples taken during Antarctic surveys. The latter species and *Cryptococcus albidus* were the most common species isolated. Genera of yeasts found in soil and other habitats are listed in Table 3.3.

Yeast species isolated from soils in other regions included two species of *Saccharomyces*, one of *Torulaspora*, four of *Kluyveromyces*, five of *Hansenula*, eight of *Pichia*, thirteen of *Candida* (including five originally described as *Torulopsis*), one of *Nadsonia* (*N. commutata*), one of *Debaryomyces* (*D. nepalensis*), and one of *Trichosporon* (*Tr. terrestre*).

The microbial flora of a prairie soil was investigated under the auspices of the International Biological Program in western Canada in 1971. Spencer and colleagues (in Phaff and Starmer 1987) found that the numbers of yeasts in the semiarid prairie soils ranged from 1.4×10^4 cells/g soil in the surface layers, and from 600 to 1800 cells/g in the subsurface layers. The yeast species found were all basidiomycetes; *Cryptococcus albidus, Cryptococcus laurentii*, and *Cryptococcus terreus*, which constituted the majority of the species isolated, and some examples of *Cryptococcus dimennae, Rhodotorula minuta, Sporobolomyces pararoseus*, and *Sporobolomyces salmonicolor* were also isolated. Most of these yeast could utilize the flavonoid rutin, and sometimes other flavonoids. The numbers of yeasts were greater in summer. The presence or absence of capsules on yeast species inhabiting soils, especially of the arid and semiarid types, may influence the ability of the yeast cells to survive low moisture conditions.

In the soils from citrus waste dumps, in Florida and California, total numbers of viable yeast cells reached 7.64×10^4 in the Florida samples, and 84×10^4 in the Californian samples. The numbers of yeast cells in soils from an onion field in California and from lemon and orange groves in Florida, which were used as controls, were less than 1×10^4 cells/g in California and somewhat lower in the Florida samples. The total numbers of yeasts in the Florida samples were generally lower than in those from California. The number of yeasts in soil depends greatly on the amount of available nutrients, and is increased by the addition of metabolizable substances. Most of the yeasts found in soil were nonfermentative (aerobic) species, and so were present in greater numbers in the surface layers of the soil Phaff and Starmer 1987).

The yeast species found were similar in all of the soils investigated. The Matador soils of Saskatchewan, semiarid, lower in organic matter and moisture, were populated almost exclusively by species of *Cryptococcus* and other species of nonfermentative, basidiomycetous yeasts. The Californian and Florida soils, having higher moisture and nutrient contents, supported a wider range of species, though the numbers of each species were generally smaller. Numbers of fermentative yeasts generally amounted to a quarter to half of the numbers of nonfermentative species. We isolated numerous cultures of the nonfermentative species, *Trichosporon cutaneum*, *Cryptococcus albidus*, and *Cryptococcus laurentii*, with smaller numbers of various *Rhodotorula* species also being present. Fermentative species were *Hansenula californica*, *H. mrakii*, and *H. saturnus*, *Pichia kluyveri* (notably), *Saccharomyces cerevisiae*, *Saccharomyces rosei* (*Torulaspora delbrueckii*), and some *Candida* species, which were present in insignificant numbers.

The yeasts we found in all three soils (Matador prairie, Florida and California orchard, and onion field soils) were similar. We did not find *Kluyveromyces* species. We did not investigate the yeasts of tropical soils, which might cover a wider range of species.

4.2.2 Waters – Natural and Polluted

"Natural" waters may not exist today. There may be slightly polluted, moderately polluted, and heavily polluted waters, depending on what pollutants are introduced. Yeasts can usually be found in all three types of waters.

Numbers of yeasts found in relatively unpolluted waters are quite similar in salt waters and fresh waters, and depend on the amount of nutrients washed into both from terrestrial sources, as well as the numbers of yeast cells brought to the lake or ocean directly. The numbers of yeasts/ml in waters decreases as the distance from the shore increases, the availability of nutrients, and the numbers of yeasts washed into the water decreases.

The numbers of yeasts may vary quite widely, from a few cells/ml in relatively unpolluted waters, to more than a million/ml in effluents. The distribution of species may also be different. Such a variation was observed in the numbers and species of yeasts in lakes, rivers, sewage systems, and waste treatment basins of a paper pulp mill in the province of Saskatchewan, Canada (Hagler and Ahearn 1987).

In lakes and rivers, yeast numbers ranged from 20 cells/l in remote areas, and 300–400 cells/ml in lakes near recreational facilities, to 4000 cells/l near towns. The numbers of yeasts in flowing water decreased rapidly downstream from towns and cities, falling to the same level as those in water in uninhabited areas within a few miles downstream. Just downstream from Saskatoon, a city of approximately 100000 at the time, the numbers of yeasts were about 126000 cells/l, which was the same as at a sampling point just below the discharge from a pulp mill near Prince Albert, Saskatchewan. A few miles downriver, the numbers had fallen to 20–30 cells/ml.

The numbers of red yeasts have been suggested as an index of pollution of such waters. However, these values proved to be too variable for this, as they ranged from 0 to 50% of the total yeast population.

The species distribution in the rivers and lakes in general was similar, and included four species of *Hansenula*, three of *Pichia*, a few isolates of *Saccharomyces cerevisiae*, seventeen of *Candida*, two of *Cryptococcus*, five of *Rhodotorula*, five of *Torulopsis*. and one of *Trichosporon cutaneum*.

In the pulp mill treatment basins themselves, and in the Saskatoon sewage treatment plant, the numbers of yeasts were much higher, reaching 115×10^4 cells/l in the pulp mill treatment basins, and 9800×10^4 cells/l in the raw domestic sewage entering the Saskatoon treatment plant. The numbers in the pulp mill basins increased steadily later in the year, from February to July, and then decreased. The numbers were highest in the upper end of the primary settling basins, and decreased to a very small number at the point of discharge into the North Saskatchewan river. The numbers of yeasts in the municipal sewage treatment plant were highest in March. The numbers of yeasts were lower in the primary treatment basin. There were no yeasts in the final treatment basin, which was chlorinated.

There was a very wide range of species isolated from the pulp mill treatment basins, including five species of *Hansenula*, five of *Pichia*, one of *Kluyveromyces*, ten of *Candida*, four of *Cryptococcus*, three of *Rhodotorula*, three of *Torulopsis*, two of *Trichosporon* (*Tr. cutaneum* and *Tr. penicillatum*), and one of *Sporobolomyces*. Species isolated from the sewage disposal plant were one of *Debaryomyces*, two of *Kluyveromyces*, one of *Pichia*, one of *Saccharomyces* (*S. cerevisiae*), two of *Candida*, one of *Cryptococcus*, two of *Trichosporon* (the same species as were found in the pulp mill settling basins), and one of *Aureobasidium* sp. *Trichosporon cutaneum* is sometimes considered an indicator of pollution from human sources, and is also known to metabolize a rather wide range of aromatic compounds. It was isolated from both the sewage plant and the pulp mill basins.

The numbers of yeasts were greater in the municipal sewage treatment plant than in the pulp mill treatment basins, but the numbers of species isolated were much higher in the later site.

Numbers of yeasts in water everywhere where low in fresh and marine habitats in remote areas at some distance from the influence of the littoral, and increased in waters near human habitation (Hagler and Ahearn 1987). Numbers were particularly high in estuaries, which may be related to the proximity of neighboring cities. The effect of general urban pollution was evident in an investigation in Brazil (Hagler and Ahearn 1987), where yeast numbers reached 48 000, 1700, and 2800 cells/l, respectively, where pollution levels were estimated as high, medium, and low. Similar low levels of yeast populations and pollution were found in estuarine waters in the southeasten USA.

The yeast species found in water vary greatly. Hagler and Ahearn (1987) list twelve species of *Candida*, three of *Cryptococcus*, one of *Debaryomyces hansenii*, (a salt-tolerant yeast), two of *Hanseniaspora*, one of *Metschnikowiapulcherrima*, five of *Rhodotorula*, two of *Saccharomyces* species, and one each of *Aureobasidium pullulans*, *Geotrichum candidum Prototheca* sp., *Sporobolomyces salmonicolor*, and *Trichosporon cutaneum*.

In marine environments, yeasts are often numerous in zones of algal blooms, giant kelps, and plants and rhizosphere of oyster grass (*Spartina alterniflora*). The yeast *Pichia spartinae*, was associated with the plant stems, and *Kluyveromyces drosophilarum* with the rhizosphere sediments (10^4–10^8 cells/g).

During our survey of yeasts in aquatic environments in Saskatchewan, we isolated a number of strains of a yeast resembling *Pichia spartinae*, having the same assimilatory and fermentative phenotype and the same NMR spectrum of the cell wall mannans. However, the genomic DNAs were not homologous. The groups of different species of yeasts having identical phenotypes, found in rotting cactus, were not then known. However, *Pichia spartinae* and our unknown isolates may form part of another such group.

Yeast numbers increased when the kelps, plankton, and other marine plants decayed and released more nutrients for the yeasts. Yeasts may utilize nutrients (carbon and nitrogen) secreted by the plants. The plants may also excrete toxic substances (polyphenols), reducing the yeast population, especially on phaeo-phyceae (brown algae). The highest levels of yeasts were found on Rhodophyceae and Chlorophyceae (red and green algae). These sea weeds supported the largest populations of yeasts, which reached 6.8×10^4 cells/g. dry weight, compared to less than 100 cells/g dry weight on Phaeophyceae.

Species of *Rhodotorula* and *Cryptococcus* dominated the yeast population in the early stages, but *Candida* species predominated after about 3 weeks. Yeasts on kelps along coastal Japan were the same as in the littoral; *Candida famata* (imperfect form of *Debaryomyces hansenii*), *Candida* (*Pichia*) *guilliermondii*, *Candida tropicalis*, *Candida parapsilosis*, *Cryptococcus albidus*, *Cryptococcus laurentii*, *Rhodotorula glutinis*, and *Rhodotorula rubra*. Yeasts on marine algae on the east coast of the United States were not the same as those of the littoral (Hagler and Ahearn 1987).

Large numbers of yeasts occur in sediments, under both fresh and marine waters, especially in polluted regions. Yeasts in the upper layers are mostly nonfermentative, including *Cryptococcus albidus*, *Cr. laurentii*, *Rhodotorula glutinis*, *Rh. graminis*, *Rh. marina*, *Rh. minuta*, *Rh. rubra*, and *Trichosporon cutaneum*. In heavily populated subtropical estuaries, the *Candida krusei* complex (*C. krusei*, *C. sorbosa*, *C. valida*, *C. lambica*, *P. membranaefaciens*, *P. kluyveri*, and *P. terricola*, some of which are fermentative), the fermentative species *Debaryomyces hansenii* and *Hansenula saturnus* occurred. The mechanical and chemical compositions of the sediments is strongly influenced by the materials, including pollutants, passing over them in the water, and the depth of the water strongly influences the resident yeast population.

The total yeast number in polluted waters increases with the degree of pollution. Yeasts suggested as indices of pollution include red yeasts (easy to count), *Candida albicans*, (association with human disease), total fermentative yeasts, (easy to determine by the most probable number technique), and *Trichosporon cutaneum*, (occurring in polluted waters). Total yeast numbers may be the best indicator. Red yeasts occur in "blooms" and are unreliable indicators. Yeasts of the *Candida krusei* complex tolerate alkylbenzenesulfonate detergents, and may indicate pollution by sewage. High water temperatures may favor one or more species of yeast.

The nonyeast microorganisms *Geotrichum penicillatum* and *Geotrichum capitatum*, and *Prototheca zopfii* are potential pollution indicators. They also occur in slime fluxes of trees and may give doubtful results.

Oil pollution is a widespread and pressing problem. Some yeasts can metabolize hydrocarbons, and utilize them after an oil spill. Numbers may increase to 10^4 cells/ml from 30–200 colony-forming units/ml previously. The yeasts *Yarrowia (Candida) lipolytica*, *Candida guilliermondii*, *C. tropicalis*, *C. maltosa*, *Debaryomyces hansenii*, and *Rhodosporidium* spp. are found in marine environments, and can metabolize a wide range of aliphatic and aromatic hydrocarbons. Yeasts as well as bacteria may be useful in devouring oil and cleaning up spills.

References and Further Reading

Barker JSF, Starmer WT (eds) (1982) Ecological genetics and evolution. The cactus/yeast/*Drosophila* model system. Academic Press, New York

Bostock A, Khattack MN, Matthews R, Burnie J (1993) Comparison of PCR fingerprinting, by random amplification of polymorphic DNA, with other molecular typing methods for *Candida albicans*. J Gen Microbiol 139:2179–2184

Do Carmo Sousa L (1969) Distribution of yeasts in nature. In: Rose AH, Harrison JS (eds) The yeasts, vol 1, 1st edn. Academic Press, New York, pp 79–105

Fell JW (1993) Rapid identification of yeast species using three primers in a polymerase chain reaction. Mol Mar Biol Biotechnol 2(3):174–180

Gould SJ (1985) The flamingo's smile. Penguin Books Canada, Markham, Ontario

Hagler AN, Ahearn DG (1987) Ecology of aquatic yeasts. In: Rose AH, Harrison JS (eds) The yeasts, vol 1, 2nd edn. Academic Press, New York, pp 181–205

Kreger-van Rij NJW (1984) The yeasts, a taxonomic study, 3rd edn. Elsevier, Amsterdam

Kurtzman CP, Phaff HJ (1987) Molecular taxonomy. In: Rose AH, Harrison JS (eds) The yeasts, vol 1, 2nd edn. Academic Press, New York, pp 63–94

Lachance M-A (1990) *Metschnikowia hawaiiensis* sp. nov., a heterothallic haploid yeast from Hawaiian morning glory and associated drosophilids. Int J Syst Bacteriol 40(4): 415–420

Lachance M-A, Starmer WT, Phaff HJ (1986) Identification of yeasts found in decaying cactus tissue. Can J Microbiol 34:1025–1036

Last FT, Price D (1969) Yeasts associated with living plants and their environs. In: Rose AH, Harrison JS (eds) The yeasts, vol 1, 1st edn. Academic Press, New York, pp 183–218

Phaff HJ, Starmer WT (1987) Yeasts associated with plants, insects and soil. In: Rose AH, Harrison JS, (eds) The yeasts, vol 1, 2nd edn. Academic Press, New York, pp 123–180

Pitt JI, Miller MW (1968) Sporulation in *Candida pulcherrima*, *Candida reukaufii*, and *Chlamydozyma* species: their relationship with *Metschnikowia*. Mycologia 60:663–685

Shen R, Lachance M-A (1993) Phylogenetic study of ribosomal DNA of cactophilic *Pichia* species by restriction mapping. Yeast 9:315–330

Spencer JFT, Gorin PAJ, Hobbs GA, Cooke DA (1970) Yeasts isolated from bumblebee honey from Western Canada: identification with the aid of proton magnetic resonance spectra of their mannose-containing polysaccharides. Can J Microbiol 16:117–119, NRCC No 11150

Spencer DM, Spencer JFT, de Figueroa LIC, Heluane H (1992) Yeasts associated with rotting citrus fruits in Tucumán, Argentina. Mycol Res 96:891–892

Sucs, Meyer SA (1991) Characterization of mitochondrial DNA in various *Candida* species: isolation, restriction endonyucolease analysis, size and base composition. Int J Syst Bacteriol 41(1):6–14

Vaughan Martini A, Kurtzman CP (1985) Deoxyribonucleic acid relatedness among species of the genus *Saccharomyces sensu strict*. Int J Syst Bacteriol 35:508–511

CHAPTER 4

Ecology: The Bad Guys: Pathogens of Humans and Other Animals

J.F.T. Spencer and D.M. Spencer

1 Introduction

Yeasts may be classified as those useful for food for all animals, including invertebrates, small marine animals, birds, and mammals including man; those which live as commensals, and those which are actively pathogenic to man and other animals. Among these latter species are the virulent and widespread pathogens, *Candida albicans* and *Cryptococcus neoformans*. A number of other yeast species cause infections of the skin and mucous membranes, hair and beard, fingernails, and similar parts of the body.

For many years, yeasts have usually been regarded as relatively harmless invaders, as compared to the causal agents of bubonic plague, smallpox, diphtheria, tetanus, scarlet fever, typhoid fever, or the dreaded "white plague", tuberculosis. Yeast infections, by comparison, were seldom fatal.

This is no longer true. The use of immunosuppressive drugs for treatment of cancers, and other conditions to which humans are subject, became common. From being secondary invaders and mild pathogens at best, yeasts became a virulent menace, pathogens of the first rank, an area of study in their own right.

Living creatures other than humans suffer from yeast infections. These include mammals (dogs, cats, horses, and many others), birds, invertebrates, and even the lowly planktonic crustaceans suffer from yeast infections. The unusual yeast genus, *Metschnikowia*, contains species (*M. bicuspidata*, *M. krissii*), which infect crustaceans such as the brine shrimp, *Artemia*, and spores of these species, on release, may penetrate the intestinal wall and cause infections, probably fatal. *Metschnikowia* spp. have been isolated from other invertebrates as well. These animals also suffer from infections caused by *Candida albicans*, *Candida tropicalis*, and *Rhodotorula minuta*.

Yeasts commonly considered as pathogens of warm-blooded animals, incuding humans, include, generally, *Candida* species, *Cryptococcus*, *Rhodotorula*, *Pityrosporum* (*Malassezia*), *Trichosporon*, *Geotrichum*, and the yeast-like fungi, (*Sporothrix* (*Sporotrichum*) *schenkii*, *Histoplasma capsulatum* and *H. duboisii*, *Blastomyces dermatitidis*, *Paracoccidioides brasiliensis* and *Coccidioides immitis*) which, by a quirk of the taxonomists' imagination, are not considered true yeasts. Yeasts of the genus *Candida* are the most numerous, widespread, versatile, and troublesome.

J.F.T. Spencer/D.M. Spencer (eds)
Yeasts in Natural and Artificial Habitats
© Springer-Verlag Berlin Heidelberg 1997

At one time any yeast which did not form spores and which formed pseudomycelium was called *Candida*. Any yeast isolated from an infected patient is still called "*Candida*" by the medical profession. The predominance of yeasts of this genus, in infections of warm-blooded animals, probably lies in the nature of yeast-like fungi in general, since they tend to form pseudomycelium when placed under conditions of reduced aeration. Originally, the standard taxonomic test for production of pseudomycelium was to place a sterile coverslip on a couple of fine streaks of the strain, inoculated on to a medium such as corn meal agar or Wickerham's morphology agar, which created microaerophilic conditions. Many yeast species formed "mycelial" strands under these conditions. However, strains which were alike otherwise in their fermentative and assimilative characteristics differed in their propensity to form pseudomycelium. Thus, yeasts which had become aggressive pathogens and could invade the host tissues frequently also produced pseudomycelium under microaerophilic conditions. Some species, of course, maintained their unicellular habit of growth.

This does not hold for a number of the fungal pathogens which have a yeast-like phase, but are not considered true yeasts. While *Malassezia furfur* (*Pityrosporum orbiculare*) and numerous species of *Candida* grow in the mycelial phase in the host tissues, and are yeast-like in laboratory culture, some pathogens such as *Histoplasma capsulatum*, *Sporotrichum schenkii*, *Paracoccidioides brasiliensis*, *Blastomyces dermatitidis*, and *Coccidiodes immitis* are yeast-like in the host tissues and adopt the mycelial phase in culture.

Meyer et al. (1984) list 196 species as *Candida*. Hurley tabulates 32 as being principally or occasionally isolated from warm-blooded animals, including man. Thus, less than 20% of the known species of *Candida* have been found in association with pathogenic conditions, actual or potential, of man and other animals.

2 Diseases Caused by Species of *Candida*

Professor R. Hurley (Hurley et al. 1987) lists 23 species of *Candida* which have been isolated principally from man and other warm-blooded animals, out of 54 species of such yeasts which are not imperfect (asporogenous) forms of ascosporogenous yeasts. Nine of these are occasionally isolated from this source, and 42 have not been reported as being isolated from this source. These lists do not include species originally classified as *Torulopsis* (for instance, *Torulopsis glabrata*). Nearly half of the species of *Candida* mentioned by Hurley, then, are associated with diseases of man and other animals.

The antigenic patterns of the pathogenic yeasts do not seem to be consistent with their habits of life or their pathogenicity. Of the yeasts in Hurley's three lists, *C. albicans*, *C. tropicalis*, *C. stellatoidea*, and *C. claussenii* shared 80–100% of their antigens. The first three species were included in Hurley's first group; *C. claussenii* was in the third. *C. parapsilosis*, *C. guilliermondii*, and *C. sake* shared 50–70% of their antigens with *C. albicans* and *C. tropicalis*, but *C. parapsilosis* was said to be a member of the first group, isolated principally from warm-blooded animals, including man (it can be isolated from other sources, as we isolated this species repeatedly from water samples and samples of slime fluxes of trees taken near Davis, California; H.J. Phaff, pers. comm.); *C. guilliermondii* was also in the first group, and only *C. sake* fell in the third group. *C. utilis, C. pseudotropicalis, C. krusei,* and *C. zeylanoides* shared only 30% of their antigens with *C. albicans* and *C. tropicalis,* but *C. pseudotropicalis, C. krusei,* and *C. zeylanoides* are in Hurley's first group, and only *C. utilis* is in the second group. It should be remembered that the ascosporogenous form of *C. utilis, Hansenula jadinii,* is commonly regarded as a possible pathogen, in spite of its long history of use in industry as a food and fodder yeast. Apparently, the overall antigenic structure of a yeast has little relation to its pathogenicity. The virulence of a yeast is under closer control than that.

2.1 *Candida albicans*

This species is one of the most frequently occurring yeasts in human infections. There is no part of the body which has not been mentioned as the focus of an infection with *Candida albicans.* It causes infections of the mouth and throat in children and adults, vaginitis in women, skin infections in people of all ages whenever moisture and temperature are favorable, and in poultry, pigs, cattle, dogs, various primates, and wild animals. It is a ubiquitous pathogen of most warm-blooded animals. However, even though *C. albicans* is the most important and wide-spread yeast species causing infections in man and animals, the infections arise only in hosts that are damaged or compromised in some way. Otherwise, it is associated with normal, healthy humans as a harmless and often symptomless commensal.

Nevertheless, systemic infections of humans by *C. albicans* are dangerous and often fatal. The prognosis for systemic candidosis is poor, and prompt treatment is extremely important.

The same is true of other species of *Candida,* causing infections of man and warm-blooded animals. These species can be listed in descending order of virulence as follows (Table 4.1): The list includes yeasts involved in such conditions as endocarditis and acute disseminated candidosis.

Table 4.1. Species of *Candida* and other yeasts pathogenic to humans[a]

Species	
Candida albicans	C. pseudotropicalis
C. tropicalis	C. guilliermondii
C. stellatoidea	C. viswanathii
C. glabrata	C. lusitaniae (Clavispora lusitaniae)
C. krusei	
C. parapsilosis	Rhodotorula rubra (mucilaginosa)

[a] In probable descending order of virulence.

Conditions caused by species of *Candida* include superficial and deep-seated candidosis, the former being widespread and taking the form of oral thrush (caused by *C. albicans*), denture stomatitis, vulvovaginitis (causal agents *C. albicans* and *C. glabrata*), napkin (diaper) rash, onychia, and paronychia (predominant causal agents *C. parapsilosis* and *C. guilliermondii* rather than *C. albicans*). These often occur in housewives, fruit canners, workers who handle fish or otherwise have wet hands for long periods, and congenital cutaneous candidosis. The rare condition, chronic mucocutaneous candidosis, usually occurs in children less than 2 years old. It is frequently associated with endocrinological disorders and genetic defects, and based on defects of cellular immunity. *Candida albicans* is the causative agent. Recently, it has been found that patients receiving long-term treatment with aspirin may be more susceptible to infections of the mouth and throat with *Candida albicans*.

Deep-seated candidosis may involve the spread of the organism from the mouth and throat to the gastrointestinal tract; esophagus, stomach, and intestines. Endocarditis, renal candidosis, and acute disseminated candidosis (septicemic candidosis) are other serious conditions resulting from infections with *Candida* species. Endocarditis resulting from such infections may follow open-heart surgery, and main-lining drugs (by injection) accounts for a significant fraction of cases of endocarditis caused by *Candida* infections. Aproximately 80% of cases of endocarditis are associated with infections by *C. albicans* (55%) and *C. parapsilosis* (26%).

Candida albicans infections of the kidney may lead to renal septicemia and hydronephrosis, usually fatal. Fortunately, the condition is rare. Nevertheless, *C. albicans* has earned its reputation as a most dangerous and ubiquitous pathogenic yeast species, to man and other warm-blooded animals.

3 *Cryptococcus neoformans*

This species, and its perfect forms, *Filobasidiella neoformans* and *Filobasidiella bacillispora*, are basidiomycetous rather than ascomycetous yeasts, and in the imperfect stage do not form mycelium or pseudomycelium. The cells are capsulated; it grows at 37 °C and utilizes inositol as sole source of carbon.

The species is found generally in association with man and other warm-blooded animals, and can be isolated in considerable numbers from dried bird droppings, though the birds themselves do not develop cryptococcosis.

Cr. neoformans causes deep mycoses, sometimes systemic, visceral, or generalized. The yeast may show a preference for nervous tissue, especially brain or meninges, which contain relatively high concentrations of thiamin, an essential growth factor. Lethal forms of infection with *Cr. neoformans* are often associated with debilitating diseases such as Hodgkin's disease, collagen disease, sarcoidosis, carcinoma, tuberculosis, systemic lupus erythmatosis, and conditions requiring corticosteroid therapy and possibly other circumstances in which the immune system of the body is disturbed. Neither virulence nor resistance to phagocytosis seems to be correlated with capsule formation.

Infection probably begins as a result of inhalation of material (such as pigeon droppings) in the nasal and pharyngeal cavities, so that pigeon fanciers are especially at risk. Tourists visiting cities where there are large numbers of pigeons in the public squares (London, New York, etc.), may also be a high risk group.

Cr. neoformans has a sexual stage, as mentioned previously, produces mating pheromones, and has mating types (Moore and Edman 1993).

Sheep, monkeys, dogs, cats, horses, ferrets, cheetahs, pigs, and foxes are also subject to *Cryptococcus* infections, mammary infections (probably mastitis) and infections of the lymph nodes being most common in cattle, and fatal generalized infections in other species. Infections of the central nervous system occur in man and other animals.

Presumptive identification of a yeast associated with an infection in humans or other animals as *Cr. neoformans* can be done with only a few tests: for growth at 37 °C, using inositol as sole source of carbon, presence of capsules (by the India ink or nigrosin method), absence of pseudomycelium in most cases, cells spherical, budding. Colonies are white to pale cream-colored, turning brown on yeast extract-peptone-glucose (or similar) media containing thistle extract, potato or carrot extract, caffeic acid or dihydroxyphenylalanine (o-diphenols are converted to melanin by phenoloxidase). Urease is produced, nitrate is not assimilated. Further confirmation is obtained by determination of carbon assimilation according to the procedures described in *The Yeasts, a Taxonomic Study*.

Benign and transient infections, pulmonary in nature, may be more frequent that was thought previously, and bone lesions may occur in about 5–10% of cases.

4 *Pityrosporum ovale, Pityrosporum pachydermatis,* and *Pityrosporum orbiculare* (also Classified as *Malassezia furfur* and *Malassezia pachydermatis*)

These species are lipophilic yeasts, requiring lipids in the culture medium, when grown in the laboratory, which are normal flora of human skin and may cause skin infections in man and other animals. *P. ovale* multiplies by budding on a broad base, and was at one time called the bottle bacillus, and alleged, incorrectly, to be the causal agent of dandruff. This species will utilize oleic acid as the lipid component of the medium.

P. orbiculare buds on a narrow base, unlike *P. ovale*. Oleic acid will not satisfy the lipid requirement of *P. orbiculare*, although this species requires fatty acids having chain lengths of C_{12}–C_{24}. The requirements can be met by addition of olive oil, linseed oil, myristic acid, or stearic acid to the medium. The optimum temperatures for growth of both species is 35–37 °C. The nature of the other components of the medium is not critical, any medium from wort agar to a synthetic medium being suitable. *P. pachydermatis* does not require lipids for growth. It is seldom found on human skin, but occurs mainly on animals.

Neither species is normally infectious, though *P. orbiculare* (*Malassezia furfur*) may undergo a transition to a mycelial stage and become infectious. It is the causal agent in tinea versicolor, a condition in which flat or slightly raised, scaly spots, light brown in color, develop, most commonly on the back, upper chest, arms, and neck. Itching occurs in severe cases. Usually it is of cosmetic importance only. It can be treated with topical applications of 20% sodium hyposulfate or 2.5% sodium selenate, or ketoconazole, 200 mg, given once a day for 4–6 weeks.

5 *Trichosporon cutaneum* (*Tr. beigelii*), *Trichosporon capitatum* and *Trichosporon inkin. Trichosporon beemeri*

The genus *Trichosporon* contains both basidiomycetous and ascomycetous species. Those species, known to be pathogenic to warm-blooded animals, are basidiomycetes.

Yeasts of this genus form true mycelium as well as budding cells and pseudomycelium. Colonies of *Tr. cutaneum* are moist and cream-colored when young, becoming tough, wrinkled, and "hairy" (due to the presence of aerial mycelium). The mycelium breaks up into rectangular arthrospores which are visible in a zigzag formation, and clusters of blastospores are sometimes formed, a feature which distinguishes this species from *Geotrichum candidum*. Sugars are not fermented and nitrate is not utilized, but numerous sugars and aromatic compounds are utilized.

Tr. capitatum likewise does not ferment any sugar, and utilizes only glucose but not galactose, maltose, sucrose, or lactose. Nitrate is not utilized.

Tr. inkin and *Tr. beemeri* are similar physiologically but differ somewhat morphologically, *Tr. inkin* forming fine hyphae and clusters of blastospores resembling bunches of grapes. *Tr. beemeri* resembles *Tr. cutaneum* but assimilates inulin.

Tr. cutaneum has been isolated from soil, marine environments, river water which may have been polluted from human sources, untreated sewage, and other habitats. *Tr. capitatum* has been isolated from sewage and wood pulp, and has been found in sputum of patients having bronchopulmonary infections. *Tr. inkin* has only been found on or in human skin, and *Tr. beemeri* has been found in infections of crocodiles.

Tr. cutaneum infects human hair (where it causes micronodules), toe- and fingernails, and all *Trichosporon* species can cause infections of the skin of man and other animals. Species of this genus may cause deep-seated infections in patients with other underlying pathogenic conditions (leukemia, brain abcesses, bronchial carcinoma, aplastic anaemia) or under treatment with cytotoxic drugs.

In cattle and buffaloes, *Trichosporon* species may cause mastitis, infertility, and galactophoritis.

Infections caused by *Trichosporon* species may not be as common as those associated with infections by *Candida* species, but are as serious when they occur.

6 *Geotrichum candidum*

Geotrichum candidum is the single species in this genus. It occurs in dairy plants and may participate in ripening of cheeses and the development of typical cheese flavors. It may also be isolated from soil and sewage, sour-milk products, spoiled wine, decaying citrus fruits, and similar habitats. It produces true mycelium with coarse hyphae, pseudomycelium, and arthrospores, which may be rectangular, oval, or round. The mycelium may spread over the medium, but does not form aerial conidiophores or conidia.

Its pathogenicity is uncertain, though it has been isolated from patients whose resistance has been lowered for any reason. It has been isolated from the alimentary tract and respiratory tract, and from the oral mucosa (oral thrush). Disseminated infections have been reported, but seem to be rare. It appears to be a harmless commensal in most cases, except in debilitated patients.

7 Yeast-Like Fungi

A number of fungi, filamentous on laboratory media, are yeast-like in the host tissues. These include *Sporothrix* (*Sporotrichum*) *schenckii*, *Histoplasma capsulatum*, *Blastomyces dermatitidis*, *Paracoccidioides brasiliensis*, and *Coccidioides immitis*. These will not be discussed further in this volume. (Hurley et al. 1987).

8 Chemotherapy in Control of Yeast Infections

The following groups of antibiotics are used:

8.1 Polyenes

These include nystatin (Fungicidin, Mycostatin, Nystan); natamycin (Natamycin, Pimafucin), candicidin, and amphotericin (Fungizone, Fungitin). All contain a macrolide ring, closed by an internal ester (lactone). The chemical structure and absolute configuration is known only for amphotericin B. They are active against yeasts, dimorphic fungi, dermatophytes, and molds, but not bacteria, and act on sterol-containing membranes. Amphotericin B enhances the action of flucytosine. Nystatin is toxic to experimental animals when given parenterally, but is used for topical application in infections of skin and mucous membranes. It is used for vaginal candidosis during pregnancy, and for bladder washouts and to lower the yeast population of the gastrointestinal tract. It is not absorbed from the intestine in its normal form, which allows it to be used against infections in the mouth. Amphotericin B is used against systemic fungal infections. Resistance to polyene antibiotics has not been a clinical problem, though resistant mutants can be obtained in the laboratory.

8.2 Flucytosine (4-amino-5-fluoro-2-pyrimidine)

This compound is most effective against species of *Candida*, *Cryptococcus*, and *Geotrichum*, and somewhat less effective against *Aspergillus* species. Resistance to flucytosine can develop in a number of ways.

8.3 Imidazoles

Chlormidazole   (Myco-Polycid;   1-p-chlorbenzyl-2-methyl-benzimidazole); Clotrimazole [Canestan; bisphenyl-(2-chlorophenyl)-1-imidazolyl-methane]; Miconazole [Daktarin; 1-2,4-dichloro-β-(2,4-dichlorobenzyloxy]-phenethyl imidazole nitrate); Econazole [Pevaryl; 1-2,4-dichloro-β-(p-chlorobenzyloxy) phenethyl imidazole nitrate]; Ketoconazole Nizoral; *cis*-1-acetyl-4-[4[2-(2,4-dichlorophenyl)-2-(1H-1-imidazolyl-methyl)-1,3-dioxolan-4-yl]methoxy phenyl piperazing. These are the only imidazoles so far investigated out of many which have useful clinical activity against yeasts. They inhibit synthesis of proteins and RNA, by inducing membrane leakage and therefore loss of phosphate and K ions. Clotrimazole may also stimulate myeloperoxidase activity in the host cells. There are other effects which may be important. The inhibitory effect of imidazoles is not dependent on the presence of sterols in the cell membrane.

8.4 Development of Resistance

Development of resistance in yeasts and yeast-like fungi to imidazoles is rare. The imidazoles can be used to treat systemic mycoses. Undesirable side effects include liver damage (Hurley et al. 1987).

To sum up, there are a number of yeast species, mostly imperfect forms, which can cause infections of the skin and mucous membranes of warm-blooded animals, including man. Deep-seated, systemic mycoses can also occur, but nearly always in patients having other debilitating conditions. Patients under treatment with immunosuppressive drugs, after organ transplants, are particularly at risk.

References

Barker JSF, Starmer WT (eds) (1982) Ecological genetics and evolution. The cactus/yeast/*Drosophila* model system. Academic Press, New York

Cutler JE (1991) Putative virulence factors of *Candida albicans*. Annu Rev Microbiol 45:187–218

Do Carmo Sousa L (1969) Distribution of yeasts in nature. In: Rose AH, Harrison JS (eds) The yeasts, vol 1, 1st edn. Academic Press, New York, pp 79–105

Gentles JC, La Touche CJ (1969) Yeasts as human and animal pathogens. In: Rose AH, Harrison JS (eds) The yeasts, vol 1. Academic Press, New York, pp 107–182

Gould SJ (1985) The flamingo's smile. Penguin Books Canada, Markham, Ontario

Hagler AN, Ahearn DG (1987) Ecology of aquatic yeasts. In: Rose AH, Harrison JS (eds) The yeasts, vol 1, 2nd edn. Academic Press, New York, pp 181–205

Hurley R, de Louvois J, Mulhall A (1987) Yeasts as human and animal pathogens. In: Rose AH, Harrison JS (eds) The yeasts, vol 1, 2nd edn. Academic Press, New York, pp 207–281

Kurtzman CP, Phaff HJ (1987) Molecular taxonomy. In: Rose AH, Harrison JS (eds) The yeasts, vol 1, 2nd edn. Academic Press, New York, pp 63–94

Lachance M-A (1990) *Metschnikowia hawaiiensis* sp. nov., a heterothallic haploid yeast from Hawaiian morning glory and associated drosophilids. Int J Syst Bacteriol 40(4):415–420

Lachance M-A, Starmer WT, Phaff HJ (1988) Identification of yeasts found in decaying cactus tissue. Can J Microbiol 34:1025–1036

Last FT, Price D (1969) Yeasts associated with living plants and their environs. In: Rose AH, Harrison JS (eds) The yeasts, vol 1, 1st edn. Academic Press, New York, pp 183/218

Meyer SA, Ahearn DG, Yarrow D (1984) In: Kreger-van Rij NJW (ed) The yeasts, a taxonomic study. Elsevier, Amsterdam, pp 585–844

Moore TDE, Edman JC (1993) The α-mating type locus of *Cryptotoccus neoformans* contains a peptide pheromone gene. Mol Cell Biol 13(3):1962–1970

Phaff HJ, Starmer WT (1987) Yeasts associated with plants, insects and soil. In: Rose AH, Harrison JS (eds) The yeasts, vol 1, 2nd edn. Academic Press, New York, pp 123–180

Pitt JI, Miller MW (1968) Mycologia 60:663

Shen R, Lachance M-A (1993) Phylogenetic study of ribosomal DNA of catophilic *Pichia* species by restriction mapping. Yeast 9:315–330

Spencer JFT, Gorin PAJ, Hobbs GA, Cooke DA (1970) Yeasts isolated from bumblebee honey from Western Canada: identification with the aid of proton magnetic resonance spectra of their mannose-containing polysaccharides. Can J Microbiol 16:117–119

Spencer DM, Spencer JFT, de Figueroa L, Heluane H (1992) Yeasts associated with rotting citrus fruits in Tucumán, Argentina. Mycol Res 96:891–892

Vaughan Martini A, Kurtzman CP (1985) Deoxyribonucleic acid relatedness among species of the genus *Saccharomyces sensu strict*. Int J Syst Bacteriol 35:508–511

CHAPTER 5

Yeasts as Living Objects: Yeast Nutrition

J.F.T. Spencer, D.M. Spencer, and L.I.C. de Figueroa

1 Introduction

Yeasts, like other living organisms, require sources of carbon, nitrogen, phosphorus, trace elements, and growth factors (Phaff et al. 1979). Yeasts cannot grow anaerobically, and so require oxygen. All wild-type yeasts utilize glucose, mannose, and fructose. Different species may also utilize numerous other sugars and organic acids. Some yeasts can utilize nitrates, others only ammonium salts. They can utilize a wide range of organic nitrogen compounds, including both L- and D-amino acids (LaRue and Spencer 1966). All yeasts require trace elements. Some species can synthesize the growth factors they require, and many cannot, and these must be supplied. *Cyniclomyces guttulatus* probably requires the most amino acids and vitamins, and an atmosphere high in Co_2 (10–15%); its normal habitat is the cecum of the rabbit, and it is not easy to cultivate in the laboratory.

2 Yeasts: Nutrient Requirements

2.1 Carbon Sources

The best-known yeast, *Saccharomyces cerevisiae*, and its close relatives assimilate and ferment hexoses and their dimers, trimers, and tetramers; glucose, galactose, maltose, sucrose, melibiose (sometimes), raffinose, melezitose, and, rarely, dextrins and starch. It does not assimilate or ferment the disaccharide, lactose. Molecular biologists, attempting to modify strains of this yeast, have attempted to construct strains capable of metabolizing this sugar. Strains capable of growth on lactose have been constructed, but they have been too unstable to be usable in industry. Recently, there have been a few reports of the construction of stable strains of this type.

Strains unable to metabolize both galactose and melibiose can be isolated readily, and are useful in fundamental research and in breeding of industrial yeast strains. The pathway for metabolism of galactose (the Leloir pathway) is well known.

Yeasts able to metabolize melibiose appear to belong to a closely related species, *Saccharomyces pastorianus* (*Saccharomyces carlsbergensis*). It appears to be a natural cross of *S. cerevisiae* and *S. bayanus*, which do not hybridize readily.

J.F.T. Spencer/D.M. Spencer (eds)
Yeasts in Natural and Artificial Habitats
© Springer-Verlag Berlin Heidelberg 1997

Metabolism of melibiose appears to be encoded in at least ten homologous genes (Naumov et al. 199•). We have transferred this gene by protoplast fusion to a baker's yeast, though the strains were not stable.

Aside from the hexoses and their dimers and oligomers, *S. cerevisiae* appears to metabolize a few nonfermentable compounds including lactic acid and some other organic acids and polyhydroxy alcohols. It does not metabolize pentoses, but will ferment xylulose even though it does not ferment xylose.

Other yeast species will metabolize n-alkanes (Tanaka and Fukui 1989), fatty acids and their esters (Spencer et al. 1979), triglycerides (Spencer et al. 1979), methanol (Phaff et al. 1979), aromatic hydrocarbons, flavonoids, and a remarkable number of other carbon compounds.

2 Nitrogen Sources

All yeasts but *Cyniclomyces guttulatus* can utilize ammonium salts as sole source of nitrogen. This species has absolute requirements for numerous amino acids (Rose 1987). Yeasts utilize ammonium nitrogen at different rates, depending on the anion present. Diammonium phosphate is utilized most efficiently, and ammonium chloride, least. Yeasts cannot fix molecular nitrogen.

Nitrate nitrogen is not used by members of the genera *Saccharomyces*, *Pichia*, *Metschnikowia*, *Zygosaccharomyces*, and species of a number of smaller genera. Brewer's yeast grew well using aspartic acid, asparagine, and glutamic acid as nitrogen sources if adequate growth factors were present, used tryptophan and oxyproline poorly, and did not utilize histidine, glycine, cystine, or lysine significantly. Different strains of *S. cerevisiae* differ in their ability to utilize various amino acids.

The differences in the ability of strains of wine yeasts to use free amino acids may account for the occurrence of "stuck" wine fermentations. It may also be responsible for differences in ethanol tolerance in strains of winery and brewery yeasts, as some of these yeasts will accumulate concentrations of ethanol of more than 20% if they are supplied with the correct concentrations of required amino acids (Ingledew 1987).

Although some yeasts do not utilize nitrate nitrogen, many can use not only L-amino acids such as L-lysine, but also many D-amino acids and purines, pyrimidines, and other organic nitrogen compounds (LaRue and Spencer 1967). Another unusual yeast, *Trigonospis variabilis*, produces D-amino acid oxidases, though the fate of the nitrogen after the amino acid has been oxidized is not known.

Some *S. cerevisiae* strains can utilize urea as a source of nitrogen as well as ammonium ions. They can also utilize many purine and pyrimidine bases as sole sources of nitrogen, including allantoin, allantoic acid, adenine, guanine or cytosine, histidine (slowly), but not other imidazole compounds. *Candida utilis* utilizes both L- and D-histidine and histamine. *S. cerevisiae* grows poorly on peptides, and not at all on proteins.

Brewer's and baker's yeasts, *C. utilis*, excrete amino acids, oligopeptides, and amides, which may be reabsorbed, depending on the energy source present. In addition, nucleotides may be excreted from yeast cells when they are suspended in water or glucose solutions, and, unlike the amino acids, are not reabsorbed.

3 Phosphorus

All yeasts, as far as is known, utilize inorganic phosphates for growth. It is taken up as the monovalent anion, $H_2PO_4^-$, and more is taken up of the monobasic potassium salt than the dibasic sodium form. Inorganic phosphate is stored as metaphosphate (volutin granules) in the vacuole.

Rapid growth of yeast requires metaphosphate synthesis in the cells. Yeasts grown in a phosphate-deficient medium may have an increased content of lipid and increased acid phosphatase activity. This enzyme hydrolyzes phosphate esters, so that the inorganic phosphate is available to yeast in an otherwise phosphate-free environment.

4 Sulfur

Saccharomyces species can utilize inorganic sulfate, sulfite, and thiosulfate. They cannot utilize sulfur from sulfur-containing amino acids (cysteine or cystine), nor from the growth factors biotin or thiamine, though some can utilize S from methionine and from glutathione. *Candida utilis* can utilize all of the above compounds and sulfide and taurine as well.

Uptake of sulfate by yeasts requires energy, so the medium must contain both glucose (or other metabolizable compounds) and available nitrogen. The cell can take up sulfate under either aerobic or anaerobic (or microaerophilic) conditions. Selenate, sulfite, and thiosulfate inhibit the utilization of sulfate, and ethionine and methionine sulfone inhibit the uptake of sulfur-containing amino acids.

5 Other Inorganic Elements

About 50 of the common elements are toxic to yeasts, ranging in effect from slightly to extremely toxic. Magnesium, potassium, strontium, calcium, sulfur, and phosphorus either are nontoxic at low concentrations or actively required by yeasts.

Potassium is required for yeast growth and fermentation, and can be partially replaced by sodium and ammonium ions. Yeasts containing abnormal amounts of potassium or having the potassium ions replaced by Na^+ or ammonium ferment more slowly than normal yeasts. Potassium can also be replaced by Rb, Li, and Cs ions, with similar effects.

Magnesium, an activator of a wide range of enzymes, is required as a growth factor by yeast. If potassium is replaced by magnesium, the growth of yeasts is inhibited and rates of fermentation and oxygen uptake are low. Calcium stimulates the growth of yeast and fermentation of sugars, if magnesium is also present.

Copper, iron, and zinc are essential to yeast. Manganese may be required in small amounts but is toxic in excess, and if *S. cerevisiae* is cultivated in the presence of 5–10 mM of manganese, the yeast undergoes mutations to the petite form and to resistance to high levels of antibiotics and inhibitors (chloramphenicol, erythromycin, oligomycin, paromomycin, daunomycin, diuron, and others). Most other heavy metals are either nontoxic or inhibitory. Chloride is apparently required for growth, but is seldom a limiting factor for yeasts.

6 Growth Factors

Biotin is required by *S. cerevisiae*, but not by some yeasts, including species of *Hansenula, Brettanomyces*, or by *Candida utilis* or *Hansenula anomala*. Yeasts may require pantothenic acid, inositol, thiamin, nicotinic acid, pyridoxin, riboflavin, p-aminobenzoic acid, and folic acid (Umezawa and Kishi 1989). Pantothenic acid is a component of coenzyme A, and participates in transfer of acyl groups, especially in fatty acid metabolism. Inositol is required by *S. cerevisiae* and related species, though its function is obscure. *Cryptococcus* species can use inositol as sole carbon source. This was once used as a taxonomic character, but an ascomycetous yeast has been discovered which also utilizes inositol.

Thiamin stimulates growth in baker's yeasts, and some strains of *S. cerevisiae* require it. Brewer's "top" yeasts require either thiamin or pyridoxin, but "bottom" yeasts do not. Thiamin, as thiamin diphosphate, is the coenzyme for cocarboxylase.

Many yeasts require nicotinic acid, including *Kloeckera, Candida, Zygosaccharomyces, Saccharomyces*, and *Schizosaccharomyces* species, but brewer's and baker's yeasts do not. *Kluyveromyces* spp., which ferment lactose, require nicotinic acid, which is a component of the coenzymes NAD and NADP.

Pyridoxine (Vitamin B_6) is required by some species of *Saccharomyces, Kloeckera, Candida, Pichia*, and *Brettanomyces*. Pyridoxal phosphate is a coenzyme for aminotransferases.

Riboflavin and folic acid are synthesized by all yeasts. *para*-Aminobenzoic acid is required by *Rhodotorula* species. In high concentrations, it inhibits invertase activity and growth of yeast.

Under anaerobic conditions, yeasts require ergosterol and oleic, linoleic, or linolenic acid.

7 Nutrient Reserves

Reserve carbohydrates in the yeast cell include glycogen and trehalose. Glycogen accumulates when yeast (*Saccharomyces cerevisiae*) is grown under conditions of nitrogen deficiency. The usual glycogen content of baker's yeast is about

12%, and a high glycogen content is correlated with resistance of the yeast to autolysis.

The importance of trehalose has been realized more and more with time. Good-quality baker's yeast contains a relatively high content (about 14%) of trehalose, which accumulates when the yeast is grown aerobically and decreases under anaerobic conditions. The role of trehalose in baker's yeasts is not known.

Dulcitol accumulates in *Candida utilis*, *Pichia farinosa*, and *Torulopsis versatilis* as a metabolic product of dissimilation of galactose. Its role in the yeast cell is not known.

Lipids (Ratledge and Evans 1987) are accumulated in some yeasts (*Candida (Metschnikowia) pulcherrima*, *Candida utilis* (*Hansenula jadinii*) *Hansenula anomala*, and some *Rhodotorula* species). Fat content may reach 60% when the yeast is grown under conditions of nitrogen and phosphorus deficiency. Xylose may sometimes be used as a carbon source.

8 Oxygen and Carbon Dioxide

Oxygen is also an essential nutrient for yeasts. *Saccharomyces* species can grow under microaerophilic conditions, but there are no strict anaerobes among the yeasts. Many species are strict aerobes, and must have oxygen for growth. Oxygen is the terminal electron acceptor at the end of the chain through which energy is released from carbohydrates taken up by the cell. It is also required for the synthesis of heme, a key component of the cytochromes of the electron transport chain. Oxygen may affect the sensitivity of the yeast to X-irradiation (James and Nasim 1987).

Oxygen differs from other essential nutrients in being only slightly soluble in water, in relation to the amount required by the organism. It also increases the difficulty of maintaining an adequate supply to the yeast cell. The pathway of transport of oxygen is the longest traversed by the yeast's essential nutrients, and the oxygen molecule must cross the gas-liquid interface to the medium. When it is in solution in the liquid phase, oxygen meets much less resistance to transport through the liquid phase to the cell wall and the cytoplasmic membrane. It is not known whether there are transport systems for oxygen in the cell.

The oxygen concentration in the gas phase also affects the oxygen transfer rate (OTR). The concentration decreases as oxygen passes out of the gas phase (probably a bubble). The concentration of other gases present also alters the rate of oxygen transfer to the cell.

Measurement of dissolved oxygen in the medium is insufficient for determining its effect on cell metabolism. The amount of dissolved oxygen is very low in relation to the cell's requirements, so the oxygen uptake rate must also be measured. (Measurement of the heat output by the yeast, by microcalorimeters, is also an accurate and rapid measure of metabolic activity.)

Carbon dioxide is produced in stoichiometric amounts during yeast growth and metabolic activity. It is also metabolized, though in smaller amounts, by yeasts, and is an absolute requirement for some yeast species such as *Cyniclomyces guttulatus*. Normally, it does not have to be supplied to the yeast culture.

9 Commercial Sources of Yeast Nutrients

Molasses, malt wort, grape must, bread doughs, spent sulfite liquors, and hydrocarbons are probably the most important commercial substrates for yeasts at present. Cane juice or cane chips are also used for the production of fuel alcohol in Brazil and neighboring South American countries. Hydrolyzed starch and cheese whey can be used for yeast and ethanol production, though their use has not yet been economically feasible. The lactose in cheese whey (a disposal problem) can be used directly, as a substrate for an engineered *Saccharomyces* strain, metabolizing lactose, or after hydrolysis to glucose and galactose. It may be converted to lactic acid and then to *Saccharomyces* biomass.

9.1 Cane and Beet Molasses

These are widely used for production of baker's yeasts, and probably will be for some time.

Cane molasses (high test, refinery cane, and blackstrap) differ in the efficiency of recovery of the sucrose from the cane juice. Cane and beet molasses contain about 50% residual sugar, invert sugar in cane molasses sugar, and sucrose in beet molasses. Both types contain nitrogenous compounds: amino acids, purines, and pyrimidines, and the nonmetabolizable compound, betaine, in beet molasses, plus 5-methyl cytosine in blackstrap. Total nitrogen content in sugar beet molasses is 1–2%, and in cane blackstrap somewhat lower, 0.4–1.4%.

The level of biotin is 2.7–3.2 ppm in blackstrap molasses, and 0.04–0.13 ppm in sugar beet molasses. Biotin must be added to media containing only beet molasses. Usually, mixtures of cane and beet molasses are used, the cane molasses supplying enough biotin for good growth.

Calcium pantothenate content is 50–110 ppm in beet molasses and 54 ppm in cane blackstrap, and *myo*-inositol levels are 8000 and 6000 ppm, respectively. Nicotinic acid, riboflavin, thiamin, pyridoxin, and folic acid are present in both molasses. The amount of nicotinic acid is not sufficient for good yeast growth.

The ash content of beet molasses, about 10%, is greater than in cane molasses, 80% of the mineral residue being composed of potassium and sodium carbonates. The amount and composition of molasses ash depends greatly on the soil type and composition of the fertilizer applied. Most of the chemical elements are present.

9.2 Malt Worts

These are produced by amylolytic hydrolysis of polysaccharides of the grain and added starchy adjuncts. They include glucose, fructose, sucrose, maltose, maltotriose, and maltotetraose, and small amounts of galactose plus traces of pentoses (arabinose, xylose, and ribose). Maltotriose and maltotetraose are the major oligosaccharides. Yeasts utilize the glucose first, from mixtures of hexoses. Fructose and sucrose are fermented next, and lastly, maltose. Only the diastatic forms of *S. cerevisiae*, and *Schizosaccharomyces pombe*, and *Brettanomyces bruxellensis* ferment the various oligosaccharides.

Malt worts contain ammonium nitrogen and α-amino nitrogen, though not enough for the best growth and fermentation by some strains of the yeasts. There is enough biotin, pantothenic acid, inositol, riboflavin, thiamin, pyridoxine, nicotinic acid, and minerals for yeast growth.

9.3 Grape Must

This contains 10–25% sugar, mostly glucose and fructose, which are fermented during winemaking. The finished wine contains traces of hexoses, small amounts of rhamnose, L-arabinose, and xylose and organic acids (malic, tartaric, citric, succinic, glycolic, oxalic, and tannic acids). The must contains ammonium compounds, amino acids, polypeptides, and albumin, which are utilized during fermentation. The must and the wine contain several mineral elements, phosphates of potassium, calcium, and magnesium being most important.

9.4 Bread Doughs

Doughs contain maltose, glucose, fructose, sucrose, and levosin, a fructose polymer. The yeast ferments glucose and sucrose, and to some extent, levosin. Maltose is fermented at the end of the process. Pentosans in the flour (xylose and arabinose residues) are not fermented, but affect loaf volume.

Nitrogenous compounds in the dough are mainly amino acids, which are taken up the yeast. There are sufficient minerals in the flour (phosphate, sulfur, magnesium, and potassium) for yeast growth. Thiamin and pyridoxine are present in the flour.

9.5 Spent Sulfite Liquor

This material contains lignosulfonic acids, aldonic acids, uronic acids, polysaccharides, and monosaccharides (glucose, mannose, galactose, L-arabinose, and monosaccharides (glucose, mannose, galactose, L-arabinose, and xylose). Mannose is the major hexose present, and xylose, the major pentose. *Candida utilis* grows well on this carbon source if a nitrogen source (diammonium phosphate) and potassium chloride are added. *Candida utilis* is grown for a food or fodder yeast on spent sulfite liquor. Baker's yeast may be produced using spent sulfite liquor, but does not utilize pentoses.

9.6 Hydrocarbons

Yarrowia (Candida) lipolytica, Candida tropicalis, and *Candida maltosa* grow on paraffin oil and paraffin wax (long-chain n-alkanes; Tanaka and Fukui 1989). Even-numbered n-alkanes (C_{14} to C_{18}) are used readily by *Candida intermedia,* giving yields of approximately 82%. Hydrocarbons of chain length C_{12}, and odd-numbered chains gave poorer yields. Short-chain alkanes and solid long-chain ones (C_{20} and longer) are not utilized. Unsaturated hydrocarbons and fatty acids which are liquid at room temperatures, are also readily utilized. So the melting point is important in their utilization (Watson 1987). Saturated hydrocarbons of chain lengths C_{20} and C_{22} are readily utilized if they are first dissolved in pristane (2,6,10,14-tetramethyl pentadecane). Branched-chain hydrocarbons are not utilized.

Some smuts (*Ustilago*), basidiomycetous yeasts (*Candida bogoriensis*), and an ascomycete, *Candida bombicola,* utilize n-alkanes and alkenes, fatty acids and esters, triglycerides, and related compounds with formation of sophorosides in which the sophoroside residue is attached to the terminal or penultimate carbon atom of the chain (Spencer et al. 1979). The bridging oxygen between the carbon chain and the sophoroside moiety is derived from molecular oxygen rather than the sugar or the ambient water. The added alkane, fatty acid ester, or triglyceride is converted to glycolipid in yields up to 90%.

Candida antarctica, a basidiomycetous yeast isolated in the Antarctic, produces mannosylerythritol lipids, having surfactant activity, from soybean oil, peanut oil, corn oil, olive oil, triolein, hexadecanol, tetradecanol, and lauryl alcohol, and probably from other vegetable oils and long-chain alcohols.

9.6.1 Aromatic Hydrocarbons

These and other compounds having an aromatic nucleus in the molecule are not utilized as readily by yeasts as the alkanes and alkenes. Benzene is broken down by way of catechol. Yeasts of the genera *Oospora, Candida., Debaryomyces, Pichia,* and *Saccharomyces* metabolize catechol. Other yeasts will metabolize flavonoids, benzoates, and other compounds having aromatic nuclei. Many *Rhodotorula* species will metabolize hydroxybenzoates. Strains which metabolized p-hydroxybenzoates did not metabolize meta-hydroxybenzoates, and vice versa. Many basidiomycetous yeasts, including *Trichosporon cutaneum,* will metabolize or coozidize aromatic hydrocarbons, and related compounds, including some, such as cinnamic acid, which are normally toxic to fungi. This species is frequently isolated from polluted waters and sewage lagoons.

9.7 Methanol as a Substrate. Methylotrophic Yeasts

Methylotrophic yeasts were discovered quite recently. They form a narrow group, including *Candida boidinii, Candida sonoriensis, Hansenula polymorpha, Pichia pastoris*, *Pichia pinus*, and *Pichia methanolica*. Other species may be discovered. They were originally grown as food yeasts utilizing a relatively cheap substrate which did not leave undesirable residues in the product. They have now been found to be useful for production of heterologous proteins of high therapeutic value (Veenhuis and Harder 1992?).

10 Extracellular Metabolites of Yeast

Yeasts produce alcohols, lipids, and, occasionally, enzymes. They can also convert steroids and other compounds into other forms. Filamentous fungi and bacteria produce a greater variety of antibiotics and related compounds. However, yeasts produce alcohols, lipids, and pigments.

10.1 Ethanol

Ethanol is produced in high yields, as beverage alcohol and for industrial use, for fuel, and in the chemical industries. Industrial alcohol is mainly produced from petroleum, but recently, ethanol produced by fermentation has become more competitive.

Production of ethanol by yeast fermentation is simple and efficient. Recently, production by fermentation using *Zymomonas mobilis* has been tested successfully. Nevertheless, most of the world's fermentation alcohol is produced by yeast.

Schwanniomyces occidentalis produces α-amylases, unlike *Saccharomyces diastaticus*, which produces glucoamylases and cannot ferment starch completely, and has considerable potential for construction of yeast strains able to ferment starch directly (Ingledew 1987).

10.2 Polyhydroxy Alcohols

10.2.1 Glycerol

Glycerol is produced by yeast in small quantities during winemaking, and can be recovered from the lees after distillation of brandies and similar beverages. It is formed via the Embden Meyerhof pathway by reduction of a 3-carbon fragment, formed by cleavage of fructose-1,6 diphosphate, to glycerol. The yield of glycerol can be improved by addition of bisulfite to the fermentation, which binds acetaldehyde and traps it as the bisulfite complex. This forces the reduction of much more of the dihydroxyacetone phosphate to glycerol, which is then recovered by standard methods. The same result could be obtained by addition of sodium hydroxide, to raise the pH of the fermentation. This induces the formation of acetic acid and ethanol from two molecules of acetaldehyde, which steers the fermentation towards glycerol formation. From 10–24% of the sugar fermented is converted to glycerol; however, these processes are not economic.

The search for an organism that produced glycerol continued during World War II and for a number of years after. Nickerson (in Phaff et al. 1979), in 1947, investigated a fermentative process using an osmotolerant yeast, *Zygosaccharomyces baillii*) and found that it apparently produced glycerol. The unaerated fermentations were slow, incomplete, and still uneconomic.

In 1951, Spencer (Spencer and Spencer 1978) showed that the osmotolerant yeast, *Zygosaccharomyces rouxii*, isolated from honeycomb, produced glycerol and D-arabitol. In aerated culture, using adequate nitrogen sources, yields of glycerol + arabitol reached 60% of the sugar utilized and concentrations of 80–100 mg/ml (8–10%).

Subsequent investigations showed that the other osmotolerant yeast species (*Endomycopsis capsularis*, *Endomycopsis chodatii*, *Torulopsis magnoliae*, *Trichosporonoides oedocephalus*) produced all of the known polyhydroxy alcohols. Onishi (in Spencer and Spencer 1978) showed that *Pichia farinosa* formed xylitol and two heptitols from xylose, and confirmed that transketolase was a key enzyme in the formation of polyhydroxy alcohols.

Investigations using radiolabeled glucose showed that though glycerol was formed by *Z. rouxii* via the EMP pathway, arabitol was not. Carbon-1 could be removed as CO_2, leaving D-arabitol as a residue, and transketolase transferred carbons 1 and 2 of glucose to one of the 3-carbon residues formed by cleavage of another glucose molecule. Erythritol was formed by some yeast species, by de-phosphorylation and reduction of the residue remaining after the action of transketolase on glucose-1-phosphate.

Compatible solutes such as glycerol protect the osmotolerant yeasts against damage by high concentrations of sugars or salts, probably by protecting key enzymes. Any explanation must consider the conditions under which the yeast lives, in its natural habital – fermenting "green" honey in beehives or nests of wild bees; in the nectar of flowers, spoiled jams and similar sweet food, and even in fondant centers of some candies, which are subject to bursting due to gas production by the yeasts. Yeasts growing under these conditions are not subject to osmotic shock in these habitats, but are under some kind of continuous osmotic stress during its normal existence. (Osmotic-sensitive yeasts produce proteins analogous to heat-shock proteins, when subjected to osmotic shock.) The natural habitats of osmotolerant yeasts are usually low in available nitrogen for yeast growth and biomass production.

The stress is due to the influx of sugars from the medium, probably by diffusion. The cell must neutralize the deleterious effects, by the production of compatible solutes. The cell produces these compounds from the sugars themselves, which reduces the intracellular concentration of sugars while the compatible solutes are being produced.

In addition, the energy-rich compound, ATP, is produced during alcoholic fermentation and glycerol production, via the Embden-Meyerhof pathway. During the process, there is a net gain in ATP content, and the cell can utilize this compound in numerous metabolic reactions. Since the level of available nitrogen in the natural habitats is low, the yield of yeast cells is not great.

Therefore, the "overproduction" of compatible solutes by osmotolerant yeasts may not be as wasteful as some investigators have suggested. In fact, the system may be a very efficient one, under the conditions of the life of osmotolerant yeasts. The yeast accomplishes three things: it reduces the concentration of sugar(s) in the cell, it produces the compatible solute, glycerol, at the same time protecting the cell components, and, finally, obtains the energy-rich compound, ATP, during the process.

The origin of salt tolerance during evolution may be simply a matter of the further adaptation of sugar-tolerant yeasts to conditions of high salt concentrations in the medium, though the mechanism of the adaptation is not known.

10.3 Lipids

10.3.1 Neutral Lipids – Triglycerides

These are formed by a number of yeast species, including *Rhodotorula* species, usually under conditions of nitrogen near-starvation. Commercial production has often been suggested, but not undertaken.

10.3.2 Sphingolipids, Tetramethylsphingosines

These are formed by some species of *Hansenula*. Their nature was investigated by Wickerham and others at the NRRL, Peoria, Illinois, USA.

10.3.3 Esters of Polyhydroxy Alcohols and Long-Chain Fatty Acids

They are produced by *Rhodotorula* species, in submerged culture, and are visible as oil droplets, in the cells, and free in the medium.

10.3.4 Glycolipids

These are formed by heterobasidiomycetes (cereal smuts), a basidiomycetous yeast, *Candida bogoriensis*, and an ascomycetous yeast, *Candida bombicola* (Spencer et al. 1979). They were first observed by Haskins (in Spencer et al. 1979), at the Prairie Regional Laboratory in Saskatoon, as slender, needle-like crystals, in submerged cultures of the yeast-like phase of *Ustilago maydis* (corn smut), and were identified as a glycolipid composed of a disaccharide and a hydroxy fatty acid, linked at the terminal or the penultimate carbon atom to the disaccharide. Later, Deinema (in Spencer et al. 1979) isolated similar crystals from cultures of the basidiomycetous yeast, *Candida bogoriensis*. This compound was identified as a similar fatty acid glycoside, having sophorose as the disaccharide moiety, by Tulloch and Gorin (in Spencer et al. 1979) of the Prairie Regional Laboratory, Saskatoon, Saskatchewan, Canada.

Spencer isolated a yeast from flowers which was identified as *Candida bombicola* and utilized fatty acid esters, long-chain n-alkanes, and the fatty acid moieties of triglycerides, and attached them to the disaccharide, sophorose, through an oxygen bridge. The source of the oxygen atom was molecular oxygen, demonstrated by Tulloch, using O^{18} as a tracer (Spencer et al. 1979).

This glycolipid was produced in large amounts and is a possible starting material for synthesis of synthetic musks (as a perfume base). They are also surface-active agents with variety of potential uses. At present, these glycolipids are not produced commercially, but there is increasing interest in them, particularly in Japan.

Yeasts also produce the enzymes invertase, inulinase (by *Kluyveromyces* spp.), proteases (*Yarrowia lipolytica*, some *Cryptococcus* species), lipases (*Y. lipolytica*, *Rhodotorula* spp.), and, possibly, pectinases or hemicellulases (*Kloeckera* sp.).

Many enzymes can be produced as heterologous proteins using methylotrophic yeasts as a vector. This offers especial advantages, since the heterologous proteins can be either packaged in peroxisomes and kept separate from the rest of the cell, or the yeast can be engineered to secrete the enzyme into the culture medium. This will be discussed in Chapter 14.

References

Ingledew WM (1987) *Schwanniomyces* – a potential super-yeast? CRC Crit Rev Biotechnol 5:159–176

James AP, Nasim A (1987) Effects of radiation on yeast. In: Rose AH, Harrison JS (eds) The yeasts, vol 2, 2nd edn. Academic Press, New York, pp 73–97

LaRue TA, Spencer JFT (1968) Utilization of organic nitrogen compounds by yeasts of the genus *Saccharomyces*. Antorie Leeuwenhoek 34:153–158

Phaff HJ, Miller MW, Mrak EM (1979) The life of yeasts, 2nd edn. Harvard University Press, Cambridge

Ratledge C, Evans CT (1989) Lipids and their metabolism. In: Rose AH, Harrison JS (eds) The yeasts, vol 3, 2nd edn. Academic Press, New York, pp 367–455

Rose AH (1987) Responses to the chemical environment. In: Rose AH, Harrison JS (eds) The yeasts, vol 2, 2nd edn. Academic Press, New York, pp 5–40

Spencer JFT, Spencer DM (1978) Production of polyhydroxy alcohols by osmotolerant yeasts. In: Rose AH (ed) Economic microbiology, vol 2, Primary products of metabolism. Academic Press, New York, pp 393–425

Spencer JFT, Spencer DM, Tulloch AP (1979) Extracellular glycolipids of yeasts. In: Rose AH (ed) Economic microbiology, vol 3. Secondary products of metabolism. Academic Press, New York, pp 523–540

Tanaka A, Fukui S (1989) Metabolism of n-alkanes. In: Rose AH, Harrison JS (eds) The yeasts, vol 3, 2nd edn. Academic Press, New York, pp 261–287

Umezawa C, Kishi T (1989) Vitamin metabolism. In: Rose AH, Harrison JS (eds) The yeasts, vol 3, 2nd edn. Academic Press, New York, pp 457–488

Watson KG (1987) Temperature relations. In: Rose AH, Harrison JS (eds) The yeasts, vol 2, 2nd edn. Academic Press, New York, pp 41–71

See also *Notes Added in Proof* on p. 353.

CHAPTER 6

Outside and Inside: The Morphology and Cytology of the Yeast Cell

J.F.T. Spencer and D.M. Spencer

1 Introduction

The appearance of a yeast cell is not exciting. The noted authority on yeast cytology, C.F. Robinow (Robinow and Johnson 1991), has said of the yeast cell wall and the underlying plasmalemma that "Yeasts seem drab and featureless, and are clothed more for respectability than for beauty". Nevertheless, Robinow has filled more than 100 pages on the cytology of yeasts. If Robinow finds the cytology and ultrastructure of yeasts interesting, others will also. We cannot, here, go into the same amount of detail as Robinow, nor clothe them in such entrancing language. We can only summarize the present knowledge of the structure of the yeast cell, and recommend that the interested reader then consult Robinow's writings, both for the more detailed information within them, and for the incomparable enjoyment of the language in which it is described.

2 External Aspects

Yeast cells range in size from $1–2\,\mu M$ in diameter to $7–8 \times 10–12\,\mu M$; from small to large. *Saccharomyces cerevisiae* is one of the largest, especially when hexaploid.

Under the light microscope, most such cells are round, ellipsoidal, elongated cylindrical forms, or are variations on these themes. *Trigonopsis variabilis* is an exception, having triangular cells, budding at the corners, under some nutritional conditions.

The fission yeasts multiply by fission and form distinctive crosswalls, which can be observed microscopically. The morphology of the separating cells after crosswall formation is characteristic.

The form of the vegetative cells, pseudomycelium, and, where present, true mycelium, blastospores, and arthrospores found in some yeast species is shown in Fig. 6.1.

2.1 Outside the Cell Wall

Scanning electron microscopy reveals features of the outer surface of the cell wall (Fleet 1991). In *Saccharomyces cerevisiae* the bud scars, where buds have been

J.F.T. Spencer/D.M. Spencer (eds)
Yeasts in Natural and Artificial Habitats
© Springer-Verlag Berlin Heidelberg 1997

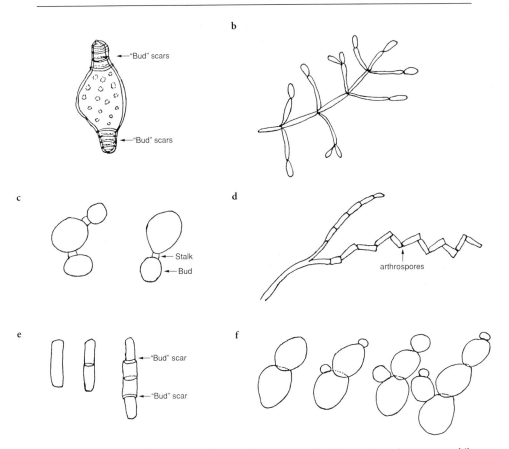

Fig. 6.1a–f. Vegetative cells, pseudomycelium, arthrospores, and budding. **a** *Hanseniaspora osmophila* – thin section showing bipolar budding with multiple bud scars. **b** *Candida parapsilosus* – pseudomycelium. **c** *Sterigmatomyces halophilus* – buds on stalks. **d** *Trichosporon* arthrospores. **e** *Schizosaccharomyces pombe* – "bud scars". **f** *Saccharomyces cerevisiae* – "budding"

detached, are visible, and there may be 12–15 per mother cell. Few such cells are seen, since the mother cell becomes senescent and dies after producing a few buds. However, if the daughter cells are removed by micromanipulation as they arise, up to 43 daughter cells may be produced per mother cell. The cell surface of one of these "grandmother" cells resembles nothing so much as a series of lunar craters, since the rim of the bud scars is raised and prominent. By comparison, the birth scar (one only to a cell) is smaller and inconspicuous (Fig. 6.2).

In apiculate yeasts, where budding is bipolar, the bud scars form easily visible rings of scar tissue, their polar extensions increasing in length as more buds are produced by the mother cell. Fission yeasts divide by formation of a crosswall arising centripetally, and the cells eventually separate. New growth occurs by outgrowth of the scar plugs, so that a ring remains between the original cell and the

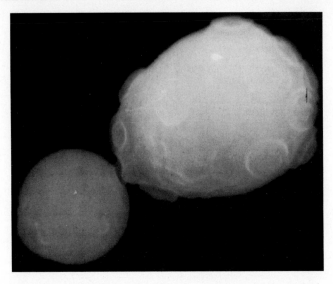

Fig. 6.2. *Saccharomyces cerevisiae.* Vegetative cells, Scanning electron micrograph showing multiple bud scars (grandmother cell)

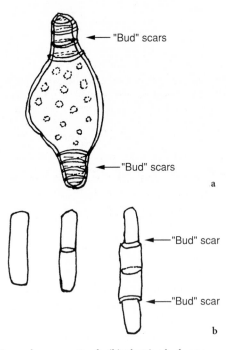

Fig. 6.3a,b. *Hanseniaspora osmophila* (**a**) and *Schizosaccharomyces pombe* (**b**), showing bud scars

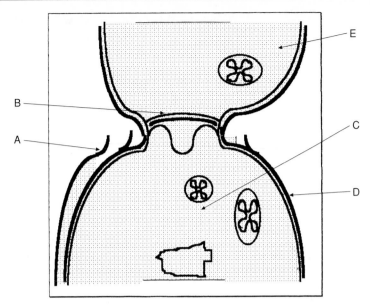

Fig. 6.4. *Phaffia rhodozyma.* Thin section through a budding cell of a basidomycetous yeast showing collar-like remnants of cell wall formed by repeated budding, surrounding the point of origin of the current bud (*A*), *B* septum, *C* mitochondria, *D* mother cell, *E* bud

new one. The "mother" cell is difficult to identify. A cell which has divided several times has a number of scar rings (plug-wall bands). Features of the cell wall can be visualized by light microscopy after differential staining, to reveal new (freshly laid down) and old wall material, with a fluorescent dye (primulin), and are more clearly visible by scanning electron microscopy (Fig. 6.3).

In the basidiomycetous yeasts *Rhodotorula* and *Phaffia*, budding resembles that in the apiculate, ascomycetous yeasts. Budding takes place through the first bud scar, and the new wall forms inside the old, so that collar-like structures surround the neck of the new bud. These are remnants of the walls remaining from previous rounds of budding (Fig. 6.4).

2.2 Spores

The morphology of yeast sexual spores is more varied, and interesting to the taxonomist. The shape varies from round to fat ovals, to bean- and kidney shapes and to hat-, helmet-, Saturn-, crescent- and other angular shapes of spores (Fig. 6.5). A few yeasts produce needle-shaped spores, which are distinctive. The asci containing the spores may be club-shaped, as in *Metschnikowia*, or complex

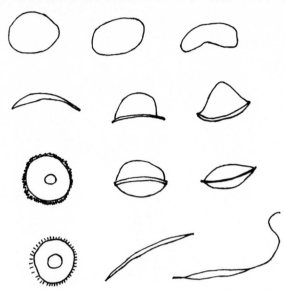

Fig. 6.5. Ascospores in yeasts. Spores round, oval, kidney- or bean-shaped, crescent-shaped, hat-shaped, helmet-shaped, round with warty surface, walnut-shaped, Saturn-shaped, round with spiny surface, needle-shaped, needle-shaped with appendage (whip)

(*Nadsonia*). The club-shaped asci of *Metschnikowia* contain one or two needle-shaped spores. In *Kluyveromyces*, *Hansenula*, and *Pichia* the spores are released spontaneously from the asci. The ascus may seldom or never be visible. Some intergeneric hybrids obtained by protoplast fusion form combination asci consisting of normal diamond-shaped asci attached to a linear ascus usually containing four, but sometimes five or six ascospores. In addition, the fission yeasts form cylindrical asci which contain four to eight or more spores. The *Zygosaccharomyces* species, and others where the ascus is formed by mating of two haploid vegetative cells, form dumbbell-shaped asci, usually with two spores in each end of the dumbbell. During mating, most species form conjugation tubes, ranging in form from the minor protuberances of *Saccharomyces cerevisiae*, to the long, sometimes bent conjugation tubes of some *Hansenula* and *Pichia* species. In *Pachysolen tannophilus*, the original mother cell forms a very long (30 μM), thick-walled tube (sporophore). The spores are formed in a thin-walled ascus at the tip (Fig. 6.6). They are released from the ascus soon after maturity.

Although the morphology of the ascospores (hat-shaped, warty, etc.) has been used as a taxonomic criterion, spore morphology is not always consistent; the genus *Pichia* in particular contains species which form several types of spores. In *Pichia ohmeri* and *Yarrowia lipolytica*, spore morphology is dependent on mating type and the particular strains involved. The taxonomic value of spore morphology has decreased recently.

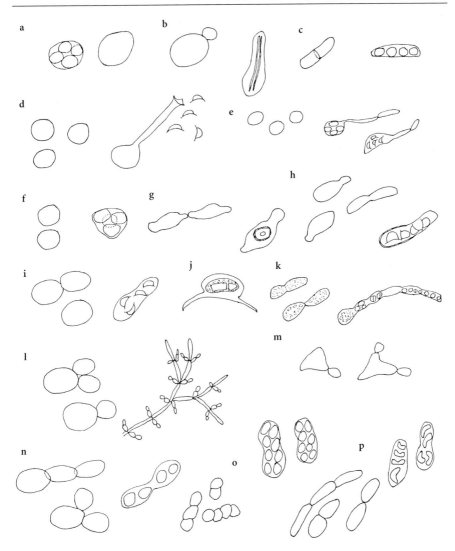

Fig. 6.6a–p. Vegetative cells and spores of: **a** *Saccharomyces cerevisiae*. **b** *Metschnikowia bicuspidata*. **c** *Schizosaccharomyces pombe*. **d** *Pachysolen tannophilus*. **e** *Pichia ohmeri*. **f** *Saccharomyces kluyveri*. **g** *Nadsonia* spp. **h** *Kloeckera vini*. **i** *Pichia membranaefaciens*. **j** Ascospore of *Pichia* spp. (TEM section). **k** *Saccharomyces capsularis*. **l** *Candida tropicalis*. **m** *Trigonopsis variabilis*. **n** *Zygosaccharomyces bailii*. **o** *Schizosaccharomyces octosporus*. **p** *Kluyveromyces* spp.

A few species form asexual arthrospores (*Trichosporon* spp.) or blastospores (*Candida albicans*). These are characteristic of the species and are very useful in identification and classification. The sexual spore, will be discussed in more detail in Chapter 8.

3 Inside the Yeast Cell

3.1 The Cell Wall

The internal morphology of the yeast cell, its cytology, is more interesting and more complex than its external appearance (Byers 1981). In *Saccharomyces cerevisiae*, the cell wall has a complex structure (Fleet 1991). There is an outer amorphous layer of mannan, phosphorylated to different degrees in different yeast species, a middle layer of alkali-soluble β-glucan, and an inner rigid layer of alkali-insoluble β-glucan. The latter gives shape and rigidity to the cell. The wall contains various protein inclusions, though its structure is more complex than Lampen (1968) originally postulated. The wall bears the scars of the cell's history: the original birth scar, and bud scars varying in number and location according to the species.

The chemical structure of the outer mannan layer is characteristic of the yeast species, and can be used as a taxonomic aid. Determination of the nature of a mannan is a long process. Most of the ascomycetous yeasts have a cell wall of the type described. The exact structure of both glucans and mannan differs with the species. Basidiomycetous yeasts have walls containing chitin, whereas in ascomycetous yeasts, especially *Saccharomyces cerevisiae*, chitin occurs only in the bud scars. *Schizosaccharomyces* species do not contain chitin, but contain an α-(1->3)-linked glucan, pseudonigeran.

3.1.1 Capsules

Many yeasts produce capsular material as well, usually phosphomannans, β-linked mannans, heteropolysaccharides containing pentoses and glucuronic acid residues, D-galactose, and a few, such as *Hansenula ciferrii*, produce the hydrophobic compounds tetraacetyl phytosphingosine and triacetyl dihydrosphingosine. Some *Cryptococcus* species produce extracellular starch (amylose). The function of these compounds is not known. The phosphomannans produced by *Hansenula* and *Pichia* species associated with bark beetles attacking coniferous trees may increase the number of yeast cells carried by the beetles and deposited in the tunnels made by the pests. The yeasts are not pathogenic for the trees, but may metabolize sugars and other compounds in the sap.

3.1.2 Activities of the Cell Wall. Signaling

The cell wall of yeasts is not an inert object, a fortress made by yeasts for themselves. The office of a wall is not a passive one; nutrients pass in and products pass out. Although the plasma membrane controls the passage of materials into and out of the cell, the wall decides whether a product reaches the environment, or is retained in the periplasmic space.

3.1.3 Agglutinins

The agglutinins are glycoproteins which bind to each other at the cell surface and hold the cells in contact. They are encoded in the genome and translated from a specific mRNA by the ribosomes, and secreted normally through the endoplasmic reticulum and the Golgi body, enclosed in the secretory vesicles, and eventually released into the periplasmic space. They are passed through the cell wall (the mechanism is not understood), and finally are anchored in the outer layers of the wall. There, they make contact with an agglutinin from another cell and begin the manufacture of a new diploid cell.

3.2 The Plasmalemma

Inside the cell wall lies the plasmalemma or cytoplasmic membrane, but before this structure is reached, one must cross the periplasmic space, the region where such enzymes as invertase and acid phosphatase are normally located.

The cytoplasmic membrane itself (Henschke and Rose 1991) is no simple structure. Although it consists basically of a unit membrane, which is a lipid bilayer visible in electron micrographs as a structure in which two electron-dense layers (dark) are separated by an electron-transparent, light layer. The composition of the membrane can be altered to some extent by use of appropriate mutants, and by subjecting the cell to rather unusual cultural conditions (anaerobic growth, in the presence of particular fatty acids); it contains, as well, protein inclusions, which probably are largely enzymes such as permeases involved in transfer of sugars and other nutrients across the membrane and into the cell. There are the receptors for the mating hormones, which are triggered when the hormones bind to them, and pass the signals down the chain of command until the reactions that lead to fusion of the haploid cell to its opposite mating type, then to fusion of the nuclei, and finally, to the production of a diploid cell. The membrane is essential to the life of the cell, and will be described in detail later in this chapter.

3.3 The Cytoplasm and Organelles. Ribosome, Endoplasmic Reticulum, Golgi Body

Inside the cell, the cytoplasm, or cytoplasmic ground substance, is packed with myriads of very minute structures, the ribosomes, the basic machinery of protein synthesis (Lee 1991). These are associated with components of the cytoskeleton, the endoplasmic reticulum. This structure, with its associated ribosomes, is concerned with the synthesis and secretion of proteins. The Golgi body sorts and packages the proteins into the secretory vesicles, which afterwards migrate to the cytoplasmic membrane or the vacuoles. The vesicles fuse with the membrane and release their contents into the periplasmic space or the vacuolar lumen.

3.3.1 Ribosomes

The ribosomes consist of ribosomal RNA (rRNA) and protein. During protein synthesis they are complexed as polysomes in chains with messenger RNA (mRNA), which encodes the information for synthesis of the proteins. Transfer RNA (tRNA) carries activated amino acids to the site of protein synthesis on the ribosome. Ribosomes in *Saccharomyces cerevisiae* are of two types: cytoplasmic ribosomes, consisting of two subunits with sedimentation coefficients of 60S and 40S, which combine to form the 80S ribosomes, and mitochondrial ribosomes, which have a sedimentation coefficient of 70S and resemble prokaryotic rather than eukaryotic ribosomes. Mitochondria may represent an ancestral prokaryote which lived as a symbiont in an archaic eukaryote and eventually became irrevocably associated with it to form the modern eukaryotic cell.

3.3.2 The Golgi Body

The Golgi body (Schwencke 1991) in *Saccharomyces cerevisiae* consists of a central pile of three flattened structures (cisterna) surrounded by numerous small vesicles, or sometimes of a single disk, perforated by a few pores. Transport from one cisterna of the Golgi body to the next occurs by a complex process. The material is enclosed in a coated vesicle, detached from the donor cisterna, transferred to the acceptor cisterna, attached to it, the coat removed, and the contents of the mature transport vesicle is deposited in the receiving cisterna (see Chap. 5.2).

   The cytoplasm also contains glycogen, the other storage carbohydrate trehalose, some polyphosphates, and various glycolytic and other enzymes. Numerous lipid globules may be present, depending on the yeast species and growth conditions.

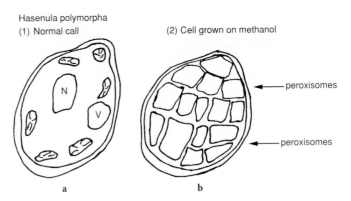

Fig. 6.7a,b. *Hansenula polymorpha.* **a** Normal cell growth on glucose. **b** Cell grown on methanol, showing peroxisomes. *V* Vacuale; *N* nucleus

3.4 Methylotrophic Yeasts and Microbodies. Peroxisomes

Methylotrophic yeasts, grown on methanol as sole carbon source, have a much reduced growth rate. The cytoplasm becomes packed with microbodies (Fig. 6.7), the peroxisomes, which contain the enzymes, alcohol oxidase, and catalase (Veenhuis and Harder 1991). The gene encoding alcohol oxidase may be disrupted and a sequence encoding another protein of technological or pharmaceutical interest introduced. This gene is then expressed and the protein may be produced in quantity. *Candida tropicalis*, when grown on n-alkanes, forms peroxisomes which are filled with catalase. Peroxisomes have a homogenous matrix and are bounded by a single unit membrane.

3.5 The Vacuole

The vacuole (Schwencke 1991), or vacuoles, is normally clearly visible in the light microscope, and is the most conspicuous organelle in the cell. The vacuole(s) are bounded by a single membrane, carrying numerous small particles, which may be ribosomes. The vacuole may contain polymetaphosphate granules (volutin), lipid droplets, numerous hydrolytic enzymes: numerous proteinases having many functions, ribonuclease, and esterases. Many of these enzymes may be localized only in the vacuole. Acid phosphatase, α-glycerophosphatase, and α-glucosidase are absent. S-adenosylmethionine, lysine, other amino acids, purines such as isoguanine and uric acid, and potassium ions may accumulate in the vacuole.

3.6 The Nucleus

There are two other important organelles; the nucleus (Williamson 1991) and the mitochondria (Guerin 1991). The nucleus comprises a crescent-shaped, electron-dense body, the nucleolus, plus a more translucent part containing the chromatin. The nucleus houses the major genome of the cell. The messenger RNA is synthesized there, on the chromosomal DNA, and transported to the cytoplasm and its ribosomes, where it complexes with amino acids from the transfer RNA, and protein synthesis occurs. The chromosomes in most yeasts, especially in *Saccharomyces cerevisiae*, are small and difficult to visualize. Lindegren (see Robinow and Johnson 1991) believed that he had stained the chromosomes and could see them under the light microscope, but the objects were artifacts. Recently, microphotographs of stained preparations of chromosomes have been published. The quality of the preparations, however, was much poorer than those obtainable with plant cells or *Neurospora*.

The nucleus is bounded by a membrane resembling the endoplasmic reticulum, and is penetrated by numerous pores. In *S. cerevisiae* and other ascomycetous yeasts, the nuclear membrane does not break down during cell division. Part of the nucleus enters the bud and separates from the nucleus of the mother cell, and forms the nucleus of the bud. For a description of the nucleus during mitosis and meiosis, see Chapter 8.

In *Schizosaccharomyces pombe*, the nucleus consists of two hemispherical regions. One is DNA-rich and contains the chromosomes (chromatin), and the other is rich in RNA and includes the nucleolus. The three chromosomes of *S. pombe* can be easily seen by fluorescence microscopy.

3.7 Mitochondria

The mitochondria (Stevens 1981) are visible in the cytoplasm, in thin section, by electron microscopy, as round, oval, or elongated objects, often with little visible internal structure as in *Saccharomyces cerevisiae* grown under conditions of limited aeration and/or glucose excess, or containing cristae, well or poorly developed, depending on the yeast species and cultural conditions. Yeasts having an active oxidative metabolism have more highly developed mitochondria, since these organelles are the site of oxidative phosphorylation. They contain mitochondrial DNA (mtDNA), visible under the fluorescence microscope after staining with DAPI, which stains DAN preferentially (Figs. 6.8, 6.9, 6.10). In thick section, electron micrographs often show the mitochondrion as a object resembling a length of hosepipe, and reconstruction of the entire organelle from serial sections of this nature yields a single convoluted structure, which breaks up into smaller units at

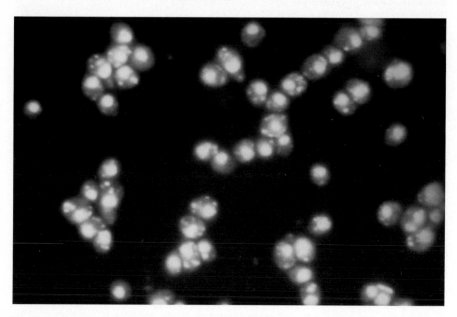

Fig. 6.8. *Saccharomyces cerevisiae.* Cells stained with DAPI (fluorescent stain) to show mitochondrial and nuclear DNA

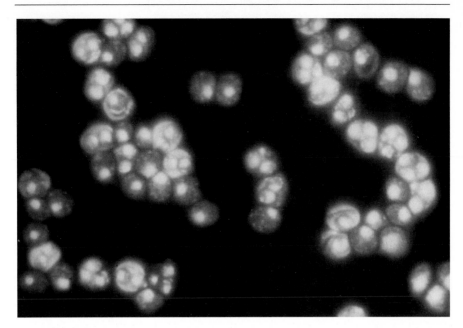

Fig. 6.9. *Saccharomyces cerevisiae.* Normal cells and asci with ascospores, stained with DAPI fluorescent stain

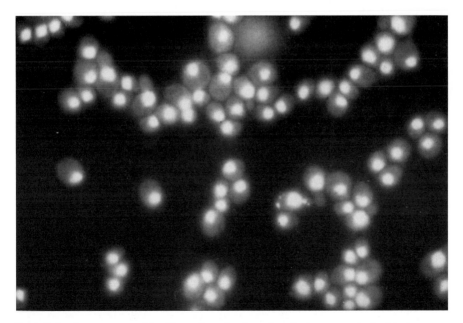

Fig. 6.10. *Saccharomyces cerevisiae.* Vegetative cells of a petite colonie mutant, obtained by treatment of the cells with ethidium bromide. Stained with DAPI (fluorescent stain). Note absence of mitochondrial DNA in most cells

certain stages of growth. The nature of the association of the mtDNA with the physical framework of the mitochondrion has not been determined, nor has the mode of replication of the mtDNA been discovered. When it is known, it may help explain the manner of formation of petite mutants, where a small fragment of mtDNA is replicated several times, yielding a complete circle of mtDNA, but lacking most of the mitochondrial gemone, and is nonfunctional (Fig. 6.10). In *S. cerevisiae*, the cell can reproduce without functional mitochondria, if fermentable sugars are present. Few other yeast species can do this.

The mitochondrial DNA in *Saccharomyces cerevisiae* is a circle approximately $21-25\,\mu m$ in length (molecular weight $46-52 \times 10^{-6}$), carrying some of the genes for the electron transport chain, particularly cytochromes a and b. Cytochrome c is encoded in the nucleus. Resistance to high levels of the antibiotics chloramphenicol, erythromycin, oligomycin, paromomycin, and other drugs is also encoded in the mitochondrial genome. The size of the mtDNA in yeasts varies, from $6.0\,\mu m$ in *Torulopsis glabrata* (molecular weight 12.8×10^{-6}) to approximately $34\,\mu m$ (molecular weight 71×10^{-6}) in *Brettanomyces custersii*. These figures are larger than those for various Metazoa ($4.5-6.2\,\mu m$) but similar to those for filamentous fungi, being $10-106\,\mu m$ for *Aspergillus nidulans* (molecular weight $20-21 \times 10^{-6}$) and approximately 31 (molecular weight $61-63 \times 10^{-6}$) for *Podospora anserina*. These are small, compared to plant mtDNA, which can reach molecular weights of 300×10^{-6} for corn (*Zea mays*) and 1600×10^{-6} for *Cucumis melo*.

The size of the mitochondrial genome in *Saccharomyces cerevisiae* is usually determined by restriction enzyme analysis, since it is extremely difficult to isolate intact mtDNA circles directly for contour measurement. The actual size of the mtDNA molecules is variable (68.0–78.3 kbp). The differences are probably caused by differences in the size and number of introns within mitochondrial genes. The differences are more pronounced in industrial yeast strains.

Yeast mitochondrial DNA carries genes for approximately 24 mt tRNAs, two rRNAs, three subunits of cytochrome c oxidase (subunits COI, II and III of complex IV), and of cytochrome b apoprotein (including the *cob-box* locus), and two subunits of the ATP-synthesizing complex (subunits 6 and 9 of the mitochondrial "ATPase"). It also contains several unassigned reading frames (URFs), and determines the size of the *var I* protein that is associated with the mitoribosomes.

The nature of the interactions between the nucleus and the mitochondria is more subtle than was realized originally, and they are often reciprocal. Fermentation of carbohydrates, flocculation, sporulation, germination, and other metabolic functions, once thought to be controlled entirely by the nuclear genome, are often influenced by mitochondrial function as well. Mitochondria have been shown to affect such diverse matters as the performance of baker's yeast and the maintenance of a foreign killer plasmid, from *Kluyveromyces lactis*. Substitution of mitochondria from a different strain of the same species, in *S. cerevisiae*, influences the metabolic processes (Spencer et al. 1988).

3.8 Other Organelles

Other cytoplasmic organelles include the 2-micron circle, a 3-micron circle, the psi factor, "killer" VLPs in *S. cerevisiae*, composed of dsRNA within a protein coat, linear DNA plasmids in *Kluyveromyces lactis*, also encoding a protein toxin and plasmids known by their effects in other species of yeasts showing "killer" activity (see Chap. 9.2).

3.8.1 The 2-Micron "Circle"

The 2-micron "circle" is a well-studied non-"killer" organelles. Its structure and base sequence have been determined and the designation of circle is incorrect. The particle is a more or less the shape of a dumbbell, in which one end is slightly larger than the other. It seems to have to function in the yeast cell, encodes no essential proteins, and has no adverse effect on the cell. It carries ARS sequences allowing its own replication, and is maintained in copy numbers of approximately 20–30 per cell. Its location has not been determined exactly, but it may be intranuclear, on the inner surface of the nuclear membrane. It segregates as a cytoplasmic particle. Cells can be "cured" of 2-micron circles by various treatments.

It is widely used in biotechnology and genetic engineering, as its ARS sequences are usually included in chimeric plasmids, for production of heterologous proteins (see Chap. 12.2).

The cytoplasmic membrane carries receptors for signals, and gated channels through which nutrients, organic and inorganic, may pass. Even more interesting, these elements are not visible, but their behavior and function can be investigated by genetic, biochemical, and biophysical methods.

References

Byers B (1981) Cytology of the yeast life cycle. In: Srathern JN, Jones EW, Broach JR (eds) The Molecular biology of the yeast *Sacchaaromyces*. Life cycle and inheritance. Cold Spring Harbor Laboratory, Cold Spring Harbor, pp 59–96

Fleet GH (1991) Cell walls. In: Rose AH, Harrison JS (eds) The yeasts, vol 4, 2nd edn. Academic Press, New York, pp 199–277

Guerin B (1991) Mitochondria. In: Rose AH, Harrison JS (eds) The yeasts, vol 4, 2nd edn. Academic Press, New York, pp 541–600

Henschke PA, Rose AH (1991) Plasma membranes. In: Rose AH, Harrison JS (eds) The yeasts, vol 4, 2nd edn. Academic Press, New York, pp 297–346

Lampen JO (1968) External enzymes of yeast; their nature and formation. Antonie Leeuwenhoek J Microbiol Serol 34:1–18

Lee JC (1991) Ribosomes. In: Rose AH, Harrison JS (eds) The yeasts, vol 4, 2nd edn. Academic Press, New York, pp 489–540

Robinow CF, Johnson BF (1991) Yeast cytology, an overview. In: Rose AH, Harrison JS (eds) The yeasts, vol 4, 2nd edn. Academic Press, New York, pp 7–120

Schwencke J (1991) Vacuoles, internal membranous systems and vesicles. In: Rose AH, Harrison JS (eds) The yeasts, vol 4, 2nd edn. Yeast Organelles. Academic Press, New York, pp 347–432

Spencer JFT, Spencer DM, Reynolds N (1988) Effect of changes in the mitochondrial genome on the performance of baking yeasts. Antonie Leeuwenhoek J Microbiol Serol 55:83–93

Stevens B (1981) Mitochondrial structure. In: Strathern JN, Jones EW, Broach JR (eds) The molecular biology of the yeast *Saccharomyces*. Life cycle and inheritance. Cold Spring Harbor Laboratory, Cold Spring Harbor, pp 471–504

Veenhuis M, Harder W (1991) Microbodies. In: Rose AH, Harrison JS (eds) The yeasts, vol 4, 2nd edn. Academic Press, New York

Williamson DH (1991) Nucleus: chromosomes and plasmids. In: Rose AH, Harrison JS (eds) The yeasts, vol 4, 2nd edn. Academic Press, New York, pp 433–488

CHAPTER 7

Membranes

M. Höfer

The plasmalemma or plasma membrane surrounds the yeast cell and is visible in an electron micrograph as a dark double-layered cellular boundary (Fig. 7.1). It comprises a hydrophobic middle layer within a hydrophilic mantle (the electron-dense layers). Biological membranes separate compartments and allow a controlled exchange of solutes between them. The plasma membrane encloses the cell and controls the selective uptake of nutrients and extrusion of metabolic products. The nucleus and mitochondria are enclosed by double membranes; the vacuoles, Golgi complex, and other vesicles by single ones. An endoplasmic reticulum is present. The subcellular membranes each perform specific tasks during metabolism.

Living organisms have a coordinated metabolism. Biological membranes ensure that the substrates, cofactors, and macromolecules required for metabolism are compartmented in the cytoplasm where individual biochemical reactions can proceed without interference. The exchange of intermediates between compartments is mediated by membrane-bound transport catalysts (carriers or channels) with affinities for different chemically related compounds. These allow hydrophilic molecules to penetrate through the hydrophobic biological membrane (selective permeability).

Biological membranes also provide the structural framework for the multienzyme complexes of the respiratory chain. The spatial arrangement of the individual enzymes within the inner mitochondrial membrane results in a flow of reducing equivalents through the redox reactions involved in charge separation (electrons from protons) and vectorial translocation of H^+. Ion translocation between compartments and generation of electrochemical gradients are fundamental functions of biological membranes. This process is the molecular mechanism of energy transduction, and requires biological membranes.

In living cells, the membrane components are synthesized in the endoplasmic reticulum (ER), processed and intercalated into membranes in the Golgi apparatus, and the newly synthesized membranes are distributed in specialized vesicles to their final destination (Sect. 1.2.1).

J.F.T. Spencer/D.M. Spencer (eds)
Yeasts in Natural and Artificial Habitats
© Springer-Verlag Berlin Heidelberg 1997

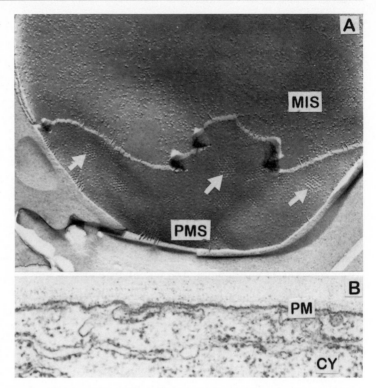

Fig. 7.1A,B. Electron micrographs of the yeast plasma membrane. **A** Protoplast of baker's yeast after freeze-etching showing the granular structure of the interior of the plasma membrane as well as the typical hexagonal assemblies of protein clusters (*arrows*) on the plasma membrane surface. *PMS* Plasma membrane surface; *MIS* membrane inner structure; magnification ×20 000. **B** Ultrathin section of a *Schizosaccharomyces pombe* cell protoplast after fixation with 3% glutaraldehyde and 2% osmium tetroxide in 50 mM phosphate buffer (pH 7), showing the "unit membrane" structure. *PM* Plasma membrane; *CY* cytoplasm; magnification ×48 000. Photographs courtesy of P. Peters and D. Volkmann, Institute of Botany, University of Bonn

1 Membrane Structure and Function

1.1 Chemical Composition of Membranes in Yeast

There are three principal components of biological membranes (Höfer 1981).

1. Lipids, mostly phospholipids. These comprise two nonpolar fatty acid chains (hydrophobic) esterified via glycerol to a polar phosphate, esterified to an organic base. The fatty acid residues are the tail and the phosphate ester is the hydrophilic head (Fig. 7.2). In aqueous milieu, phospholipids form spontaneously bilayer aggregates with the hydrophobic tails inside the bilayer and the polar hydrophilic heads facing the aqueous phases. The phospholipid bilayer is the fundamental structure of biological membranes and renders them impermeable to polar (hydrophilic) molecules.

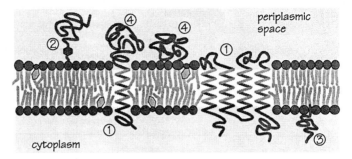

Fig. 7.2. Types of association of protein molecules with the lipid bilayer in biological membranes. *1–3* Integral membrane proteins (firmly anchored in the lipid bilayer): *1* transmembrane proteins; *2* glycoprotein; *3* lipoprotein; *4* peripheral membrane proteins (extractable by polar salt solutions). The lipid bilayer is built up of phospholipids (polar circular heads and hydrophobic tails of two fatty acids residues oriented towards the middle of the membrane) and ergosterol (smaller molecules within each phospholipid monolayer with planar rings between their heads and tails). Transmembrane proteins span the lipid bilayer one or more times with parts of their polypeptide chain, forming regular α-helices. The lipid bilayer behaves as a two-dimensional fluid in which integral membrane proteins can move laterally

2. Proteins, structural and catalytic. The latter are either transport catalysts (carriers, channels) or enzymes. Their catalytic activity depends on their interactions with the surrounding phospholipids. Membrane proteins contain proportionately more amino acids with nonpolar side chains than other, water-soluble proteins. These form hydrophobic domains of the polypeptide chain, spanning the membrane, and anchoring the protein molecule in it. The hydrophilic regions of the polypeptide chain interact with water at the membrane/water interface and stabilize the protein in the membrane further. Solute-binding sites within the molecule may be alternately exposed to one or the other side of the membrane by rearrangement of the tertiary structure of the protein. Carriers or channels operate in this way.

3. Carbohydrates. Glycolipids or glycoproteins, which amount to 3–6% of the membrane, occur at the outer face of membrane bilayers. Their sugar groups are exposed at the cell surface, where they can interact with the environment. The mode of synthesis of the glycolipids and glycoproteins in the Golgi apparatus determines their distribution in the bilayer membrane. They are glycosylated in the lumen of the Golgi apparatus.

1.1.1 Membrane Lipids

Most yeast membrane lipids are phosphoglycerolipids, phosphoesters of diacylglycerol. The great variety of phospholipids is due to the variability in length and number of double bonds of the two fatty acid residues (Prasad 1985). The organic base, also, may be choline (lecithins), ethanolamine, serine, or inositol (Table 7.1).

Table 7.1. Phospholipid composition of membranes of *Saccharomyces cerevisiae*, *Schizosaccharomyces pombe*, and *Candida albicans*

Phospholipid[a]	Yeast species		
	S. cerevisiae[b]	*S. pombe*[c]	*C. albicans*[b]
PC	41.2	36.2	38.9
LysoPC		5.6	
PE	18.5	13.5	25.8
PS + PI	25.4	33.0	19.2
CL	2.9	4.7	4.0
Noncharacterized	12.0	7.0	12.1

[a] Abbreviations: PC, phosphatidylcholine; LysoPC, lysophosphatidylcholine; PE, phosphatidylethanolamine; PS, phosphatidylserine; PI, phosphatidylinositol; CL, cardiolipin.
[b] Data compiled from Prasad (1985).
[c] Own unpublished data.

Biological membranes are dynamic structures. There is rapid lateral diffusion within a monolayer, rotation about the long axes, and flexing of the aliphatic chains, especially near the center of the bilayer. Migration of lipids from one monolayer to the other (flip-flop) is rare. The lipid bilayer behaves as a two-dimensional fluid in which the constituent molecules (including membrane proteins) can move laterally. The fluidity of a lipid bilayer depends on its composition and on the temperature. The liquid crystalline state of the bilayer undergoes phase transition at a characteristic temperature (transition point), to a rigid crystalline (or gel) state having much less freedom of lateral movement and rotation.

Ergosterol also affects membrane fluidity by interaction of the rigid planar rings with the hydrophobic tails of the fatty acid residues (Fig. 7.2). It also decreases the permeability of the yeast plasma membrane to small hydrophilic molecules and probably increases the flexibility and mechanical stability of the membrane. It migrates easily between monolayers (flip-flop).

1.1.2 Membrane Proteins

Most of the specific functions of biological membranes are carried out by a variety of membrane-associated proteins. The yeast plasma membrane is about 50% by weight protein, while the inner mitochondrial membrane may contain 75% protein by weight. There are about 50 lipid molecules for each protein molecule in the plasma membrane.

The membrane proteins are linked to the phospholipid bilayer either by covalent bonds, by extending the polypeptide chain across the bilayer (integral membrane proteins) or by polar bonding to glycolipids or integral proteins, at the membrane surface (peripheral membrane proteins), cf. Fig. 7.2. The first can be freed from the membrane only by disrupting the membrane by detergents. Proteins of the second group can often be released from membranes by gentle extraction with salt solutions which disrupt the noncovalent polar interactions without affecting the lipid bilayer.

The integral transmembrane proteins span the bilayer with parts of their polypeptide chain, forming regular α-helices. The α-helix domains are composed of amino acid residues with nonpolar side chains. A transmembrane protein may be anchored in the membrane by one or more α-helices (Fig. 7.2). The N- and C-terminals of the polypeptide chain and the hydrophilic domains between the α-helices are exposed to the aqueous phases on both sides of the membrane.

Integral membrane proteins, also, move laterally in the lipid bilayer. Protein molecules can aggregate to form functional protein complexes, as in the mitochondrial respiratory chain. Some membrane proteins require the presence of specific phospholipid head groups, as many enzymes require a particular ion for activity. Most transmembrane proteins are glycosylated. Like glycolipids, they have the oligosaccharide moiety on the outside of the membrane.

Several transmembrane signaling systems are located in the plasma membrane; the most important is the mating system. The mating factors, a and α, are peptides which initiate conjugation between haploid cells of opposite mating types. They arrest the cell cycle in G1, induce shmoo formation (copulation tubes), and induce expression of several genes essential for conjugation. Receptor proteins in the plasma membrane, encoded by the STE2 and STE3 genes, recognize the mating factors. The receptors probably have seven transmembrane domains. Binding of the mating factor(s) to the receptor activates a G protein in the membrane.

Three proteins of the cAMP system are part of the family of GTP-sensitive G proteins. They include a GTP-sensitive adenylate cyclase, which is activated by products of the RAS1 and RAS2 genes, and are regulated by the CDC25 gene product, probably a sensor responding to nutrient concentration.

1.2 Functional Membrane Morphology and Protein Topology

Davson and Danielli (1943) proposed that biological membranes have a triple-layer structure, with the main membrane components, phospholipids and proteins, the phospholipids forming a bimolecular central layer with proteins on either side. The first electron-microscopic photos of biological membranes fixed with osmium tetroxide showed a triple-layered, railroad track-like structure (Fig. 7.1). The two dark outside lines, each approx. 2 nm wide, were considered the hydrophilic regions of phospholipid heads with attached protein molecules, while the light layer separating them, approx. 3 nm thick, was probably the hydrophobic domain of phospholipid tails. Robertson (1964) developed the concept of the **unit membrane** in membrane structure. Studies using artificial phospholipid membranes supported this idea.

Use of the freeze-fracture technique gave a better picture of the nature of the bilayer membrane, supporting the current concept of **fluid mosaic membrane** structure (Singer and Nicolson 1972). This model suggests that biological membranes are two-dimensional solutions of oriented globular proteins dissolved in bilayers of lipids – phospholipids, sterols, and glycolipids. The orientation is thermodynamically controlled. Both membrane proteins and phopholipids are assumed to be amphipathic molecules; they are asymmetric, with polar and nonpolar regions on their surfaces. The hydrophilic heads of the phospholipids and the polar side chains of the proteins face the water phases (Fig. 7.2). The extent to which a protein molecule is intercalated in the membrane depends on its hydrophobic interactions with its molecular surroundings. The binding between protein and phospholipid minimizes the free enthalpy of the system as a whole. Lipids in a functional yeast membrane are fluid rather than crystalline. The amphipathic properties of the integral proteins maintain their molecular orientation and degree of intercalation in the membrane even during translational diffusion in the plane of the membrane. They occur in the hydrophobic environment of the membrane and cannot be extracted by aqueous solutions.

Ion transport channels are formed from protein molecules having the polar groups of amino acid residues lining the channel and nonpolar residues anchoring it in the membrane. In carrier-mediated transport, also, according to the fluid mosaic model, substrate molecules are translocated through the membrane by binding to and releasing from transmembrane proteins whose binding sites are alternately exposed to one or the other side of the membrane.

Membranes are impermeable to hydrophilic (polar) molecules because the protein and lipid components form a hydrophobic barrier. The lipids give the membrane elasticity. The catalytic function of integral membrane proteins depends on the fluidity of membrane lipids and thus on temperature. Below the transition point, the membrane is crystalline (solid), the transmembrane proteins are "frozen" in one conformational state, and the carriers are inactive. "Channel" proteins are less affected by temperature.

1.2.1 Membrane Biogenesis

The protein and lipid components of cellular membranes are synthesized in the endoplasmic reticulum (ER), which contains all the enzymes required for their synthesis (Van der Rest et al. 1995). The proteins are imported from the cytosol and bind to the membrane by a hydrophobic signal peptide, recognized by a signal-recognition particle. This initiates an ATP-dependent translocation of the polypeptide chain across the ER membrane. Soluble proteins to be transferred to the lysosomes or for extracellular secretion pass into the ER lumen. Transmembrane proteins for the ER or other cell membranes are anchored in the ER lipid bilayer by membrane-spanning, α-helical regions in the polypeptide chain. The polypeptide chain may cross the membrane several times.

The proteins and lipids in the ER are transferred to the Golgi apparatus and sorted for transfer to the plasma membrane, lysosomes, and secretory vesicles. The

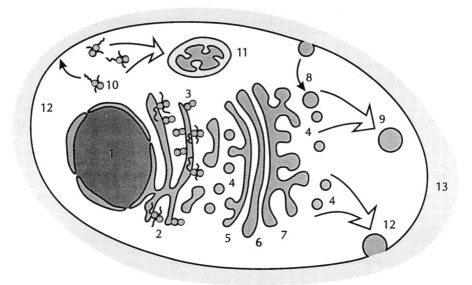

Fig. 7.3. Schematic representation of membrane biogenesis: generation of cellular membranes by membrane flux from the nucleus 1-2-4-5-6-7-4 to the plasma membrane; incorporation of specific proteins, e.g., in a mitochondrion 10–11 or in the plasma membrane 10–12; genesis of a lysosome by fusion of a transport vesicle with an endosome 4-8-9. The numbers stand for: *1* cell nucleus; *2* endoplasmic reticulum (ER); *3* membrane-bound ribosomes in the rough ER; *4* transport vesicles; *5 cis, 6 medial, 7 trans* cisternae of the Golgi apparatus; *8* endosome resulting from endocytosis; *9* lysosome formed by fusion of transport vesicle(s) with endosome(s); *10* free ribosomes synthesizing nuclear coded mitochondrial proteins or specific plasma membrane proteins; *11* mitochondrion; *12* plasma membrane; *13* cell wall

Golgi apparatus is made up of at least three distinct compartments, *cis, medial,* and *trans* cisternae, surrounded by membrane vesicles. Membranes containing specific integral proteins are exported to the *cis* face of the Golgi stack of membranes by transport vesicles. They move *cis-trans* across the stack from one cisterna to the next, and, in particular types of vesicle, to their final destination (Fig. 7.3). The molecular mechanisms involved are not fully understood. More than 25 genes and their products are required for controlling membrane biogenesis.

Inducible plasma membrane transport proteins and several proteins of subcellular organelles are encoded in the nuclear DNA and synthesized in the cytosol. They are incorporated into the target membrane separately from membrane lipids. Membrane recognition is done through a signal sequence of amino acids at the N-terminal of the polypeptide chain. Attachment of the signal sequence to the membrane initiates polypeptide translocation across or intercalation into the membrane by a mechanism of protein subunits forming a transmembrane channel along the central axis of the aggregate. Afterwards, the signal sequence is cleaved and the mature protein either released into the lumen of an organelle or anchored in the target membrane by its α-helical domains. Intercalation via the protein translocation machinery integrates the protein molecules in the correct orientation in the membrane. (Evans and Graham 1989).

2 Transport Function of Membranes

Membranes are hydrophobic boundaries, impermeable to hydrophilic molecules, to each cell and its subcellular compartments. The transport proteins in the membrane allow exchange of material between the cell and its environment, and between the subcellular compartments and the cytosol. The polar surface on each side of the membrane is the interface between the aqueous phases and the hydrophobic region of the membrane, and facilitates the passage of hydrophilic solute molecules.

The permeability of membranes to various substances is variable. It depends on the properties of both solute and membrane. Many transport processes occur spontaneously. Some proceed because of interaction between the dissolved molecules and the membrane and are analogous to noncatalyzed chemical reactions. Most transport processes are mediated by specific carriers or channels, and resemble enzyme-catalyzed reactions. **Passive transport** is independent of metabolic energy. **Active transport** requires metabolic energy and is always mediated by a carrier or a channel. Passive, carrier-mediated transport is **facilitated diffusion**.

2.1 Simple (Physical) Diffusion

Gases (O_2, CO_2), alcohols (C_2H_5OH), undissociated organic acids (CH_3COOH) and bases (NH_3), and other small uncharged molecules migrate through the membrane by simple (physical) diffusion driven by thermal molecular motion. The driving force is the free energy of the concentration gradient of the solute molecules. Diffusion of each molecular species is independent of all others. The rate of transport obeys Fick's first law and, at a given temperature, is determined only by the concentration difference, Δc. Transport by simple (physical) diffusion does not exhibit saturation kinetics.

2.1.1 Membrane Permeability to Water

Water penetrates the membrane rapidly because of its very small molecular size, as it can pass through the spaces between the lipid fatty acid chains. Few water molecules are present within the lipid bilayer because water solubility in lipids is low and the time of their passage is very short. In addition, the concentration of water in aqueous solutions is very high (approx. 55 M) and consequently, aggregates of water molecules form by hydrogen bonding. These aggregates are still small enough to penetrate easily through the membrane.

Water transport is: (1) a passive process, coupled to the active transport of solutes (including ions) and/or to osmotic pressure differences on each side of the membrane. (2) It apparently requires metabolic energy, because of its link with active solute transport. (3) The permeability of biological membranes to water depends on their physical state (e.g., cold resistance of erythrocytes to hemolysis at 4 °C).

2.1.2 Diffusion of Ions

The principle of electroneutrality requires that the sum of positive charges on each side of the membrane equals the sum of negative charges. Otherwise, a large amount of energy is required to separate the positive from the negative charges across the membrane. The charge separation is powered by either the metabolic energy (via ion pumps) or the free energy of a concentration gradient of ions (diffusion potentials).

The diffusion of ions, via membrane channels, is controlled by the electrical potential difference across the membrane, the **membrane potential** $\Delta \psi$ (Fig. 7.6B). The channel is open for a certain range of $\Delta \psi$ values only, the membrane being impermeable to ions at other voltages, (channel gating). The channel may be activated by agonists (Ca^{2+}, cGMP) and inhibited by antagonists (tetraethylammonium, quinine). Some channels are specific for a particular ion species, others are rather unspecific. Electric currents through individual ion channels can be studied by patch clamping (Sect. 5.3.2).

Biological membranes are also slightly permeable to small ions (H^+, Na^+, K^+, Cl^-), which pass through without interacting with a particular membrane component (passive or nonspecific membrane permeability). The nonspecific permeation of small ions through membranes is much slower than that mediated by channels. It leads to continuous dissipation of the electrochemical ion gradients. Cells expend energy to maintain a constant ionic composition of the cytosol. Maintenance uses up to 25% of energy released in metabolic reactions in resting cells.

Each ion passing through the membrane must, in general, be accompanied by an ion of the opposite sign, or an ion of the same sign must cross the membrane in the opposite direction. In this way, ion transport is electrically neutral (no net charge is translocated), and therefore, no additional electric work is required. A bulk flow of ions occurs in **electroneutral transport**. However, during charge separation when the ion translocation is not electroneutral, the membrane potential will be generated, eventually preventing further translocation of ions across the membrane. Transport of ions without electric compensation is called **electrogenic transport** (Fig. 7.4). To generate membrane potentials of the usual magnitude, only a small number of ions needs to cross the membrane. It is too small to be determined by chemical methods. The electrogenicity of ion transport can be deduced only from the formation of the membrane potential. If an ion transport leads to measurable changes in ion concentrations, we must be dealing with electroneutral transport. It may be difficult to decide whether an ion pump operates in an electrogenic or electroneutral mode.

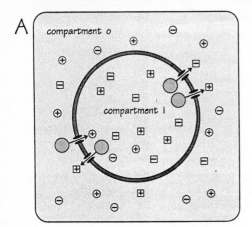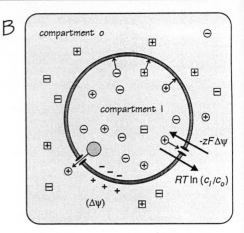

Fig. 7.4. Schematic visualization of electroneutral (**A**) and electrogenic (**B**) ion transport. In the course of the electroneutral transport, the charge translocation is compensated for by either a flow of oppositely charged ions in the same direction or of other ions of the same charge in the opposite direction. The ion transport is electrogenic when there is no compensation for charge translocation (the membrane is permeable for only one ion species). Consequently, the charges (on ions) become separated as the ions pass through the membrane (via channels), driven by their chemical potential difference (concentration gradient) $RTln(c_i/c_o)$. The resulting membrane potential, $\Delta\psi$, soon counterbalances the concentration gradient, thus preventing further flow of ions. This equilibrium of the chemical and electrical potential differences is described by the Nernst equation: $-zF\Delta\psi = RTln(c_i/c_o)$

2.2 Carrier-Mediated Transport

Most solutes cross biological membranes more rapidly than expected from their estimated permeability coefficients, and their transport kinetics resembles that of an enzyme-catalyzed reaction. The rate of transport does not obey Fick's first law, but reaches the maximum transport velocity with increasing concentration gradients. This is **saturation kinetics** and is characteristic of carrier-mediated transport (Fig. 7.5). Carrier-mediated transport further differs from simple diffusion in: **substrate specificity, competition** for the carrier when multiple substrates are present, and **sensitivity to specific inhibitors.**

These properties are determined by transmembrane proteins, (carriers), which, like enzymes, bind a substrate (solute) molecule on one side of the membrane and translocate it to the other side, where the bound molecule is released. The unoccupied (free) carriers then resume the previous conformation with the substrate binding site exposed to the original compartment, and the next translocation cycle can start (Fig. 7.6A). The velocity of mediated transport is determined by the carrier-substrate complex concentration and is limited by the number of carrier molecules present in the membrane.

Solute molecules in a mixture compete for the binding site on the carrier, leading to competitive inhibition of transport of a particular solute. Because bind-

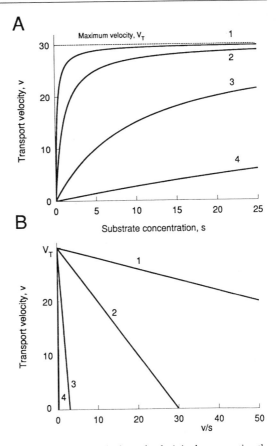

Fig. 7.5A,B. Kinetics of transport of three sugars in the yeast *Rhodotorula glutinis*, demonstrating the characteristic features of carrier-mediated transport. **A** The plot of *v* against *s* shows the saturation kinetics for: *1* D-glucose; *2* D-xylose; *3* L-xylose; the substrate specificity (different half-saturation constants, K_T, especially for the two optical isomers of xylose); and the competition for the carrier: *4* strong inhibition of D-xylose uptake in the presence of D-glucose (author's results). The saturation curves obey the Michaelis-Menten (1913) kinetics $v = V_T \, s/(K_T + s)$ (see text for more details). **B** Linear transformation of the saturation kinetics (noninverted plot according to Hofstee 1959), $v = -K_T(v/s) + V_T$, facilitating the analysis of kinetic data. The resulting straight lines of slopes $-K_T$ can be drawn from fewer experimental points and analyzed more precisely using simple statistics (linear regression). The coincidence of the V_T values is in accord with the fact that the three sugars are translocated by the same carrier

ing of a solute to its carrier is specific, molecules having similar physicochemical properties bind to the carrier with different affinities. The transport is substrate-specific. Transport specificity distinguishes even between optical isomers: D- and L-monosaccharides or L- and D-amino acids. The former isomers are the natural substrates of transport (cf. Fig. 7.5). Carrier-mediated transport is sensitive to specific inhibitors. Transport inhibitors have played a crucial role in identifying and characterizing carrier proteins.

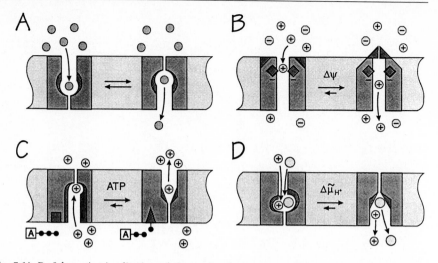

Fig. 7.6A–D. Schematic visualization of the mode of transport catalysis by a symmetric carrier in facilitated diffusion **A**, by a membrane potential-regulated ion channel **B**, by a transport ATPase **C**, and by a symporter (H⁺ cotransport) **D** in active transport. The *dark areas* indicate the binding site on the carrier (the recognition site in the channel) for either a substrate (○), an ion (⊕ ⊖), or a phosphate group (–●)

2.2.1 Ionic Channels

Channels catalyze the passage of ions through membranes and are either open or closed, switching being regulated by the membrane potential (Fig. 7.6B). When the potential is high, the channels are closed; when the membrane is depolarized, the channels open and ions can pass through the membrane, regenerating the membrane potential. One function of ion channels is to form a membrane-potential buffer in yeast cells, at high intracellular K^+ concentrations. Channels can distinguish between ions solely according to their hydration shells. The interaction between the channel and the ions is weak, so they exhibit a lower substrate specificity than carriers.

Carriers and channels may work with or without metabolic energy. In the latter, carrier-mediated transport of uncharged molecules is driven by the chemical concentration gradient of the solute. The electrochemical gradient drives the flow of ions through channels. Facilitated diffusion is mediated transport down an (electro)chemical gradient, independent of metabolic energy. Carrier-mediated transport, which requires metabolic energy and generates (electro)chemical gradients of solute, is active transport. Solute transport, associated with hydrolysis of ATP, is by **transport ATPases** (in yeast, H^+-pump, Fig. 7.6C). This coupled to an electrochemical ion gradient (in yeast, H^+ gradient) is H^+ cotransport or **H⁺ symport** (Fig. 7.6D).

Differences between enzyme- and carrier/channel catalysis include: (1) Enzymes catalyze chemical transformations of a substrate into product. Carriers and channels do not alter the substrate during translocation. (2) Enzyme catalysis is not spatially oriented. Carriers and channels catalyze directional translocation of a solute across a membrane structure. The catalytic activity of isolated transport proteins can be tested only in reconstituted proteoliposomes (Sect. 5.1).

2.2.2 Kinetics of Carrier-Mediated Transport

Carriers accelerate solute translocation in a system towards equilibrium, but cannot alter its position. Before translocation, solute molecules, while acquiring enough internal energy by collisions, reach an activated transition state. The translocation rate is proportional to the number of activated molecules in the transition state. It depends on the energy difference between the mean energy level of the molecular population and that of the transition state (activation energy E_S, Fig. 7.7). The higher the activation energy, the slower the reaction.

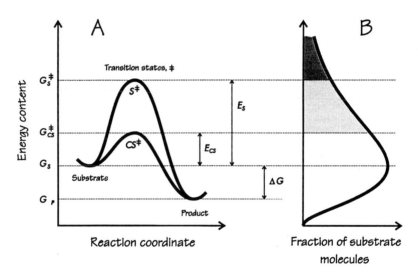

Fig. 7.7. **A** Energy relations in the course of a noncatalyzed and a catalyzed reaction. G_S Mean energy content of the substrate molecules; $G_S^{\ddagger}$ energy content of the activated substrate molecules in the transition state‡; exclusively the $S^{\ddagger}$ molecules are capable of reacting to make the product, with a mean energy content G_P; the difference between G_P and G_S gives the free enthalpy change of the reaction, ΔG, which determines its spontaneity. E_S is the activation energy (the free enthalpy of activation), an energy barrier $G_S^{\ddagger} - G_S$, which must be overcome by each molecule before reacting to product (crossing the membrane). Only a small fraction of substrate molecules gained enough "activation" energy, by collisions with other molecules, to reach the transition state, $S^{\ddagger}$. A catalyst (carrier) forms a transient complex with the substrate, $CS^{\ddagger}$, with a distinctly lower transition energy, $G_{CS}^{\ddagger}$, thus reducing the activation energy E_S to E_{CS}, without affecting the free enthalpy change of the reaction (translocation), i.e., the equilibrium. **B** Boltzmann distribution of energy content among a population of substrate molecules. A very small fraction of molecules possesses energy higher than $G_S^{\ddagger}$ (*very dark area*); many more molecules possess energy higher then $G_{CS}^{\ddagger}$ when an enzyme (carrier) decreased the activation energy. Consequently, the reaction (transport) velocity is significantly accelerated

Two ways of accelerating a translocation process are: by increasing the temperature, increasing the mean energy level of all molecules in a system, or by decreasing the activation energy of transport via carrier catalysis. Formation of a carrier-substrate complex lowers the activation energy for solute translocation. Carrier-mediated transport may be regulated metabolically, by modulation of carrier activity and de novo synthesis or breakdown of carrier molecules. Changes from the liquid to the crystalline state of the membrane also affect the lipid-protein interactions and the activation energy of the fundamental translocation step, carrier conformational change.

The catalytic cycle of a carrier is more complicated than that of an enzyme-catalyzed reaction. However, transport, too, exhibits saturation kinetics, described by the Michaelis-Menten rate equation. The half saturation constant, K_T, measures effective carrier affinity, giving the solute concentration at half-maximal transport velocity, where half the carrier population is occupied by bound solute molecules.

In transport molecular mechanisms, the carrier facilitates the passage of solute molecules in both directions. Competition between two structurally similar solute species for the carrier determines the kinetics of countertransport (Höfer 1981). The solutes compete for the carrier outside the membrane, which reduces the influx of the first solute. The outflow of the first solute is unimpeded until the second solute accumulates inside the cell. The flows of the two solutes are coupled only if both are translocated by a common carrier. Countertransport demonstrates the operation of a membrane-bound carrier, and becomes more pronounced if one of the two solutes is metabolized while the other is not. The metabolized solute does not compete for the carrier during the outflow.

3 Energetics of Membrane Transport

Transport of most solutes across biological membranes requires metabolic energy. In yeasts, energy is released only by chemical oxidation, which occurs mainly in the respiratory chain of the inner mitochondrial membrane. Yeasts can also release energy under anaerobic conditions by fermentation.

Redox energy is eventually stored as adenosine triphosphate, ATP, which supplies energy for all endergonic (energy-requiring) cellular reactions. The conversion of redox energy into ATP in biological membranes, is associated with the generation in the respiratory chain and use (by the F_0,F_1-ATPase) of the electrochemical proton gradient ($\Delta\bar{\mu}_{H^+}$). Uphill transport of solutes into cells as well as from the cytosol into cell organelles proceeds in symport with H^+. ATP is hydrolyzed by membrane-bound H^+-translocating ATPase in either the plasma membrane (P-type) or tonoplast (V-type), whereby $\Delta\bar{\mu}_{H^+}$ is generated. H^+-Symport couples the solute transport to proton flow down the electrochemical proton gradient.

The electrochemical proton gradient, $\Delta\bar{\mu}_{H^+}$, consists of the electrical and chemical potential differences, the membrane potential $\Delta\psi$ and the H^+ concentration gradient ΔpH, respectively. $\Delta\tilde{\mu}_{H^+}$ has the dimension of energy, its driving force is the proton motive force, pmf,

$$\Delta\tilde{\mu}_{H^+} = \left(\Delta\psi + Z\Delta pH\right)F$$

$$pmf = \Delta\psi + Z\Delta pH,$$

where Z is the conversion factor of chemical to electrical energy (Z = 2.303 RT/F), R is the gas constant (R = 8.314 J K^{-1} mol^{-1}), T is the absolute temperature (K), and F the Faraday constant (F = 96480 C mol^{-1} of monovalent ion); the factor 2.303 converts the natural to the common logarithm. At 30 °C, Z is 0.06 V (or 60 mV).

3.1 Energetics of Transport Processes

The first law of thermodynamics states that energy can neither be created nor destroyed, but only converted from one form to another. The first law, when applied to transport, does not indicate the direction of translocation of a solute between compartments.

The second law of thermodynamics states that all physical and chemical changes proceed toward disorder of the system, (increasing entropy). **Entropy**, S, indicates the degree of randomness or disorder on a molecular scale. A system containing gradients is in a state of high order (low entropy). When the gradients dissipate, the entropy increases. Energy is required to generate gradients and decrease the entropy of a system.

At constant temperature and pressure, the entropy change of the surroundings is equivalent to the change of enthalpy, ΔH, of the system and ΔH = ΔG + TΔS; where ΔG and ΔS are the free enthalpy and entropy of the system, respectively, and T is the absolute temperature. **Enthalpy** (heat content) is the internal energy (the sum of all energy forms) of the system, under constant pressure. According to the above equation, only part of the total energy change (ΔH) in a system can be used to perform work (ΔG), while another part is lost as an increase in entropy (ΔS). **Free enthalpy** (Gibb's free energy) is defined by the above equation rewritten as ΔG = ΔH − TΔS.

Processes showing a decrease in free enthalpy (ΔG < 0; exergonic) proceed spontaneously. Endergonic processes are accompanied by an increase in free enthalpy (ΔG > 0) and require energy to proceed in the given direction. At G = 0, the system is in equilibrium. No net change in state of the system occurs; free enthalpy is at a minimum (maximum entropy).

The equations consider only changes in, not absolute values of, the entropy, enthalpy, and free enthalpy. Thermodynamic quantities depend only on the state of the system, not the path traversed or its history. A system generally proceeds towards equilibrium. Living cells are open systems, exchanging energy and matter with their surroundings. Maintenance of the "equilibrium state" (steady state) of a cell requires a continuous release of free enthalpy from nutrients during metabolism. Part is lost as heat, increasing the entropy of the surroundings. Thus, biological systems satisfy the second law of thermodynamics (for more information, e.g., Harold 1986; Smith and Wood 1991).

3.1.1 (Electro)chemical Potentials and Free Enthalpy Change

Chemical potential, μ, determines the readiness of a compound to react. It is defined as the partial derivative of the free enthalpy of a system by the concentration of the compound. The chemical potential μ_i is proportional to the concentration c_i (in mol l^{-1}) of the i-component of a system, $\mu_i = \mu_i^0 + RT \ln c_i$. μ_i^0 is defined as the standard chemical potential of the i-component when $c_i = 1$ M, R, and T have their usual meanings. μ_i^0, and thus, μ_i cannot be determined as an absolute value. However, the change in chemical potential and the change in free enthalpy ΔG can be calculated when c_i changes during a biochemical or transport process.

Passage of a solute between compartments separated by a membrane along its concentration gradient leads to a decrease in free enthalpy (increase of entropy). The free enthalpy is given by the sum of the chemical potentials of the solutes in the compartments. At equilibrium, the free enthalpy is in minimum and the gradients dissipated. If the molecules are charged, an electric term must be added to the chemical potential, which is then called **electrochemical potential** $\mu_i = \mu_i^0 + RT \ln c_i + zF\psi$; where in addition to the above defined symbols, z is the number of charges of the ion species i, F is the Faraday constant, and ψ the electrical potential of the locus of i. The Nernst equation relates the free enthalpy of a chemical (concentration) gradient to that of an electrical potential difference (membrane potential) $\Delta\psi = -(RT/zF)\ln(c_{in}/c_{out})$.

The number of ions necessary to generate a significant membrane potential is very small, too small to be determined chemically. Hence, generation of diffusion potentials does not lead to dissipation of ion concentration gradient.

No natural membrane is perfectly impermeable to ions. The permeability of membranes to various ions differs only by the values of permeability coefficients of the individual ion species. Calculation of the diffusion potential resulting from multiple concurrent ion fluxes can be solved using the Goldman-Hodgkin-Katz equation (Nobel 1992), which simplifies to the Nernst equation if the permeability of any ion is considerably greater than that of the others. This may be induced by the use of valinomycin or other ionophores, which confer on the membrane a specific permeability for a single ion species. In systems such as planar black membranes or phospholipid vesicles, a concentration gradient of K^+ in the presence of valinomycin is used to generate an electrical potential difference across the membrane which becomes negative on the side of the higher K^+ concentration. It can be calculated for a given K^+ concentration gradient using the Nernst equation.

3.1.2 Transport Work

The energy requirement for the generation of an electrochemical gradient is determined by the amount of osmotic work necessary to overcome the concentration gradient and is increased by the amount of electric work necessary to overcome the membrane potential. The electric work depends on the charge of the transported species, and the magnitude and direction of the membrane potential. A considerably larger amount of energy can be stored in an electrochemical ion gradient and used to do work.

Thermodynamic equations indicate only the minimum energy requirement for the generation of, or the maximum work achievable by, the dissipation of a given gradient. Any change in real systems is accompanied by a loss of energy as increased entropy. On the molecular level, energy loss during transport is due to the inefficiency of coupling between the endergonic (transport) and exergonic (metabolism) reactions and the progressive breakdown of the gradient by backdiffusion of the accumulated solute down its (electro)chemical gradient.

3.1.3 Closing Remarks

Classical equilibrium thermodynamics do not apply in biological (open) systems. These can be treated by rules of nonequilibrium or irreversible thermodynamics. However, at the molecular level, under ideal conditions, the classical approach using equilibrium thermodynamics may be adequate.

Thermodynamic considerations do not involve the term of time; they determine only the direction of a process and its equilibrium state. Rates of reactions can be determined through study of reaction kinetics, not from thermodynamic considerations. In biological systems, the action of enzymes which decrease the activation energy (lowering the potential barrier which must be surmounted before the reaction can take place) becomes important in order to maintain the proper velocity of cellular processes.

3.2 Active Transport

Transport processes described so far are independent of metabolic energy. The transport-driving force is the free enthalpy of the concentration gradient of the transported species, so transport proceeds towards dissipation of the gradient. By active transport, gradients are built and continuously regenerated on expenditure of metabolic energy. These transport systems must be coupled to the cell metabolism, and are called active, energy-dependent, accumulative, or uphill transport systems.

In a yeast cell, active transport supplies nutrients and maintains the intracellular concentrations of ions and metabolites, so metabolic fluxes remain constant regardless of environmental variations. Optimum concentrations of inorganic ions (K^+, Mg^{2+}, Ca^{2+}, and HPO_4^{2-}) are required for the regulation of many cellular processes. Active transport maintains the osmotic balance between cells and their environment, and participates in the transduction of the free enthalpy from electron transport in the respiratory chain to its chemical form in ATP.

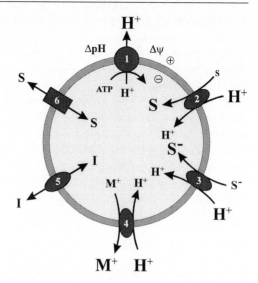

Fig. 7.8. Schematic representation of various transport systems in yeast plasma membrane: *1* Primary active transport by H⁺-ATPase, the proton pump, generating the electrochemical H⁺-gradient, $\Delta\bar{\mu}_{H^+}$. *2–5* Secondary active transport: symport of uncharged (*2*, electrogenic) or negatively charged (*3*, electroneutral) solute molecules *S*, driven by $\Delta\bar{\mu}_{H^+}$ or by ΔpH, respectively; ΔpH drives also the electroneutral antiport (*4*) of positively charged solute molecules *M*, whereas the electrogenic uniport (*5*) of either positively or negatively charged ions *I* is driven by the membrane potential $\Delta\psi$. *6* Passive transport of uncharged molecules *S* down their concentration gradient catalyzed by a carrier

There are two modes of active transport, depending on the mode of coupling to metabolic energy (Fig. 7.8). **Primary active transport** is a direct interaction of the carrier molecules with metabolic energy, provided by ATP in yeast cells. The carrier proteins have two catalytic activities, ATP-hydrolyzing and ion-translocating; they are called **transport ATPases** or **ion pumps**. The known transport ATPases translocate only small ions (H⁺, Na⁺ and K⁺, Ca²⁺, or Cl⁻). Yeast cells have only transport ATPases which translocate H⁺ and Ca²⁺ ions. The H⁺ pump generates an electrochemical H⁺ gradient, which drives numerous **secondary active transport** processes not coupled directly to metabolic energy. Their driving force is the free enthalpy of the H⁺-gradient generated by a primary active transport system, and they are coupled to a downhill flow of H⁺ (H⁺ cotransport). The process is either **symport** (both in the same direction) or **antiport** (opposite directions); **uniport** is a translocation of ions, mostly through channels, driven solely by the membrane potential (Fig. 7.8).

Energy coupling in biological processes requires a common intermediate between the driven endergonic and the driving exergonic reactions. In primary active transport, the common intermediate is mostly the phosphorylated carrier-protein (the product of ATP hydrolysis), while in secondary active transport (in yeast) it is the electrochemical gradient of H⁺.

4 Transport Systems in Membranes of Yeast

Most transport processes are catalyzed by transmembrane proteins (carriers, per-meases, translocators, channels, ion pumps, and ion symporters). Most transport processes are functionally asymmetric, so that solute accumulates on one side of the membrane. This may be important if microorganisms inhabit biotopes where nutrient levels are low. Solute accumulation results in generation of an (electro)chemical gradient of the solute, which requires metabolic energy. Active transport requires both the symmetrical operation of a carrier basically indepen-dent of metabolic energy, and superimposed, energy-dependent conformational changes of the carrier causing its asymmetrical function in the membrane. The two

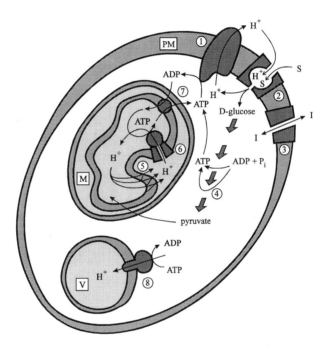

Fig. 7.9. Schematic representation of the variety of transport systems in a yeast cell. Note that various transport systems are compartmentalized into subcellular organelles: *1* the plasma membrane-bound H^+-ATPase (H^+-pump) driven by the free energy of ATP hydrolysis to generate $\Delta\tilde{\mu}_{H^+}$ across the plasma membrane (this is a P-type ATPase); *2* the H^+/substrate symporter using the free energy of $\Delta\tilde{\mu}_{H^+}$ to accumulate the substrate, e.g. D-glucose, inside the cell; *3* an ionic channel mediating the charge compensation during the generation of $\Delta\tilde{\mu}_{H^+}$ as well as its dissipation by H^+ symport; D-glucose is catabolized to pyruvate in the glycolysis, whereby ATP is produced (*4*); pyruvate enters the mitochon-drion, becomes decarboxylated, and eventually is oxidized to CO_2 in the citric acid cycle and to H_2O in the respiratory chain *5*; the H^+-ATPase of the inner mitochondrial membrane (the F-type ATPase) *6* transduces the free energy of $\Delta\tilde{\mu}_{H^+}$ generated by the respiratory chain in the chemical form of ATP; *7* ATP/ADP translocator catalyzing the exchange of adenine nucleotides between the mitochondrion and the cytosol; the vacuolar membrane contains still another H^+-ATPase *8* of the V-type which generates $\Delta\tilde{\mu}_{H^+}$ across the tonoplast, positive and acidic inside the vacuole. PM Plasma membrane; M mitochon-drion; V vacuole

resulting conformational forms of the carrier exhibit different affinities for the solute and differences in the rate constants of their translocation across the membrane, the two essential features of solute accumulation.

The plasma membrane must be energized for efficient uptake of nutrients (carbohydrates, amino acids). In yeasts and other eukaryotes, the plasma membrane H^+-ATPase generates an electrochemical H^+ gradient across the plasma membrane (Sect. 4.1.2) by hydrolyzing cytosolic ATP, which originates either from glycolysis or from mitochondrial oxidative phosphorylation. This plasma membrane-bound ATPase pumps protons out of the cell and the plasma membrane becomes polarized, with the cytoplasmic side of the membrane being more negative and alkaline than the outside. The resulting membrane potential is defined as negative.

Another essential compartment within yeast, fungal, and plant cells participating in cellular solute transport is the vacuole. Its membrane, the tonoplast, has a vacuolar H^+-ATPase generating an electrochemical H^+ gradient across the tonoplast, which is more positive and acidic inside the vacuole than the cytosol. The energized tonoplast is necessary for effective solute translocation into and out of the vacuole. The basic cellular energy-releasing and transport processes are summarized in Fig. 7.9.

The schemes in Figs. 7.8 and 7.9 show that the membrane of each compartment contains only one primary active ion pump, its H^+-ATPase, which energizes all other membrane transport systems of that compartment. The latter, secondary systems are driven by the free enthalpy of the electrochemical proton gradient generated by the former pumps. The ATPase is an integral membrane protein, having an ATP-hydrolyzing activity in which the protein changes conformation, and a specific H^+-translocating activity. Three types of electrogenic ion-translocating ATPases developed during evolution. The variety of ions translocated by the ATPases is quite limited. H^+-ATPases are found in most organisms, including yeasts.

4.1 Proton-ATPases – the Primary Active Ion Pumps in Yeast

Three different types of membrane-bound proton-ATPases are found in eukaryotic cells. In yeast, the plasma membrane is energized by electrogenic H^+ extrusion via the P-type ATPase (Fig. 7.9, transport system 1); a V-type H^+-ATPase generates an electrochemical H^+ gradient across the tonoplast (Fig. 7.9, transport system 8); the F-type ATPase in the inner mitochondrial membrane (Fig. 7.9, transport system 6) is primarily an ATP synthase.

4.1.1 The F_0F_1-ATPase of the Inner Mitochondrial Membrane

This is the universal ion-translocating ATPase. It occurs in bacterial plasma membranes and the inner membrane of mitochondria and chloroplasts of eukaryotic cells. It transduces energy released by the electron transport in the respiratory chain (or in the photosynthetic electron transport chain) into ATP. It catalyzes the final step in the *oxidative phosphorylation* of mitochondria or in *photophosphorylation* of chloroplasts. Its action is reversible: if the electrochemical proton gradient is high enough, F_0F_1-ATPase catalyzes ATP synthesis using free enthalpy of the electrochemical H^+-gradient. However, under anaerobic conditions (or in the dark), the energy-transdusing membrane becomes depolarized. Then, the same enzyme catalyzes the reversed reaction and the membrane is energized using energy from hydrolysis of ATP formed in glycolysis. This process is essential in facultative anaerobes.

The F_0F_1-ATPase molecule has two domains: a hydrophilic, peripheral coupling factor, F_1, having the complete catalytic activity and being composed of five subunits, and a hydrophobic, integral membrane protein F_0, a specific H^+ channel. The F_1 catalytic sector is on the inside of the particular compartment and can be easily dissociated from the membrane sector, F_0. The isolated F_1 contains all the structural elements for ATP hydrolysis. There are three ATP-binding sites on each F_1 sector. Isolated F_0 is a membrane channel specifically permeable to protons.

Only the intact F_0F_1-ATPase can catalyze ATP synthesis, the isolated coupling factor F_1 catalyzes only the hydrolysis of ATP. Reconstituted vesicles from bacterial plasma membranes depleted of F_1 are unable to maintain an electrochemical proton gradient (Fig. 7.10). The mitochondrial ATPase inhibitor, another component of the enzyme, renders the enzyme able to synthesize ATP. The F_0F_1-ATPase

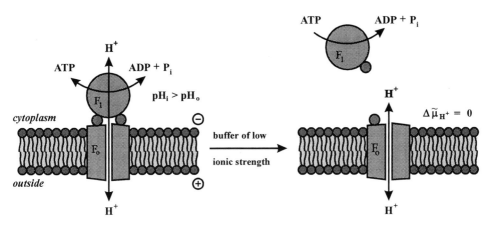

Fig. 7.10. Schematic construction of a bacterial H^+-ATPase from the coupling factor F_1 and the membrane H^+-channel F_0. The integral enzyme (F_0F_1 complex) operates as an ATP-synthase if there is sufficiently high $\Delta\tilde{\mu}_{H^+}$ across the membrane (visualized by $\oplus$ and $\ominus$ at the membrane as well as by pH_i > pH_o). A solubilized coupling factor F_1 subunits (by washing it off with a buffer of low ionic strength containing EDTA) catalyzes ATP hydrolysis, whereby the stored energy is released as heat ($\Delta\tilde{\mu}_{H^+}$ = 0). The residual membranes are permeable to H^+ due to the uncapped F_0 channels. The small subunits act in binding the F_1 to the membrane-bound sector, F_0. Corresponding yeast H^+-ATPase of the F-type is found in the mitochondrial inner membrane

is inhibited by the antibiotic oligomycin and by sodium azide. The enzyme is not phosphorylated during its catalytic cycle (synthesis or hydrolysis of ATP).

4.1.2 The Plasma Membrane H⁺-ATPase

This enzyme is a P-type ATPase, composed of one catalytic subunit of approximately 100 kDa, which forms a transient, acid-stable β-aspartyl phosphate intermediate as the product of ATP hydrolysis. It is inhibited by ortho-vanadate, which substitutes for phosphate during the reaction cycle. The phosphate-binding site faces the cytoplasmic side of the plasma membrane, so inhibition by ortho-vanadate of the ATPase is reduced by the low permeability of plasma membranes to the anion. Fungal plasma membrane ATPase, like F-type ATPases, is an electrogenic proton pump, except that the action of P-type ATPases is irreversible. They generate electrochemical ion gradients using energy from ATP hydrolysis.

The plasma membrane H⁺-ATPases from *S. cerevisiae* and *Schizosaccharomyces pombe* and the fungus *Neurospora crassa* have been isolated, purified, and reconstituted in proteoliposomes. Addition of ATP to these generates a membrane potential (inside positive) and a H⁺ concentration gradient (inside acidic). The highest membrane potential was observed in the absence of permeant anions, but with them, ΔpH was at a maximum. With the *Neurospora crassa* ATPase, monomers of the 100-kDa polypeptide can carry out efficient proton translocation. In native membranes the monomer subunits aggregate in oligomers. There are eight in the *S. pombe* plasma membrane.

The monomer polypeptide is the product of the *PMA1* gene. The N terminus is extremely hydrophilic, followed by four hydrophobic sequences 20–30 amino acids long. Their hydropathy index averages more than +1.5, so they probably span the membrane. The central part of the molecule (one third of it) is hydrophilic. It contains residues involved in ATP binding and phosphorylation, so is probably in the cytoplasm. The C terminus has six putative membrane-spanning segments and a short hydrophilic end-sequence exposed at the cytoplasmic surface of the membrane (Goffeau and Green 1990).

The H⁺-ATPase is the primary active transport system which energizes the plasma membrane in yeast. The uptake mechanisms for nutrients and inorganic ions are H⁺-symport, H⁺-antiport, and uniport (Fig. 7.8). The ATPase maintains the high intracellular K⁺ concentration, which is an effective membrane potential buffer. When the plasma membrane becomes depolarized during H⁺ symport, K⁺ flows out of the cell through K⁺-specific membrane channels, regenerating the negative membrane potential.

4.1.3 The Vacuolar H⁺-ATPase

The V-type ATPases are also composed of a hydrophilic sector catalyzing ATP hydrolysis, and a hydrophobic, membrane-bound H⁺ channel. These and the F-type ATPases probably evolved from a common ancestral precursor and are distinct from the P-type ATPases of eukaryotic plasma membranes. V-type ATPases are found also in clathrin-coated vesicles, synaptic vesicles, chromaffin granules, endosomes, lysosomes, ER, and the Golgi membranes. Here, these H^+ pumps can build up an interorganelle pH gradient which may be important for targeting proteins to cellular membranes. The vacuolar H^+-ATPase is a nitrate-sensitive enzyme which is not inhibited by azide (inhibitor of F-type ATPases) or orthovanadate (inhibitor of P-type ATPases). Probably V-type ATPases, like the F-type, do not form phosphorylated intermediates.

Yeast vacuolar ATPases have a peripheral, hydrophilic catalytic sector, which is exposed to the cytoplasmic side of the organelle, and a membrane sector composed of hydrophobic polypeptide(s). The yeast enzyme is composed of only three major subunits: the catalytic subunit a, subunit c, the proton channel, and subunit b, function unknown. The b subunit of the vacuolar ATPase shows homology with the a subunit of the F-type ATPases.

The H^+-ATPase performs the chemiosmotic work in the vacuole by energizing the tonoplast (ΔpH acidic and $\Delta\psi$ positive inside the vacuole). Yeast vacuoles are metabolically active organelles with low internal pH (pH_v), and can regulate ionic homeostasis in the cytosol (Anraku et al. 1989). There may be seven specific H^+/amino acid antiporters in the membrane, driven by $\Delta\tilde{\mu}_{H^+}$. The nH^+/Ca^{2+} antiport system, required for the regulation of free Ca^{2+} concentration in the cytosol, is in the tonoplast. The vacuolar Cl^- transport system consists of a saturable one and a linear one, sensitive to DIDS, a potent anion transport inhibitor. Chloride transport across energized vacuolar membranes is related to acidification inside the vacuole.

The tonoplast contains high conductance, membrane potential-dependent cation channels, which have low cation selectivity but do not conduct Cl^-, detected by the patch clamp method (Sect. 5.3.2). Their function is optimal at potentials around 30 mV (interior positive), with Ca^{2+} on the cytoplasmic side. The K^+ channel is an energy converter, coupled to the H^+-ATPase, the Cl^- transport systems, and other solute transport systems, to maintain the ionic homeostasis of the cytoplasm. Figure 7.11 shows a schematic view of the tonoplast transport systems.

4.1.4 Other Transport ATPases

Another type of transport ATPases recently identified in yeast plasma membrane, the M-type ATPases, share similarity with the multidrug-resistant p-glycoprotein in that they consist of two symmetrical polypeptide regions each possessing two nucleotide-binding cassettes (Goffeau 1993). Two of them are products of the *PDR5* and *STE6* genes; both show sequence and structure homologies to a large superfamily of membrane proteins, designated ABC transporters or traffic ATPases (Higgins 1992). In *S. cerevisiae*, STE6 is responsible for the efflux of the mating factor a; its deficiency leads to the failure of pumping a-pheromone out of

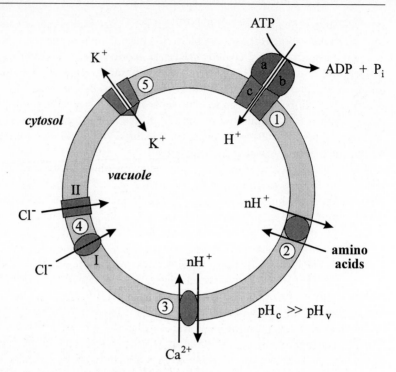

Fig. 7.11. Schematic view of the yeast vacuolar transport systems. *1* The H$^+$-ATPase of the V-type as the primary active H$^+$ pump; *2* H$^+$/amino acid antiporters of various kinds; *3* H$^+$/Ca^{2+} antiporter; *4* the two components of the Cl$^-$ transport system (*I* saturable and DIDS-insensitive; *II* nonsaturable and DIDS-sensitive); *5* $\Delta\psi$-regulated K$^+$-channel of large conductance but low cation selectivity; $\oplus$ and $\ominus$ denote the tonoplast polarization due to the electrogenic H$^+$ pumping from the cytosol into the vacuole (cytosolic pH$_c$ >> pH$_v$ inside the vacuole); see text for further explanation

the cell, and hence to sterility. The physiological substrate of PDR5 is as yet unknown; however, genetic evidence indicates that the ATPase is involved in the cellular efflux of a variety of drugs such as cycloheximide, erythromycin, and sulfonamides. The PDR5 protein (*p*leiotropic *d*rug *r*esistance) corresponds to the M-type ATPases, which are sensitive to both ortho-vanadate and oligomycin.

4.1.5 Concluding Remarks

H$^+$-ATPases of biological membranes convert the electrochemical energy of proton gradient into ATP, and vice versa. The characteristic features of yeast H$^+$-ATPases are summarized in Table 7.2. These ATPases energize the cellular membranes, acting as a proton pump.

The most powerful proton pump of aerobic yeast cells and other heterotrophic cells, however, is the respiratory chain in the inner mitochondrial membrane. The energy released by the electron transport components of the respiratory chain is

Table 7.2. Characteristic features of the three types of yeast H$^+$-ATPases

Type	F	P	V
Example	Mitochondrial F$_0$F$_1$-ATPase	Plasma membrane H$^+$-ATPase	Vacuolar H$^+$-ATPase
Molecular size (kDa)	350	100	135–410
No. of subunits	10–12	1–2	3–9
pH optimum (ATPase)	8.5	6–6.5	7–7.5
Phosphorylated intermediates	No	Yes	No
Inhibitors	Oligomycin Azide	o-Vanadate	Nitrate
Substrate specificity	ATP > GTP > NTP	ATP >>> NTP	ATP > GTP > NTP

used for charge separation in hydrogen atoms which reach the respiratory chain as reduction equivalents from redox reactions. Protons are then extruded into the intermembranous space. The two corresponding electrons remain associated with the respiratory chain in the membrane and are transported down the redox potential in the chain, eventually reducing an oxygen atom to form a molecule of water. The electrochemical proton gradient generated by the respiratory chain in an aerobic cell is used by the reversible F-type H$^+$-ATPase to synthesize ATP (Fig. 7.9). This energy transduction occurs with biological membranes separating charges across the boundary of a closed compartment. Energy transduction requires an impermeable membrane bounding a compartment. Without it, energy release is uncoupled from ATP synthesis, and the energy is dissipated as heat. An electrical potential difference across membranes is a fundamental feature of a living cell.

4.2 H$^+$/Solute Symport – the Secondary Active Uptake of Nutrients

Most solute transport systems in yeasts are coupled to the electrochemical proton gradient. In Mitchell's (1966) chemiosmotic theory of energy coupling in oxidative and photosynthetic phosphorylation, the proton gradient represents an intermediary energy form, usable for either ATP synthesis or solute accumulation. The primary active transport generates a proton motive force which drives protons across a membrane as they tend to move back into the cytosol or mitochondrion, or out of the vacuole. However, the membrane is impermeable to protons, and there are only a few specific routes by which H$^+$ can return: either by the F$_0$F$_1$-ATPase in mitochondria during ATP synthesis (cf. Fig. 7.9), or by an electroneutral antiport system in exchange for another cation or positively charged metabolite which accumulates, perhaps, in the vacuole (Fig. 7.11), or by symport systems which may be either electroneutral, when coupled to an anion, or electrogenic when an electroneutral nutrient molecule (sugar) is cotransported by (systems 3 and 2, respectively, of Fig. 7.8).

These processes depolarize the membrane. The energy to repolarize it is supplied by the respiratory chain of the mitochondrion, or by an H^+-ATPase, whereby the membrane potential and/or the pH gradient is regenerated. However, biological membranes are not completely impermeable; there is a small penetration of H^+ down their electrochemical gradient. This proton backflow steadily depolarizes the membrane and a significant amount of energy is required to keep the membranes of resting cells polarized (about 25% of the endogenous respiration). Secondary active transport includes symport, antiport, and uniport systems (Fig. 7.8). In yeast, antiport systems are important in vacuolar solute transport (Fig. 7.11). Uniports transport ions across the membrane via aqueous channels in response to the membrane potential. Antiport and uniport share many common features with symport, hence, only the last will be dealt with in detail.

4.2.1 Mechanism of H^+/Solute Symport

Coupling of ion transport to movement of H^+ along an H^+ electrochemical gradient is difficult to extend to include accumulation of neutral molecules (nutrients). Similarly, a flux of H^+ in one direction cannot be distinguished from a flux of OH^- in the other; nevertheless, some transport systems are defined as OH^-/solute antiports, e.g. the uptake of inorganic phosphate in mitochondria.

The mechanism for H^+/solute symport assumes that the carrier can bind and release protons reversibly. The alternate protonation and deprotonation at the opposite sides of the membrane alters the charge on the carrier molecule, and thus the direction of its "movement" in the electric field of the membrane potential. Further, the state of protonation of the carrier influences its affinity for the solute. If the carrier is a neutral molecule (Fig. 7.12), it becomes protonated on the external side of the membrane because of the low ambient pH. Here, it has a high affinity for the solute and forms a carrier-solute (substrate) complex. Since the carrier is protonated, the carrier-substrate complex is positively charged and because the membrane is negative inside, there is a conformational change by which the proton- and the solute-binding sites face the compartment's interior (cf. Fig. 7.6D). The carrier-H^+-solute complex is deprotonated because of the higher cytosolic pH, releasing the substrate molecule. This induces another conformational change and the proton- and substrate-binding sites face the outside of the membrane, completing the transport cycle. The carrier can be protonated and a new transport cycle begins.

The accumulation of substrate in the compartment is caused by the protonation of the carrier on the outside of the membrane due to the low external pH, and the conformational change of the carrier-H^+-substrate complex caused by the electrical field of the membrane potential in which the substrate-binding site is exposed to the inside of the membrane (Fig. 7.12). If the carrier is an anion and the protonated carrier-substrate complex is neutral, substrate accumulation is caused by increased deprotonation of the carrier on the inside of the membrane, and by the membrane potential-induced conformational change of the free (negatively

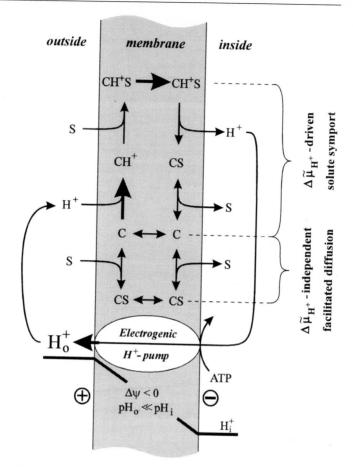

Fig. 7.12. Scheme illustrating the operation of a H⁺/solute symport. *From the bottom* The plasma membrane H⁺-ATPase operates as a proton pump, generating the membrane potential $\Delta\psi$ and the proton concentration gradient ΔpH; facilitated diffusion of the solute operates independently of $\Delta\tilde{\mu}_{H^+}$; H⁺/solute symport is driven by the entire $\Delta\tilde{\mu}_{H^+}$ (*upper part*). Transport steps which are in or very near to equilibrium are marked with *two-way arrows*, those driven by the individual components of $\Delta\tilde{\mu}_{H^+}$ are distinguished as *thick one-way arrows*. A generally accepted assumption implies that the protonated free symporter cannot traverse the membrane without a bound solute

charged) carrier back to its original state (substrate-binding site outside). Consequently, the number of carriers with substrate-binding site on the inner side of the membrane is diminished, and thus fewer carriers are available to mediate substrate efflux as compared with its influx.

H⁺ symport is a common mechanism for uptake of sugars and amino acids. H⁺ symport in the genus *Rhodotorula* was described by Höfer (1989). The H⁺/suger symporter in *Rhodotorula* is tightly coupled to the existence of the electrochemical proton gradient across the plasma membrane. In depolarized membranes the symporter becomes immobile and there is no net translocation of sugars across the membrane. In *S. pombe*, the energy coupling of the symporter can be suspended by dissipating the proton gradient (with protonophores or inhibitors of the plasma membrane H^+-ATPase). Here, the symporter protein can still catalyze facilitated diffusion and the equilibration of sugar concentrations on both side of the membrane.

H^+/solute symport exhibits the following characteristic features: (1) Alkalinization of the extracellular medium at the onset of solute transport indicates tight coupling between solute uptake and cotransport of H^+. Proton uptake is rapidly counterbalanced by an increased activity of the H^+ pump, so alkalinization is brief. (2) The half-saturation constant of the H^+ cotransport is determined by the saturation kinetics of the given solute transport. (3) The H^+/solute stoichiometry remains constant under physiological conditions which cause distinctly different rates of solute uptake. Kinetic parameters and stoichiometry of some yeast H^+/solute

Table 7.3. Kinetic parameters and the stoichiometry of various H^+/solute symports in several yeast species under different physiological conditions

Solute	Temp. (°C)	K_T (mM) solute	K_T (mM) H^+	V_T (nmol solute $min^{-1} mg^{-1}$ dry wt)	H^+/solute stoichiom.
Rhodotorula glutinis					
D-Xylose	22	0.7 ± 0.2	0.6 ± 0.2	20.5 ± 2.9	1.0 ± 0.1
	26	1.1 ± 0.3	1.1 ± 0.3	31.3 ± 1.6	1.0 ± 0.1
	28	1.2 ± 0.3	1.0 ± 0.1	43.9 ± 2.8	1.1 ± 0.1
D-Galactose	28	1.2	2.2	35.7	1.0 ± 0.1
2-Deoxy-D-glucose	22	0.3	0.3	18.9	1.2
	28	0.4	0.4	28.7	1.0
Glucuronate	28	10 (pH 4)	n.m.	11.5 (pH 4)	1.0 ± 0.05
R. glutinis M8 (glucokinase-deficient mutant)					
D-Glucose	30	n.m	n.m.	6.37 ± 0.3	1.0 ± 0.1
D-Xylose	25	n.m.	n.m.	23.9 ± 1.9	1.1 ± 0.1
Metschnikowia reukaufii					
D-Glucose	28	0.9	1.1	39.4	0.8
D-Xylose	28	2.0	2.1	41.0 ± 3.7	1.0 ± 0.1
3-O-Methyl-D-glucose	28	1.2	1.1	21.7 ± 2.7	1.0 ± 0.05
Schizosaccharomyces pombe					
2-Deoxy-D-Glucose	30	2.1 ± 0.2	3.7	23.8 ± 3.5	0.43 ± 0.05
	10	n.m.	n.m.	9.7	0.4
+ oligomycin	30	n.m.	n.m.	11.4	0.35
anaerobiosis	30	2.6	n.m.	21.2	0.45
anaerobiosis	10	n.m.	n.m.	8.7	0.45
D-Gluconate	30	3.0 ± 0.2	n.m.	10.0 ± 2.0	1.0 ± 0.05

K_T, the half saturation constant of either solute or H^+ transport. V_T, the maximum solute flux. The H^+/solute stoichiometry is calculated as the ratio of the maximum H^+ and solute fluxes. The given values are arithmetic means of at least four independent measurements $\pm$ SEM. n.m., not measured – in this case a number of single measurements (≥ 6) of H^+ and solute fluxes at a saturating solute concentration were carried out. No SEM is indicated when less than the given number of measurements has been available.

symports are given (Table 7.3). (4) H$^+$ symport is electrogenic. The translocation of the positively charged complex (or the back translocation of the negatively charged free carrier) represents a net transport of charges, depolarizing the membrane. This is also transient, as it is compensated by an opposite flow of intracellular K$^+$ and operation of the H$^+$ pump. (5) The coupling between H$^+$ and substrate transport is manifested by inhibition of substrate transport by agents affecting the permeability of membranes to H$^+$, DCCD (dicyclohexylcarbodiimide), or uncouplers (CCCP; carbonylcyanide m-chlorophenylhydrazone). Active transport parallels the generation and maintenance of an electrochemical H$^+$ gradient by the plasma membrane.

4.2.2 Glucose Transport in Yeast

Glucose is the most common yeast nutrient. *Rhodotorula* species take up glucose by a constitutive, high-affinity H$^+$ symport, In *Kluyveromyces*, *Hansenula*, and *Candida* species, the high-affinity uptake system may be induced by low glucose concentrations. Induction of H$^+$/glucose symport at low glucose concentration is common in *Candida* and *Hansenula* species. At high glucose concentrations, the high-affinity H$^+$ symport is repressed and glucose uptake occurs by facilitated diffusion. In *S. cerevisiae* glucose uptake is through a proton-independent transport by facilitated diffusion. There is also a high-affinity glucose transport system, found in cells grown on substrates other than glucose, dependent on hexokinase synthesis, and controlled by the *SNF3* gene. A nonsaturable glucose uptake system appears when the high-affinity system is repressed by high glucose concentration. Despite the lack of saturability, the glucose transport is also catalyzed by a transport protein since it is substrate-specific (no uptake of L-glucose) and can be suppressed by inhibitors of protein synthesis. Regulation of expression of the *SNF3* gene and of the low-activity glucose carrier is not yet understood.

4.2.3 Molecular Biology of Membrane Transport Proteins in Yeast

The isolation and purification of transport proteins is extremely difficult mainly because of the lack of an easy test of their transport catalytic activity. The glucose carrier has been isolated from a plasma membrane fraction from erythrocytes and reconstituted in proteoliposomes. Identification and characterization of transport proteins was made easier by using the *Escherichia coli* in vivo expression system. *E. coli* cells are transformed with an expression vector containing the given transport protein gene so that the transport protein is overexpressed, and can be identified and isolated. Similarly, the *S. cerevisiae* arginine carrier was isolated from plasma membranes and reconstituted in plasma membrane vesicles.

The spatial structure of functional transport proteins (carriers, symporters) is unknown. All attempts to crystallize transport proteins for X-ray or electron diffraction analysis have failed. Current models of spatial structure of transport proteins are based on a few known structures of integral membrane proteins, e.g., bacteriorhodopsin. Cloning the carrier gene is another method for determination of the structure and mechanism of operation of a transport protein. The cloned gene is sequenced and the derived amino acid sequence subjected to the hydropathy test, which determines the probability that the α-helix of the protein spans the membrane (Kyte and Doolittle 1982). Segments of the polypeptide chain, probably spanning the membrane, can be identified, allowing prediction of the basic protein structure.

Genes can be isolated from yeast by genetic complementation (Rose 1987). This method can be applied to cloning genes which encode transport proteins, if a suitable transport mutant is available. In *S. cerevisiae*, genes for glucose (*SNF3*, *HXT1*, and *HXT2*) and galactose transporters (*GAL2*) and maltose symporter (*MAL61*) were isolated by complementation. Two additional genes (*HXT3* and *HXT4*) were identified as suppressors of the potassium transport defect in potassium transport-deficient cells (Ko et al. 1993). The presence of any of the *HXT* genes was sufficient to allow growth in high-glucose medium. In contrast, the normal expression of SNF3 alone does not provide these cells with enough glucose for growth. Furthermore, since snf3Δhxt1Δhxt2Δhxt3Δhxt4Δ cells were able to grow on low-glucose media, it has been concluded that SNF3 may function as a negative regulator of glucose transport. Indeed, whereas there is 62 to 88% sequence identity between each of the HXT transporters and the GAL2 transporter, there is only 22% homology between HXT and SNF3.

Recently, three novel *HXT*-related genes (*HXT5* to *HXT7*) were isolated and a mutant strain deficient in all seven *HXT* genes was constructed (Reifenberger et al. 1995). Using this *hxt* null mutant, the function of the individual transporters in glucose metabolism was studied. The results demonstrated that expression of each of the *HXT1-7* genes is sufficient for aerobic growth on glucose, and that the change in apparent affinity observed in the wild-type strain is the consequence of glucose-regulated expression of the *HXT* genes (Özcan and Johnston 1995).

In *Schizosaccharomyces pombe*, glucose is taken up by an H^+-symporter. The symporter gene was recently isolated also by complementation of a glucose transport-deficient mutant (Milbradt and Höfer 1994). The *S. pombe* symporter consists of 565 amino acids and displays a significant homology to the HXT multigene family of *S. cerevisiae*. All these sugar transporter genes belong to a transporter gene superfamily, which also includes genes from mammals, bacteria, plants, as well as other yeast species and fungi (Marger and Saier 1993). All members of this superfamily contain 12 putative membrane-spanning hydrophobic domains with one larger hydrophilic region between domains 6 and 7. The membrane-spanning domains 3, 5, 7, 8, and 11 have been postulated to assemble into a pentagonal form, thus forming a channel through which a sugar molecule could be transported (Goswitz and Brooker 1995).

5 Special Techniques in Membrane Studies

These techniques make use of artificial phospholipid membrane vesicles (liposomes) in which isolated membrane fractions or individual solubilized and purified transport proteins are reconstituted (proteoliposomes), and include the use of ionophores and electrophysiological techniques, including electroporation.

5.1 Reconstituted Plasma Membrane Vesicles

Reconstituted membrane vesicles are the only biochemical system for testing catalytic properties of the transport proteins. The catalytic activity of a transport protein can be measured only by incorporating it in a membrane separating two compartments, as in proteoliposomes. Proteoliposomes offer several advantages over intact cells: (1) vesicles are devoid of metabolic activities except for those associated with the membrane; (2) the intravesicular milieu can be defined; (3) solute transport can be coupled to a defined electrical, chemical, or electrochemical potential difference; and (4) the symmetry of solute transport in vesicles of known orientation can be determined.

The transport properties of reconstituted membrane vesicles depend on how they were prepared. Yeast plasma membrane fraction can be purified by using cationic silica microbeads. These are attached to the membranes of cell protoplasts to enhance their density relative to other membranes. The microbeads are added to a purified protoplast suspension incubated in a buffer, the protoplasts separated from the unbound microbeads by centrifugation and then lysed by resuspending in a hypotonic buffer. Purified plasma membranes (less than 10% contamination with mitochondrial protein) are recovered from the lysate by low-speed centrifugation. The plasma membrane with attached silica microbeads can be reconstituted in proteoliposomes. These can generate a proton concentration gradient and/or an electrical potential difference (Gläser and Höfer 1987).

Membrane vesicles may be right-side-out or inside-out, depending on which side was originally external. This is important in the energization of membrane vesicles (Fig. 7.13). Plasma membrane vesicles are energized via the H^+-ATPase upon addition of Mg-ATP. Since the ATP-binding site of the ATPase is exposed to the cytoplasmic side, only inside-out vesicles can hydrolyze the ATP and pump H^+ into the vesicles. The right-side-out vesicles can be energized either by coreconstituted cyt c-oxidase system (Driessen and Konings 1993) or by generation of diffusion potentials.

Because concavalin A (ConA) binds to α-D-mannopyranosyl and α-D-glucopyranosyl residues only on the outside of the plasma membrane, labeled ConA can be used as a marker of the outer face of plasma membranes, and thus allows estimation of the proportion of right-side-out to inside-out vesicles.

Transport studies require tightly sealed membrane vesicles, capable of generating and maintaining the (electro)chemical H^+ gradient. Leakiness of the vesicles can be measured by determining the rate of dissipation of the pH gradient follow-

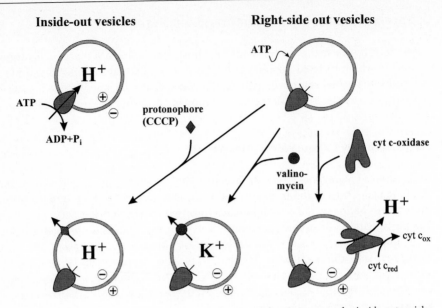

Fig. 7.13. Two possible orientations of plasma membrane vesicles. Contrary to the *inside-out* vesicles (*upper left*), the *right-side-out* vesicles cannot be energized by the nonpermeant Mg-ATP; a purified mitochondrial cyt c-oxidase complex is frequently coreconstituted to energize plasma membrane vesicles. The K^+/valinomycin system can also be used to generate $\Delta\psi$ along a preset K^+ concentration gradient (positive on the side of lower K^+ concentration, in the figure *outside* the vesicle). The protonophore system (H^+/CCCP) functions similarly. Note that the orientation of other membrane vesicles may be different, e.g., tonoplast vesicles would be oriented in the opposite way, i.e., the right-side out vesicles can be energized by added Mg-ATP (cf. Fig. 7.11)

ing the addition of hexokinase and glucose (removal of ATP). Leakiness may be reduced by an addition of 5 mol% ergosterol to the phospholipid mixture for reconstitution (Höfer et al. 1991).

Membrane vesicles may be reconstituted by freezing and thawing purified plasma membranes mixed with a phospholipid fraction (with or without ergosterol), or by solubilization of plasma membranes using nonionic detergents (n-octylglucoside), addition of phospholipids, and removal of the detergent by dialysis (Mair and Hofer 1988). By the former method, the H^+-ATPase activity of *Metschnikowia reukafii* was strongly stimulated by K^+ and less so by Na^+ (Höfer and Gläser 1989). Without cations, there was no pH change after addition of Mg-ATP. Potassium ions stimulated the H^+-ATPase and narrowed its pH activity profile to a peak at pH 6.5. With K^+ present, other nucleoside triphosphates were hydrolyzed at much lower rates than for ATP. The half-saturation constant of the ATPase for ATP was significantly lower. This plasma membrane ATPase appears to operate as an electrogenic H^+/alkali metal exchanger with a strong preference for K^+.

The symporter proteins show lower stability during the reconstitution procedure than the membrane-bound H^+-ATPases. Only the galactose symporter from *Kluyveromyces marxianus* was successfully reconstituted in plasma membrane vesicles, energized by a coreconstituted mitochondrial cytochrome c-oxidase complex (van Leeuwen et al. 1991).

5.2 Ionophores – Ion Transport-Mediating Reagents

Ionophores mediate specific translocation of particular ions across biological membranes and phospholipid bilayers. They are valuable tools enabling generation, modulation, or dissipation of electrochemical ion gradients in a defined way, in reconstituted vesicles. There are three classes of ionophores: (1) Protonophores ("uncouplers"), are lipophilic weak acids (HA) which mediate H^+/A^- symport down the pH gradient and A^- uniport in the opposite direction, driven by the membrane potential. Protonophores dissipate electrochemical H^+ gradients by rendering membranes permeable to H^+. In reconstituted membrane vesicles, they induce energization when the pH is appropriate (Fig. 7.13). (2) Valinomycin-type ionophores are electroneutral oligopeptides mediating electrogenic uniport of cations. They are carriers (valinomycin), or channels (gramicidin). Valinomycin is specific for K^+ and is frequently used to energize membrane vesicles by applying defined K^+ concentration gradients (Fig. 7.13). Gramicidin forms channels without a distinct cation specificity. (3) Nigericin-type ionophores catalyze electroneutral cation antiport for H^+. The molecules are either protonated and electroneutral, or deprotonated and negatively charged, depending on the pH. The deprotonated molecule has a ring configuration, binding the appropriate cation. Nigericin forms complexes with K^+, transporting them in one direction and H^+ in the other, down the pH gradient. Nigericin-type ionophores dissipate ion concentration gradients without interfering with the existing membrane potential. Monensin catalyzes antiports of Na^+ for H^+, and the compound A23187, Ca^{2+} for $2H^+$ (Harold 1986).

Note. We found that only protonophores were effective on H^+ and K^+ fluxes or gradients in intact yeast cells or protoplasts. Reconstituted yeast plasma membrane vesicles displayed normal ionophore sensitivity.

5.3 Electrophysiological Measurements on Yeast Cells

Since only a few ions need be translocated electrogenically across a membrane to affect the membrane potential, this potential is the most sensitive indicator of the energetic state of biological membranes. The high intracellular concentration of K^+, in combination with the membrane potential-dependent K^+ channels in the plasma membrane, represents an efficient membrane-potential buffer of yeast cells.

5.3.1 Measurements of Membrane Potential ($\Delta \psi$)
by Glass Microelectrodes in Yeast Cells

Measurements of $\Delta \psi$ with glass microelectrodes in yeast cells is difficult, because of their small size and mobility in suspensions. Nevertheless, cells of *Pichia humboldtii* (Lichtenberg et al. 1988) can be trapped in a microfunnel by suction and impaled on a microelectrode (tip diameter <0.5 μm) mounted on a micromanipulator. The values from microelectrode measurements corresponded to those calculated using the lipophilic cation tetraphenylphosphonium (Höfer 1989).

5.3.2 Measurements of Ion Channels by the Patch-Clamp Technique in Yeast Cells

This technique allows recording of ion fluxes through a single channel protein. It requires an extremely tight seal between a glass micropipette and the membrane, so that current can flow only by passing ions through channels in the patched membrane. This is possible with the whole cell attached to the micropipette (cell-attached or whole-cell mode) or with the patch separated from the cell (inside-out patch or outside-out patch; Fig. 7.14). Use of detached patches permits study of

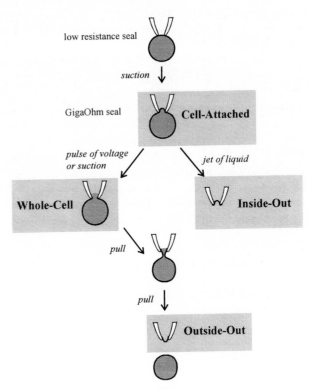

Fig. 7.14. Patch-clamp recording modes. A high resistance seal (>10^9 Ohm) will spontaneously form when a cell with a clean membrane surface is gently sucked to the tip of a micropipette – *Cell-Attached* mode. The GigaOhm seal is very stable, it is possible to break the patch covering the micropopette tip either by a short voltage pulse or by a pulse of suction – *Whole-Cell* mode. *Inside-Out* patch is obtained by shooting off the cell in the cell-attached mode with a stream of medium, whereas an *Outside-Out* patch is formed by a gentle pull of the micropipette in the whole-cell mode

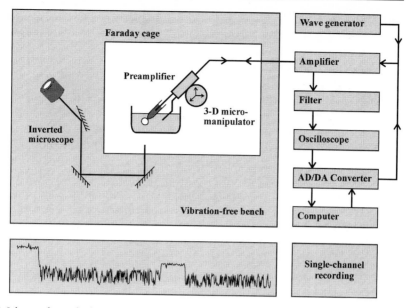

Fig. 7.15. Scheme of a patch-clamp setup. An inverted microscope is used to do the micromanipulation in a measuring chamber in which the patch-clamp micropipette (measuring electrode) and a reference electrode are submerged; both electrodes are connected to a preamplifier which can be moved by a three dimensional hydraulic micromanipulator (3-D micromanipulator). The microscope stands on a vibration-free bench, and the whole setup is shielded by a Faraday cage. After a formation of the GigaOhm seal, the current fluctuations due to opening and closing of individual membrane channels can be recorded (*bottom shadowed part* of the figure), using a voltage-clamping circuit. The small currents, in the range of pA, are measured as a voltage drop across a large resistor. An operational amplifier varies its output to keep the micropipette potential at a preset values; this proceeds very rapidly and precisely so that the micropipette potential remains virtually constant. Hence, the voltage imposed across the membrane patch can be set at any desired value and the fluctuations of current due to ions flowing through the channel(s) in the membrane patch can be monitored. The signals from the preamplifier are further amplified, filtered, and registered by an oscilloscope. Passing through an AD/DA converter, the current signals are eventually recorded, analyzed, and evaluated by a computer. At the same time, the voltage imposed over the studied membrane can be controlled by a computer software. The wave generator (*top right* of the scheme) is used to measure the resistance of sealing of the micropipette to the membrane

channel gating by recording single channel events related to the applied voltage. The composition of the solution on either side of the patch can be altered to test the effect of different solutes on the channel behavior. In the whole-cell mode, electric currents through a full complement of plasma membrane channels and/or pumps can be measured and used for detection and characterization of electrogenic transport systems having low turnover rates.

The patch-clamp system is shown in Fig. 7.15. A protoplast is sucked onto the tip of a micropipette. Electrical resistance between the electrode tip and the plasma membrane should increase rapidly from tens of megaohms to about 10 gigaohms (GigaOhm seal). Yeast, fungal, and plant cell protoplasts must be true protoplasts without traces of cell wall residues. If the protoplasts are suspended in a mildly hypoosmotic patch-clamp medium, some of the protoplasts will inflate slowly and prove suitable for patch-clamping.

Such a GigaOhm seal, using a voltage-clamping circuit, permits recording of current fluctuations due to openings and closings of individual membrane channels (Fig. 7.15). Gustin et al. (1986) found in *S. cerevisiae* TEA-sensitive channels with a unit conductance of 20 pS with selectivity for K^+.

Plasma membrane patches from *S. pombe* (Vacata et al. 1993) had voltage-dependent ionic channels, mildly selective for potassium over sodium, lithium, and chloride. They showed conductances with a maximum of 153 pS. The channels gated (opened and closed) as expected, in the region of physiologically relevant voltages, being closed at hyperpolarizing and open at depolarizing voltages. The channels are closed in metabolically active, energized cells. When the plasma membrane becomes depolarized, during H^+ symport, the channels open, intracellular K^+ ions move out of the cell and repolarize the plasma membrane. In electrogenic charge translocation it is irrelevant which ionic species carries the charge. Reconstituted plasma membrane vesicles from *S. pombe* showed little difference in ionic preferences for charge compensation during ATP-induced generation of membrane potential (Mair and Höfer 1988). *S. pombe* had approximately 35 ion channels per cell.

5.3.3 Inclusion of Nonpermeant Molecules into Yeast Cells by Electroporation

Controlled electric pulses of high charge intensity and short duration (ms) appear to cause a transient opening of the cell membrane, which spontaneously reseals after the treatment. In yeast, this technique was successfully used to introduce plasmids for cell transformation (Becker and Guarente 1991), the fluorescent dye pyranine to measure the intracellular pH_i (Peña et al. 1995) and nonpermeant agents such as ortho-vanadate and dicyclohexylcarbodiimide (Höfer et al. 1995). Pyranine included in yeast cell allows for a continuous recording of the pH_i and thus the cytosolic response to, e.g., stimulation of the plasma membrane H^+-ATPase by glucose. Including nonpermeant inhibitors intracellularly makes it possible to study the sidedness of their effect under in vivo conditions.

In a typical electroporation protocol, a 50% (fresh wt/vol) yeast cell suspension in an appropriate buffer, containing the reagent to be included, is used. A sample of cell suspension (0.7–1 ml) is placed in an electroporation cuvette and pulsed at 5.7 kV cm^{-1} field strength, 200 Ohm, and 25 μF. The elecrtroporated cell suspension is rinsed with water by centrifugation (to remove external reagent) and the cell vitality is checked by measuring cell respiration. The vital cell suspension with included reagent is used for experiments.

Acknowledgments. The work used in this chapter was supported by the Deutsche Forschungsgemeinschaft, Bonn (Grant Series Ho 555), and partially by the Volkswagen Stiftung, Hanover, the DAAD (German Academic Exchange Service), Bonn, and the Commission of European

Union, Brussels. The illustrations were designed by Dr. V. Vacata, University of Bonn, whose sound and skilful expertise is highly appreciated.

References

Anraku Y, Umemoto N, Hirata R, Wada Y (1989) Structure and function of the yeast vacuolar membrane proton ATPase. J Bioenerg Biomembr 21:589–603

Becker DM, Guarente L (1991) High efficiency transformation of yeast by electroporation. Methods Enzymol 194:182–187

Davson H, Danielli JF (1943) The permeability of natural membranes. Cambridge Univ. Press, Cambridge and New York

Driessen AJM, Konings WN (1993) Insertion of lipids and proteins into bacterial membranes by fusion with liposomes. Methods Enzymol 221:394–408

Evans WH, Graham JM (1989) Membrane structure and function. IRL Press at Oxford University Press, Oxford

Gläser H-U, Höfer M (1987) Ion-dependent generation of the electrochemical proton gradient $\Delta\bar{\mu}_{H^+}$ in reconstituted plasma membrane vesicles from the yeast *Metschnikowia reukaufii*. Biochim Biophys Acta 905:287–294

Goffeau A (1993) Transport ATPases and yeast metabolism. In: Scheffers WA, van Dijken JP (eds) Metabolic compartmentation in yeast. Pasmans, The Hague, pp 42–44

Goffeau A, Green NM (1990) The H^+-ATPase from fungal plasma membrane. In: Pasternak CA (ed) Monovalent cations in biological systems. CRC Press, Boca Raton, pp 155–169

Goswitz VC, Brooker RJ (1995) Structural features of the uniporter/symporter/antiporter superfamily. Protein Sci 4:534–537

Gustin MC, Martinac B, Saimi Y, Culbertson MR, Kung C (1986) Ion channels in yeast. Science 233:1195–1197

Harold FM (1986) The vital force: a study of bioenergetics. W.H. Freeman and Company, New York

Hille B (1992) Ionic channels of excitable membranes. Sinauer Associates Inc., Sunderland, MA

Higgins CF (1992) ABC transporters: from microorganisms to man. Annu Rev Cell Biol 8:67–113

Höfer M (1981) Transport across biological membranes. Pitman Publish Ltd, London

Höfer M (1989) Accumulation of electroneutral and charged carbohydrates by proton cotransport in *Rhodotorula*. Methods Enzymol 174:623–653

Höfer M, Gläser H-U (1989) K^+ ions stimulate both ATPase activity and ΔpH generation in reconstituted yeast plasma membrane vesicles. In: Kotyk A, Skoda J, Pačes V, Kostka V (eds) Highlights of modern biochemistry. VPS International, Utrecht, vol I, pp 753–760

Höfer M, Mair T, Wernsdörfer E (1991) Reconstituted plasma membrane vesicles: a tool to study transport in yeast. In: Prasad R (ed) Yeast molecular biology and biotechnology. Omega Sci Publ, New Delhi, pp 239–253

Höfer M, Calahorra M, Klein B, Paña A (1995) Assessment of $\Delta\bar{\mu}_{H^+}$ in *Schizosaccharomyces pombe*; intracellular inclusion of impermeable agents by electroporesis. Folia Microbiol 40: in press

Hofstee BHJ (1959) Non-inverted versus inverted plots in enzyme kinetics. Nature 184:1296–1298

Ko CH, Liang H, Gaber RF (1993) Roles of multiple glucose transporters in *Saccharomyces cerevisiae*. Mol Cell Biol 13:638–648

Kyte J, Doolittle RF (1982) A simple method for displaying the hydropathic character of a protein. J Mol Biol 157:105–132

van Leeuwen CCM, Postma E, van den Broek PJA, van Steveninck J (1991) Proton-motive force-driven D-galactose transport in plasma membrane vesicles from the yeast *Kluyveromyces marxianus*. J Biol Chem 226:12146–12151

Lichtenberg H-C, Giebeler H, Höfer M (1988) Measurements of electrical potential differences across yeast plasma membranes with microelectrodes are consistent with values from steady-state distribution of tetraphenylphosphonium in *Pichia humboldtii*. J Membr Biol 103:255–261

Mair T, Höfer M (1988) ATP-induced generation of pH gradient and/or membrane potential in reconstituted plasma membrane vesicles from *Schizosaccharomyces pombe*. Biochem Int 17:593–604

Marger MD, Saier MH Jr (1993) A major superfamily of transmembrane facilitators that catalyze uniport, symport and antiport. Trends Biochem Sci 18:13–20

Milbradt B, Höfer M (1995) Glucose-transport-deficient mutants of *Schizosaccharomyces pombe*: phenotype, genetics and use for genetic complementation. Microbiol 140:2617–2623

Michaelis L, Menten ML (1913) Kinetics of invertase action. Biochem Z 49:333–369

Mitchell P (1966) Chemiosmotic coupling in oxidative and photosynthetic phosphorylation. Biol Rev 41:445–502

Özcan S, Johnston M (1995) Three different regulatory mechanisms enable yeast hexose transporter (*HXT*) genes to be induced by different levels of glucose. Mol Cell Biol 15:1564–1572

Peña A, Ramírez J, Rosas G, Calahorra M (1995) Proton pumping and the internal pH of yeast cells, measured with pyranine introduced by electroporation. J Bacteriol 177:1017–1022

Prasad R (1985) Lipids in the structure and function of yeast membranes. Adv Lipid Res 21:187–242

Reifenberger E, Freidel K, Ciriacy M (1995) Identification of novel HXT genes in *Saccharomyces cerevisiae* reveals the impact of individual hexose transporters on glycolytic flux. Mol Microbiol 16:157–167

Robertson JD (1964) Unit membranes: a review with recent new studies of experimental alterations and a new subunit structure in synaptic membranes. In: Locke M (ed) Cellular membranes in development. Academic Press, New York, pp 1–81

Rose MD (1987) Isolation of genes by complementation in yeast. Methods Enzymol 152:481–511

Smith C, Wood EJ (1991) Energy in biological systems. Chapman & Hall, London

Singer SJ, Nicolson GL (1972) The fluid mosaic model of the structure of cell membranes. Science 175:720–731

Vacata V, Höfer M, Larsson HP, Lecar H (1993) Ionic channels in the plasma membrane of *Schizosaccharomyces pombe*: evidence from patch-clamp measurements. J Bioenerg Biomembr 24:43–53

Van der Rest ME, Kamminga AH, Nakano A, Anraku Y, Poolman B, Konings WN (1995) The plasma membrane of *Saccharomyces cerevisiae*: structure, function and biosynthesis. Microbiol Rev 59:304–322

See also *Notes Added in Proof* on p. 353.

CHAPTER 8

The Yeasts: Sex and Nonsex. Life Cycles, Sporulation and Genetics

J.F.T. Spencer and D.M. Spencer

1 Life Cycles

1.1 Introduction

The vegetative life cycle of a yeast appears simple. The cell may produce a bud, the bud matures, separates from the mother cell, and produces its first bud. The mother cell continues to produce other buds. It is a simple way of life. However, a single yeast cell, if haploid, besides budding, can find a partner, fuse with it, produce a diploid cell, sporulate, and produce other yeast cells, with different arrangements of genes. Its life is not so simple.

The sexual cycle in yeasts differs mainly in the fraction of the cycle that the yeast spends in the haplophase and how much in the diplophase. "Imperfect" yeasts may be haploid, the diplophase being suppressed. However, yeasts such as *Candida albicans* undergo mitotic recombination when irradiated with ultraviolet radiation, and auxotrophic mutants are expressed. Therefore, *C. albicans*, which has no known sexual cycle, is diploid.

In "perfect" species, the yeast alternates between the haplophase and diplophase, and forms spores. *Saccharomyces cerevisiae* is normally diploid. Reduction division occurs during sporulation. The ascus contains four haploid spores, which germinate, producing haploid cells which sometimes mate immediately to form a zygote, which produces a diploid bud after karyogamy, and the yeast returns to its normal diploid state. The zygote may be a transient dikaryotic stage, though the two haploid nuclei fuse almost immediately. If haploid cells are produced for a few generations, the diploid state is restored when the haploid cells come in contact with haploids of the opposite mating type, or mating may be delayed for a few generations. Some strains of *Saccharomyces cerevisiae* undergo mating-type switching, where some cells of a haploid strain change to the opposite mating type. These then mate with cells of the original mating type, and the diploid state is restored.

Fowell (1969) describes eight types of life cycle in yeasts.

1. *Torulopsis* type (budding of vegetative cells only, without spore formation).
2. *Schizosaccharomyces pombe* type, (vegetative cells multiply by fission, not budding. The sexual phase begins with fusion of two cells of opposite mating type, to form a zygote which functions as an ascus. The diploid state is very short and

J.F.T. Spencer/D.M. Spencer (eds)
Yeasts in Natural and Artificial Habitats
© Springer-Verlag Berlin Heidelberg 1997

meiosis takes place. Haploid spores are formed. These germinate and restore the haploid state. If the cells are homothallic they form spores. *Coccidiascus, Debaromyces, Endomyces, Lipomyces, Nematospora, Pichia, Schizosaccharomyces, Schwanniomyces, Wingea,* and *Zygosaccharomyces* belong to this group.

3. *Schwanniomyces* type, (conjugation between a mother cell and its bud). Meiosis takes place in the bud. The mother cell functions as the ascus. One (haploid) spore is normally formed. *Debaryomyces, Pichia,* and *Zygosaccharomyces* are similar, though in the latter species two to four spores are generally formed. *Nematospora, Endomyces,* and *Endomycopsis* form parthenogenic asci without previous conjugation. The spores and vegetative cells are haploid.

4. *Nadsonia* species have a very short dikaryotic stage. The cell is haploid during most of the life cycle. The nucleus of a bud associates with the nucleus of the mother cell during the dikaryotic phase. Both nuclei move into a bud at the other end of the mother cell and fuse. This bud is the ascus. Meiosis and then spore formation follow. A single spore forms, germinates, and the haplophase is restored. Classical genetic analysis of yeasts which form a single spore after nuclear events within a single cell envelope is impossible, and can only be done by using molecular genetics.

5. *Hansenula wingei* type. The diplophase is prolonged. Diploid cells may exist alongside haploids. Conjugation occurs between haploid cells of opposite mating types, nuclear fusion follows, and the diploid phase lasts for 50% of the life cycle. Haploid spores are formed in an ascus, on weak medium, and produce haploid vegetative cells of opposite mating types. These fuse and the cycle begins again.

6. In this type, there may be a short, transient dikaryotic stage. The vegetative stage is diploid.

7. In *Saccharomyces cerevisiae*, ascus formation occurs after conjugation between two haploid vegetative cells, or after diploidization of the spores, or of the vegetative cells immediately after germination. The haplophase is very short. The haploid vegetative cells may switch mating types after several generations and then conjugate.

8. In *Saccharomycodes ludwigii* the spores conjugate in the ascus after germination, restoring the diploid phase. The diploid vegetative cell forms an ascus and sporulates.

Some species of *Pichia, Hansenula,* and *Pachysolen tannophilus* are interesting because some of them ferment xylose and can produce heterologous proteins (Chap. 12). *Hansenula polymorpha*, a methyltrophic yeast, has a typical life cycle. It may be either haploid or diploid. If the culture is haploid, mating-type switching induced by nitrogen starvation occurs, the cells conjugate, and the diploid state forms. The culture may remain diploid, or sporulate and the haploid strains are isolated. Diploid strains sporulate readily on malt extract agar. The spores are difficult to separate by microdissection because they are bound together by a threadlike structure. However, tetrad analysis then shows normal segregation of markers in heteroallelic diploids.

Pichia pinus is similar and can be analyzed in this way. *Pachysolen tannophilus* has an unusual ascus, but genetic analysis is possible.

Sporidiobolus and *Rhodosporidium* are basidiomycetes with a long dikaryotic stage. They form a dikaryotic mycelium with clamp connections after conjugation of haploid vegetative cells. Chlamydospores are formed and meiosis occurs. A promycelium having two to four haploid cells is produced. These bud off haploid cells, the *Rhodotorula* phase. In *Sporidiobolus*, the chlamydospores may bud off diploid cells. These may be analogous to the teliospores of other heterobasidiomycetes.

2 The Cell Division Cycle in Yeast

(Hartwell 1974; Carter et al. 1983; Wheals 1985)

A sequence of very complex processes takes place between the initiation of the budding process and the separation of the new bud from the mother cell (Fig. 8.1). The process may not be as complex as that involved in the initiation and development of a mammalian fetus, but there is much less difference than appears at first sight. New DNA is synthesized, the cell nucleus changes shape, divides, the cytoplasm divides, a new cell membrane forms between the two parts, and where there was one cell, there are two. Besides the cytoplasmic membrane, the yeast cell must continuously form a new cell wall as the bud grows in size. The development of the bud conveniently provides an external marker showing the progress of the wonderfully complicated processes going on within.

The cell division cycle is divided into several phases: G1, the presynthesis stage; the S phase, when DNA synthesis becomes active; G2, which is another pause in synthetic activity, and M, or maturation phase, when the cytoplasm of the bud separates from that of the mother cell (cytokinesis) and the bud itself separates from the mother cell. DNA synthesis can be considered separately from cytokinesis. Since the two are closely correlated unless interfered with by a molecular biologist, there is a timing device as well. Cell division includes two cycles of synthesis of cell material and possibly a "clock". There are also several checkpoints, which prevent the cycle from proceeding unless the preceding steps have been completed (Hartwell and Kastan 1994). If the DNA of the cell cannot be repaired, the cell dies. There is an analogous set of checkpoints in human cells, which normally prevent cells with damaged DNA from developing into cancer cells.

The cell division cycle has been studied quite intensively in *Schizosaccharomyces pombe*, though this species lacks the external, visible markers found in *Saccharomyces cerevisiae*. Crosswall formation is one such marker. Cell separation is another, but these cannot be followed visually so readily.

There may be a "G0" phase, differing from the G1, but the evidence for this is equivocal. "G0" is more acceptable in mammalian cells where there is a much

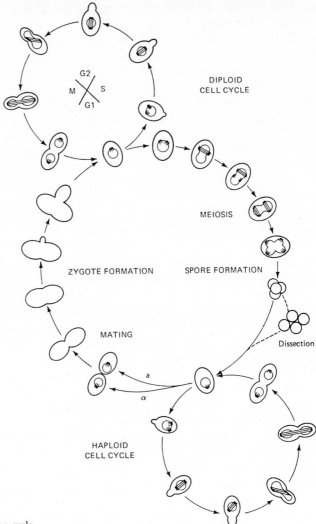

Fig. 8.1. Yeast cell division cycle

longer quiescent phase with little metabolic activity in the cell. In the G1 phase in yeast, the cell remains unbudded until there are sufficient nutrients for cell division to begin, at "Start". Cells having a conditional, temperature-sensitive mutation (cdc28; cell division cycle mutant 28) at this point, do not "Start" if held at the restrictive temperature. At "Start", a point early in interphase, the events leading to doubling of the cell mass are coordinated with DNA synthesis, centrosome duplication, and mitosis, which normally proceed faster than increase in cell mass. The bud does not reach full size until after separation from the mother cell, and cannot enter the next cycle of mitosis and bud formation until it reaches a critical size, when "Start" can take place if the concentration of nutrients in the medium is also great enough. Also, haploid cells cannot pass "Start" if the appropriate mating-type hormone is present. Cells arrested by mating hormones do not continue into mitosis, but into meiosis, and sporulate (see Sect. 4). Cells arrested before start may arrest as unbudded, nongrowing cells, as in nutrient starvation. Other mutants may arrest as growing but unbudded cells. One of the genes whose product is required for passage through "Start" (CDC28 in *S. cerevisiae* and cdc2 in *Schizosaccharomyces pombe*) encodes a protein kinase.

Hartwell (1974) used temperature-sensitive mutants to study the cell division cycle. At the restriction temperature, the mutants were readily isolated, usually by their morphology or cytology (failure of buds to separate, deficiencies in cytokinesis, failure of nuclear division, or cells in which the nuclei were blocked part way through division, so that the visible cytological marker seen by nuclear staining was a dumbbell-shaped figure lodged between the mother cell and the bud. In one mutant, none of the nucleus entered the bud, but the cellular "clock" continued to run. The cell produced an empty bud at each turn of the cycle, so the result was a mother cell with two or three empty buds attached).

Near the end of G1, the first external sign of bud formation becomes visible. Transmission electron microscopy of cells at this stage reveals the beginning of activity in the nucleus, as the spindle plaque (spindle pole body) begins to divide and microtubules begin to form in both the nucleus and the cytoplasm. Before division begins, the single spindle pole body, with its microtubules extending into the nucleus and on the opposite side into the cytoplasm, has a darkly staining region beside it in the nuclear envelope, the "half-bridge". As duplication proceeds, a densely staining satellite appears on the half-bridge, at the far side from the SPB. The SPB then duplicates, and the two bodies are connected by a bridge. They remain connected by the bridge until the bud reaches approximately one-third the size of the mother cell. The complete spindle appears very rapidly here, and the bridge is divided, half remaining with each new SPB. The two plaques separate, and move around the nuclear envelope to positions opposite each other.

One or more of the microtubules extends toward the developing bud initial, and as the latter becomes recognizable as a bud, enters the bud itself. The role that the microtubules play in bud initiation and development is not known. Approximately 15 microtubules, 15–18 nm in diameter, join the two plaques, and short sections extend into the cytoplasm. DNA replication occurs during the S phase, so that there is sufficient new DNA to form the nucleus of the developing bud. One of the SPBs is always closely associated with the bud. The S phase which follows constitutes about 25% of the cell division cycle.

2.1 Chromosome Replication Takes Place in the S Phase

The DNA and the complex architecture of the chromosomes is duplicated precisely within this phase. In eukaryotic cells, replication takes place at multiple sites, during a time characteristic of the species and the type and developmental stage of the cell. Adult cells of *Drosophila* have an S phase of 10h, while cells of early embryos have an S phase of less than 4min. Assuming that the doubling time of *Saccharomyces cerevisiae* is 2h, and the S phase is 25% of the cell division cycle, the S phase is about half an hour. Replication origins in yeast have been isolated and are autonomously replicating sequences (ARS elements) which contain a short conserved core consensus sequence, essential for replication.

DNA replication takes place by replication forks, and is mediated by DNA polymerases I and III. The replication forks move in both directions at once, polymerase I containing a primase activity and apparently participating in synthesis of Okazaki fragments on the lagging strand, while polymerase III has no primase activity and participates in synthesis of the leading strand. Numerous phosphatases participate in cell cycle control, including DNA replication. Termination of DNA replication at the ends of a chromosome appears to be by a "telomerase", which adds a simple sequence DNA repeat to the end of the chromosome.

The newly formed DNA is assembled into nucleosomes, a short distance behind the replication fork. Besides the nucleosomes, the chromosomal scaffolding proteins are duplicated, though the mechanism is not known. The ARS elements and centromeres are attached to the scaffolding at some point. The replication sites may be immobilized on a structural framework in the nucleus, and the unreplicated DNA in this organelle is passed through these fixed replication sites in the S phase.

The mechanism by which the different rounds of DNA replication are coupled to the cell cycle is not known. Replicated nuclei, for instance, will not reenter the S phase until they have first passed through mitosis, so that DNA replication occurs once and only once in each cell division cycle. However, the existence of checkpoints in the cell cycle is now better understood, and these play a great part in ensuring that cells do not proceed to the next stage in the cycle until the previous step has been completed correctly. Cells which cannot complete these steps may be instructed to die, which ensures that only normal cells complete the cycle and produce a bud (Hartwell and Kastan 1994). The mechanism and molecular biology of replication of the chromosomes is discussed in Chapter 10.

The S phase is followed by the G2 phase. The bud increases in size and the nucleus moves to the neck between the mother cell and the bud. The microtubules elongate, and the nucleus begins to change shape before nuclear division, becoming constricted at the middle and assuming a dumbbell shape.

Next is the M interval. The microtubules continue to elongate, and the constriction of the nucleus proceeds until the division is complete. Finally, the new nucleus forms in the bud, and the spindle plaques in the bud and mother cell again become quiescent. A few short microtubules project into the interior of the nucleus. Cytokinesis then occurs and the plasmalemma of mother cell and bud separate, after which the bud separates completely.

Cytokinesis begins with the appearance of a monolayer of filaments, 10nm in diameter and spaced 24nm apart, just below the plasma membrane. Also, a ring of chitin is deposited around the neck of the bud, to form the bud scar later. Several small vesicles fuse, forming a pair of membranes across the opening defined by the chitin ring. Cytokinesis is complete. Synthesis of the central part of the bud scar occurs and the bud separates. The bud is still significantly smaller than the mother cell and cannot form a bud itself until it has reached full size.

The chromosomes of *S. cerevisiae* cannot be seen in the light microscope at any stage during cell division. Structures identified as synaptonemal complexes are visible by transmission electron microscopy.

The DNA content of a haploid cell of *Saccharomyces cerevisiae* is approximately $1.2-1.4 \times 10^7$ kDa, of which 80 to 90% is located in the nucleus. *Saccharomyces cerevisiae* probably has the largest number and smallest size of chromosomes of the yeasts. There are 16 or 17 identifiable chromosomes in most strains, corresponding to the number of bands visible in pulsed-field gel electrophoretic preparations. Wild and industrial strains have distinctive chromosome banding patterns (Vezinhet et al. 1990). Other yeast species often have smaller numbers of much larger chromosomes, ranging from 3–4 to 10–12. *Schizosaccharomyces pombe* has three chromosomes visualized under the light microscope, which agrees with the number of linkage groups and the number of bands visible on pulsed-field gels and species of *Hansenula* and *Pichia* having two or three.

The cell division cycle in *Schizosaccharomyces pombe* resembles that in *Saccharomyces cerevisiae*. The major cell cycle event useable as a marker is the development of the septum, which occurs three-quarters of the way through the cell cycle.

The G1 phase is very short in the wild-type *S. pombe* cell, so growth actually begins in G2, occurring at the "old" end of the cell, which existed in the previous cycle. At a point in G2 corresponding to 0.3–0.35 of the cycle, when the cell has reached a length of 9.0–9.5 μm, the new end begins to grow. At 0.75 of the cycle, mitosis occurs and the septum begins to develop. At 1.0, cytokinesis and cell separation occur.

The distribution of F-actin changes during the cell division cycle. It is localized at the growing end of the cell in newly divided cells. When bipolar growth begins, actin appears at both ends of the cell. Actin filaments appear briefly and may extend through most of the length of the cell. At mitosis, a ring of actin appears where the septum will form, overlying the nucleus during division. After the septum forms, actin disappears from the equatorial ring and reappears at the far ends of the two new cells, where it initiates the new growth. Actin filaments also appear.

Events in the nucleus during mitosis resemble those in *S. cerevisiae*. The spindle pole body duplicates in early mitosis, not in late G1, and at this stage, the two SPBs are within the nuclear envelope. They are connected by pole-to-pole microtubules, the spindle elongating until the daughter nuclei reach the ends of the cell. The structural rearrangement of the microtubules occurs at the G2/M boundary and at mitotic telophase. The changes in the spindle of the *S. pombe* cell are more like those of higher eukaryotes than in *S. cerevisiae*.

3 Cell Division Cycle Genes of *Schizosaccharomyces pombe*

Genes involved in cell division cycle control have been identified in G1, S, G2 stages, and initiation and completion of mitosis (tubulin, nuclear division, DNA topoisomerases, chromosome separation, calmodulin formation, actin synthesis). A ras$^+$ gene has been identified and investigated.

4 Sporulation

Sporulation is a cycle of development with a different object. Sporulation and spores are important tools in yeast genetics. The spores are separable and can be grown as individual clones, each representing one of the four products of meiosis. The course of genetic events in the meiotic cycle can be followed, and segregation, recombination, and other events in the yeast determined. Cloning of genes and other DNA sequences and transforming them into yeast cells is a remarkably powerful tool for the study of the mechanism of inheritance in yeasts and other living beings.

The events occurring during meiosis and sporulation resemble those of the mitotic cycle in *Saccharomyces cerevisiae* and other sporulating yeast species. The trigger for the switch from vegetative reproduction to meiosis and sporulation is not known. In *S. cerevisiae*, starvation in a nitrogen-poor medium with a nonfermentable carbon source triggers sporulation. Instances of release of yeasts from inhibition of sporulation by glucose are known. The presporulation medium is important; cultivation on a presporulation medium containing a nonfermentable carbon source, and transfer to sporulation medium induced sporulation in poorly sporulating strains of brewing yeasts.

The temperature of incubation affects sporulation, which was better, and spore viability higher, when the incubation temperature was 21 °C, than at 27 °C. We have found that sporulation and spore viability were improved when the incubation temperature was 15–18 °C.

4.1 The External Factors Controlling Sporulation in *Saccharomyces cerevisiae* Are Well Known (Miller 1989)

4.1.1 The Presporulation Medium

A rich presporulation medium (PSM) medium improves sporogenesis on transfer to sporulation medium (SM). The carbon source in PSM may be a fermentable sugar or a nonfermentable compound: acetate, lactate, or glycerol. If it is a fermentable sugar, the cells should be grown to early stationary phase before transfer to SM containing acetate. If the carbon source is nonfermentable, cells from the logarithmic growth phase give better sporulation. The growth conditions should allow the gluconeogenic and glyoxalate bypass enzymes to be completely dere-pressed when the cells are transferred to SM.

4.2 Sporulation Media

For S. cerevisiae, the gluconeogenic enzymes and glyoxalate cycle enzymes should remain completely derepressed when transferred to the presporulation medium. Yeast cells can be sporulated on moist gypsum blocks, carrot or potato wedges, or a vegetable extract agar. Sporulation media contain nonfermentable carbon sources such as acetate, ethanol, glycerol, dihydroxyacetone, or pyruvate. Inhibi-tors (mycophenolic acid, ethyl N-phenylcarbamate, or gas-oil) may stimulate sporulation.

Low concentrations of glucose may increase sporulation. Galactose is less in-hibitory to spore formation than glucose, fructose, or mannose. Increasing the concentration of these sugars from 0.05 to 0.1% may increase the proportion of three- and four-spored asci, while the percentage sporulation is reduced.

Sporulation requirements for other yeast species may be different. *Metschnikowia* spp. usually sporulate best on dilute V8 agar or other vegetable extract media, *Debaryomyces* spp., on Gorodkowa agar containing meat extracts and NaCl, and many species (*Hansenula*, *Pichia* spp., *Kluyveromyces* spp., *Zygosaccharomyces* spp., and *Schizosaccharomyces* spp.), on malt extract agar or YM agar[1].

4.3 Population Density of the Cells in SM

The best yields of asci, on SM, are usually obtained at densities of 1×10^6 to 1×10^7 cells/ml, if the cultures are aerated adequately. Adequate sporulation of *Saccharomyces cerevisiae* in liquid McClary's medium, from 48-h cultures YEP broth, is 1:10 (v/v). Good sporulation was obtained after another 48 h growth in aerated culture. S. cerevisiae does not sporulate under anaerobic conditions, but *Schizosaccharomyces japonicus* does.

4.4 Carbon Dioxide

A little CO_2 is required by yeast during sporulation. CO_2 carboxylates phospho-enolpyruvate, forming oxaloacetate, which is converted to other TCA-cycle inter-mediates. The TCA and glyoxalate cycles are fully activated during sporulation. Their operation is required for spore formation.

---

[1] YM (yeast-malt) agar: YM agar contains yeast extract 0.3%, malt extract 0.3%, peptone 0.5%, glucose 1%, and agar 2%, and is used for maintenance and sporulation of yeasts. Dehydrated YM medium is supplied by Difco, Inc.

4.5 pH (Hydrogen Ion Concentration)

The optimum pH for sporulation in S. cerevisiae is probably neutral to alkaline, particularly in acetate-containing media. Sporulation has been obtained over the entire physiological pH range.

4.6 Temperature

Saccharomyces cerevisiae sporulates over a temperature range of 3–35 °C. The optimum temperature is about 25–30 °C. Industrial yeast strains sporulate more freely and produce more viable spores at 20–21 °C than at 30 °C, as was discovered at the Carlsberg Laboratory.

4.7 Water Activity

Sporulation is reduced in media having increased osmotic tensions. Yeast spheroplasts sporulated freely in osmotically stabilized media containing 0.6 M KCl or 1 M sorbitol. This is a useful way of obtaining ascospores free of vegetative cells in strains which form protoplasts or spheroplasts in these conditions.

4.8 Effects of Radiation on Sporulation

Sporulation in S. cerevisiae is sensitive to the effects of visible light, UV, and ionizing radiation. Premeiotic DNA formation is the most sensitive to UV and ionizing radiation. Some radiation-repair (RAD) genes are also involved in meiosis. These genes increased the sensitivity to radiation and blocked completion of normal meiosis.

Sporulation in S. cerevisiae was stimulated by red light and inhibited by blue and green light, especially by blue light. Irradiation with blue and green light altered the composition of the spore wall. The spores no longer stained with malachite green.

4.9 Inhibitory Effects of NH_4 Ions on Sporulation

NH_4^+ ions repress sporulation, in the presence of acetate or other metabolizable carbon source. Yeasts will sporulate in the presence of NH_4^+ ions in buffer or solutions containing dihydroxyacetone or lysine, and in the presence of pyruvate if its concentration is limiting. Likewise, older (large) cells may sporulate in the presence of NH_4^+ ions, where small (young) cells of the same strain do not.

The mechanism of inhibition is at present unknown. Methylamine, a nonmetabolizable analog of ammonia, suppresses sporulation, while mycophenolic acid and ethyl-N-phenylcarbamate are antagonistic to the action of

NH_4^+ ions. Finally, reduction of the intracellular concentration of NH_4^+ ions increases gluconeogenesis, which is essential for sporulation. During sporulation, the culture is influenced by multiple factors acting in combination.

5 Numbers of Spores per Ascus

S. cerevisiae normally forms four spores per ascus. However, some strains form few or no four-spored asci. A few form diploid spores in two-spored asci. Asci having six spores may be induced on media containing acetate, glucose, amino acids and glycerol. The herbicide amitrole, added to the PSM, also induced formation of multispored asci. If strains which were homozygous for mating type (a/a or α/α) were mated with haploid strains carrying the kar mutation, defective for nuclear fusion, and the resulting hybrid was sporulated, six-spored asci were obtained.

5.1 Formation of Two-Spored Asci

Sporulation in buffer or water, high cell densities, high temperature (causing reversible pachytene arrest), and addition of ethidium bromide to SM containing acetate as carbon source all increased the frequency of two-spored asci. Acetate, pantothenate, or potassium ions in the sporulation medium, and microaerophilic conditions in the PSM increased the numbers of four-spored asci. Growth and sporulation of some brewing yeasts at 21 °C in PSM and SM containing acetate increased the frequency of four-spored asci.

5.2 Apomictic Sporulation

This occurs in some strains of yeast which form two diploid spores per ascus (Miller 1989). There is a single nuclear division (meiosis I), and later the spindle apparatus and nucleus take on the characteristics of meiosis II, and two diploid nuclei are formed per ascus, which become spore nuclei. Some inhibitors of mitochondrial respiration inhibit meiotic but not apomicvtic sporulation. Caffeine interferes with meiosis II and increases apomictic sporulation.

6 Chemical Changes During Sporulation in Yeast:
   Carbohydrates, Lipids, Amino Acids, and Proteins,
   Nucleic Acids, and Polyphosphates

6.1 Carbohydrates

Carbohydrates account for about 67% of total dry weight, and include trehalose, glycogen, glucans and mannans, glucosamine, and uronic acids. Mutants which are unable to accumulate trehalose do not sporulate. Vegetative cells do not contain uronic acid, and mannan is the major polymer. Glucosamine is part of the dark outer layer of the spore wall, resistant to enzyme action. Dityrosine may also increase the resistance of the wall to enzymes.

6.2 Lipids

These include free sterols, sterol esters, and phospholipids. Inositol is incorporated into spore membranes. There appear to be two periods of active synthesis of lipids, at 10–18 h and 24–30 h.

6.3 Amino Acids and Proteins

Proline is the major amino acid present in yeast spores. Glutamate, arginine, and methionine are also present. The reserve of free amino acids and proteins in the vacuole and cytoplasm, in vegetative cells, decreases during sporulation. Protein turnover during sporulation is rapid. Levels of proteinase increase 10–25 times. At least 14 sporulation-specific proteins have been detected.

6.4 Nucleic Acids

DNA replication (S phase or premeiotic DNA synthesis) occurs during 4–12 h after placing the cells in SM. The duration of the S phase in sporulation is at least 65 min in individual cells. This is approximately twice the duration of mitosis in vegetative cells. Replication of DNA is necessary for sporulation.

Sporulation-specific RNAs can be correlated with major morphogenic events in yeast sporulation. One set of mRNAs occurs only in the epiplasm in the ascus, which may be related to synthesis of the spore wall. Free ribonucleotides, deoxyribonucleosides, and deoxyribo-nucleotides disappear during sporulation, and may be precursors for synthesis of RNA and DNA. A number of highly phosphorylated nucleotides appear in the very early stages of sporulation.

6.5 Nature and Composition of the Spore

A composite section through the yeast spore is shown in Fig. 8.2. The inner and outer layers of the spore wall, the plasmalemma, endoplasmic reticulum, cytoplasm, lipid granules, mitochondrion, vacuole, nucleus, spindle pole body, and nucleolus, all of which are present in the vegetative cell, are visible. The spore wall is much thicker than the wall of the vegetative cell (Fig. 8.3), and the plasmalemma is relatively deeply invaginated. The spore coat is relatively smooth, though it may have one or two protrusions. The surface is hydrophobic, and contains lipid or

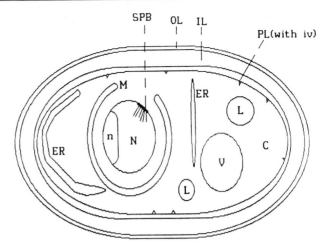

Fig. 8.2. Section through a yeast spore. *C* Cytoplasm; *ER* endoplasmic reticulum; *L* lipid globule; *M* mitochondrion; *N* nucleus; *n* nucleolus; *PL* periplasic layer; *SPB* spindle pole body; *S* spore wall (*OL* outer layer; *IL* inner layer); *iv* invaginations; *V* vacuole. (After Miller 1989)

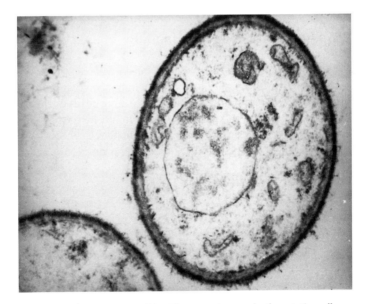

Fig. 8.3. *Saccharomyces cerevisiae.* Electron micrograph of vegetative cell

lipoprotein and a chitinous component, which is not stained with a chitin-specific dye, so that the chitin is apparently not exposed. The intact spores are resistant to enzymatic action.

Spores are less sensitive than vegetative cells to a number of types of environmental stresses, including heat, cold, mechanical abrasion, drying, UV, microwave and X-irradiation, ultrasound, starvation, action of glucanases, and various solvents such as ethanol and acetone. The high level of proline may contribute to resistance to drying. The high trehalose content also protects against desiccation.

7 Dormancy and Germination

Germination takes place in 4–65 h after the spore is placed in nutrient medium, beginning with rupture of the spore coat. Amino acid uptake and loss of acid fastness follow, and a bud appears, arising from under the wall of the mother cell. Cristae appear in the mitochondria. The cell wall becomes sensitive to glucanases. Zinc is translocated from the nucleus to the vacuole.

Germination requires the presence of an easily assimilable sugar (glucose, sucrose, mannose, or fructose. Galactose and maltose are less effective. Lactose, trehalose, and nonfermentable carbon compounds are not utilized; 2-deoxyglucose inhibits germination.

Nitrogen sources, except lysine, supporting growth of vegetative cells also support germination of S. cerevisiae spores. The pH optimum for germination is very broad, from pH 5–9.

8 Sporulation in Natural Habitats

What is the natural habitat of Saccharomyces cerevisiae? This determines the survival value of the ability to sporulate in natural habitats. It is normally found in association with human activities. It is found in breweries, distilleries, bakeries, baker's yeast plants, and wineries, where it may be the principal species occurring. The increased resistance of the spores to drying, heat, cold, abrasion, and the numerous other detrimental factors previously mentioned, must increase their chances of survival.

Other yeast species sporulate readily under natural conditions. Pichia and Hansenula species, found in association with plants and soil and natural and polluted waters, sporulate freely. Schizosaccharomyces octosporus is normally found in environments having high sugar contents. This yeast and many others also sporulate readily.

One of the advantages of being able to sporulate is the recombination of genes occurring during meiosis, with production of new strains possibly having a selective advantage in a changing environment. A second is increased resistance of the spores to adverse environments. This may preserve the newly emergent strains produced by recombination.

9 Induction and Regulation of Sporulation

Older cells, which had budded at least once, formed more three- and four-spored asci. Daughter cells (small) which had not budded formed two-spored asci, containing spores of opposite mating type.

Yeasts are redirected from the vegetative cell division cycle by a change in the environment from nutritionally rich to nutritionally poor, low in nitrogen, and having a nonfermentable carbon source.

Cells of *S. cerevisiae* do not become fully committed to sporulation immediately on transfer to a sporulation medium. Simchen et al. (1972) identified four stages leading to commitment to sporulation:

1. In the first 3 h in sporulation medium (SM), the cells do not sporulate if recovered and transferred to a water suspension.
2. If they are left in SM for from 3 to 7 h, they will sporulate if transferred to a water suspension, but if they are transferred back to presporulation medium (PSM), they revert to vegetative growth without sporulating ("readiness").
3. This is followed by a short stage after about 7 h, when the cells will sporulate in water, but if transferred back to PSM, will neither sporulate nor revert to vegetative growth ("partial commitment").
4. Following this stage, the cells will continue on to sporulation even if they are returned to presporulation or other growth medium. Visible signs of sporulation may not be immediately obvious.

9.1 Changes in the Yeast Nucleus During Sporulation
(Dawes 1983; Miller 1989; Fig. 8.4)

In *Saccharomyces cerevisiae*, a diploid cell begins sporulation when it reaches the G1 stage. If it is placed in sporulation medium at any other stage, it continues in the vegetative cycle until G1 is reached, when it may switch to the sporulation cycle.

When the cell is in G1 and sporulates, it undergoes the first meiotic division. A "polycomplex body", containing multiple synaptonemal complexes, appears in the nucleus, and the spindle pole body (spindle plaque) divides. Next, the first meiotic division occurs. The spindle plaques move to the opposite sides of the nucleus and are connected by a spindle of microtubules. The synaptonemal complexes in the polycomplex body disappear. In the next stage, the nucleus elongates and the spindle plaques divide again. The spindle plaques again move to positions opposite each other on the nuclear membrane, after disintegration of the first spindle. New spindles appear, connecting the two pairs of spindle plaques. The endoplasmic reticulum begins to form a prospore wall by forming a double membrane in the region of the four spindle plaques. The nucleus assumes a highly lobed form. The prospore membranes begin to surround the nuclear lobes, which will become the nuclei of the spores. The membranes complete the enclosure of the spindle plaque, which separates from the prospore wall. The synthesis of the spore wall continues between the membranes derived from the endoplasmic reticulum. Finally, an outer spore coat, which appears to be largely protein in nature, is formed.

The spore walls are relatively thick and are composed of glucan and mannan, but have a different composition from the vegetative cell wall. The spores are more resistant to solvents and heat, perhaps because of the nature of the outer layer of polymers of N-acetyl glucosamine (chitin) found on the spore wall.

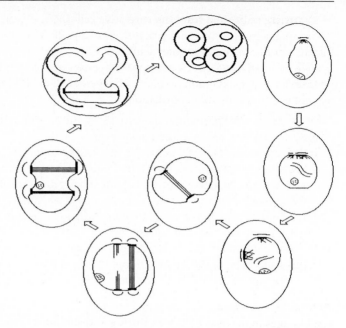

Fig. 8.4. Changes in the nucleus of *Saccharomyces cerevisiae* during sporulation

10 "Courtship" in Yeasts (Jackson and Hartwell 1990)

During courtship, haploid cells in mixtures select mating partners. The α-mating factor induces formation of the **a**-mating hormone, a signal indicating the presence of a mating partner for the α-cells. This "conversation" between cells of opposite mating types is the beginning of the "courtship" phase. It is controlled by the **a**-mating hormone, synthesized when the cells are stimulated by the presence of the α-hormone in sufficient concentration. The preferred mating partner is the cell which produces the largest amount of **a**-hormone. The cells then produce protuberances (copulation tubes), extending towards the other mating partner. When the cells make contact, the walls fuse and become contiguous, the walls and cell membranes break down at the point of contact, the cytoplasms and the nuclei fuse, and the "courtship" phase is over.

There is then an abrupt transition from the haploid to the diploid stage. The first diploid nucleus forms by fusion of the haploid SPBs along their lateral edges. At the same time, the satellites fuse. The new, diploid "conjugation SPB" is first V-shaped, but becomes flat later, and is larger than the original haploid SPBs. Soon after formation, the conjugation SPB replicates and the first bud appears on the wall nearest the SPB, showing that the SPB influences the site of bud emergence. If the nucleus and diploid SPB are displaced into one of the original haploid cell regions, the first bud arises at the corresponding pole of the zygote, again near the SPB.

11 Conjugation (Dawes 1983)

This occurs in haploid cells of opposite mating types, triggered by the mating-type hormones. The cell wall deforms, the cell assumes the characteristic shmoo[2] shape, and the two cells come into contact. Internally, the cells leave the vegetative cell division cycle in the G1 phase, with one SPB each. These are modified by the appearance of a satellite similar to the one which is observed during the later G1 stage in vegetative growth. Agglutinins accumulate in the cell walls, which adhere and fuse, the walls deforming until the zygote assumes its final form with a smooth, narrow isthmus between the two original haploid cells. The walls become perforated at the point of contact, the cytoplasmic membranes fuse and perforate, so that the cytoplasms become continuous. Nuclear fusion (karyogamy) takes place, beginning with fusion of the SPBs. Since the isthmus enlarges rather slowly, the nuclei may never pass through, with both in one of the original cells. They make contact and fusion occurs in the isthmus.

The chemical composition of the cell walls of the **a**-cells also changes, the new cell wall material laid down at this time containing more glucan and less mannan, and the mannan containing more short side chains and unsubstituted backbone than normal cell walls. The cells have higher chitinase activity than usual. As a result of these changes, the walls of **a**-cell shmoos are more susceptible to digestion by glucanases.

The α-mating-type hormone induces the formation of agglutinins in **a** cells. Agglutination of the cells as a preliminary to mating is weak at first, but increases under the influence of the α-mating hormone, until the cells are strongly bound into clumps. Some strains of **a** cells are constitutive for formation of agglutinins, at low temperatures, but must be induced at higher ones.

A knowledge of yeast life cycles, courtship, conjugation, and cell fusion is enough to permit the understanding of the methods of yeast genetics. The regulatory processes in the yeast cell, controlling mating type, mating-type switching, signal transduction, events in the cell cycle, in meiosis and sporulation and replication, transcription, and translation of the genomic DNA, require more study.

12 Yeast Genetics: a Short History

Elucidation of the genetics of *Saccharomyces cerevisiae*, and *Schizosaccharomyces pombe* is a development of the 20th century. Copulation in yeasts was observed by Hansen (in Phaff et al. 1979) in 1891 and confirmed by Schionning (in Phaff et al. 1979) in 1895. However, yeast genetics as a discipline probably began in 1935, when Winge and Lautsten (1937) demonstrated the alternation of haploid and diploid

---

[2] Shmoo is a term used in yeast genetics to describe a haploid yeast cell, of either mating type, which has been exposed to **a** or α mating pheromone, and has developed a short copulation tube. The name is derived from a fictional comic-strip character, similar in appearance, well-known in the United States in the 1930s.

generations in *S. cerevisiae*. Winge used a micromanipulator to isolate the individual spores in a yeast ascus (Johnston and Mortimer discovered the use of snail digestive juice in dissecting yeast asci in 1959, simplifying the task).

Winge and Laustsen (1937) began the study of yeast genetics with the isolation of tetrads of spores from asci of *Saccharomyces cerevisiae*. They observed differences in the morphology of giant colonies on wort gelatin, among the single-spore clones. They demonstrated Mendelian segregation in *Saccharomycodes ludwigii*, a species in which long and short cells segregate according to the Mendelian rules.

Lindegren (1949) determined Mendelian segregation of melibiose fermentation by tetrad analysis of a hybrid of *S. cerevisiae* × *S. carlsbergensis* (*S. pastorianus*). Genes for fermentation of raffinose, maltose, galactose, and sucrose, and their polymeric nature, were investigated (Lindegren 1949; Winge and Roberts 1952). They observed irregular segregations, caused by postmeiotic segregation, and linkage between several genes and with their centromeres, and established the first genetic map for *S. cerevisiae* (Lindegren 1952; Winge and Roberts 1954).

Mutants were induced in yeasts, by camphor, LiCl, brilliant green, alcohol, X-rays, UV irradiation, β-propiolatone, and other mutagens. Roman (1957) investigated the genetic basis of the adenine-deficient mutation in *S. cerevisiae*. (Ephrussi and Chimenes 1949; Ephrussi and Tavlitzki 1949) observed the formation of small colony (petite colonie) mutants, and determined some of their characteristics.

The foundations of modern yeast genetics were laid between the mid-1930s and the mid-1950s. Mapping of the yeast genome proceeded rapidly, by use of auxotrophic mutants, fermentative characteristics, and mutants to resistance to heavy metals, canavanine, and other inhibitors (Mortimer and Hawthorne 1970). Bevan and Makower (1963) discovered the "killer" character and killer particle (see Chap. 9) and showed the killer particle to be a virus-like body containing dsRNA (Bevan et al. 1973). Yeast genetics became increasingly molecular biological, as illustrated here.

13 Gene Conversion

Fogel et al. (1983) used the techniques of molecular biology to elucidate the molecular mechanisms leading to nonreciprocal exchange of genetic information during meiotic recombination. Lindegren (1955) observed aberrant segregation ratios much earlier, and named it, correctly, gene conversion. Postmeiotic segregation (PMS) is observed when single-spore colonies, obtained when yeast asci are dissected, are replica-plated directly to selective media, corresponding to all the auxotrophic markers present in the original diploid yeast strain. PMS is observed as half-colonies in those arising from spore clones in which gene coversion has taken place. Segregation ratios departing from the Mendelian ratio (2:2) occur.

The models of the mechanism of gene conversion, the Holliday (1984) structure, and the original Meselson-Radding (1975) model, postulate the heteroduplex formation in the DNA and the transfer of informational DNA sequences to one chromosome and loss of corresponding information from a homologous sequence, during repair of the heteroduplex.

Fogel et al. (1983) listed three methods for investigation of gene conversion: use of selected recombinant or prototrophic spores; use of selected tetrads containing a prototrophic spore; and use of unselected tetrads. Fogel used the third method, and he and his coworkers analyzed more than 25 000 unselected tetrads, in a system which they described as a scanning of the genetic heavens for all the meiotic events, of any kind, which they could detect. Their results confirmed their predictions based on the modified Holliday structure, and showed that gene conversions usually include a relatively large sequence of DNA, rather than only one or a few base pairs. (Holliday structures have a real existence and can be detected with the electron microscope.)

A less laborious method, in which the desired gene (*ade8-18*) was cloned on a plasmid and introduced into the recipient yeast by transformation, was later used to investigated the nature of gene conversion (Fogel et al. 1983). The yeast was then sporulated and analyzed in the normal manner, with satisfactory results. "Only" 597 tetrads were required. The methods of recombinant DNA and of molecular biology in general are now available for general use in yeast genetics and biotechnology, and are usually the methods of choice.

Further Reading

Bevan EA, Makower (1963) The physiological basis of the killer character in yeast. In: Geerts SJ (ed) Genetics today. XIth Int Congr Genet, vol 1, Pergamon Press, Oxford, pp 202–203

Bevan EA, Herring AJ, Mitchell DJ (1973) Preliminary characterization of two species of dsRNA in yeast and their relationship to the "killer" character. Nature (Lond) 245:81–86

Carter BLA, Piggott JR, Walton EF (1983) Genetic control of cell proliferation. In: Spencer JFT, Spencer DM, Smith ARW (eds) Yeast genetics: fundamental and applied aspects. Springer, Berlin Heidelberg New York, p 1

Dawes IW (1983) Genetic control and gene expression during meiosis and sporulation in *Saccharomyces cerevisiae*. In: Spencer JFT, Spencer DM, Smith ARW (eds) Yeast genetics: fundamental and applied aspects. Springer, Berlin Heidelberg New York, p 29

Ephrussi B, Chimenes AM (1949) Action de l'acriflavine sur les levures. I. La mutation "petite colonie" Ann Inst Pasteur 76:351–364

Ephrussi B, Tavlitzki J (1949) Action de l'acriflavine sur les levures. II. Etude génétique du mutant "petite colonie". Ann Inst Pasteur 76:419–50

Fogel S, Mortimer RK, Lusnak K (1983) Meiotic gene conversion in yeast: molecular and experimental perspectives. In: Spencer JFT, Spencer DM, Smith ARW (eds) Yeast genetics: fundamental and applied aspects. Springer, Berlin Heidelberg New York, pp 65–107

Fowell RR (1969) Life cycles in yeasts. In: Rose AH, Harrison JS (eds) The yeasts, vol 1, 1st edn. Academic Press, New York

Hartwell LH (1974) *Saccharomyces cerevisiae* cell cycle. Bacteriol Rev 38:164–198

Hartwell LH, Kastan MB (1994) Cell cycle control and cancer. Science 266:1821–1828

Holliday R (1984) A mechanism for gene conversion in fungi. Genet Res 5:282–304

Jackson CL, Hartwell LH (1990) Courtship in *Saccharomyces cerevisiae*. Both cell types choose partners by responding to the strongest pheromone signal. Cell 63:1039–1052

Johnston JR, Mortimer RK (1959) Use of snail digestive juice in isolation of yeast spore tetrads. J Bacteriol 78:292

Lindegren CC (1949) The yeast cell, its genetics and cytology. Educational Publishers, St Louis

Lindegren CC (1955) Non-Mendelian segregation in a single tetrad of *Saccharomyces* ascribed to gene conversion. Science 121:605–607

Meselson MS, Radding CM (1975) A general model for genetic recombination. Proc Natl Acad Sci USA 72:358–361

Miller JJ (1989) Sporulation in *Saccharomyces cerevisiae* In: Rose AH, Harrison JS (eds) The yeasts, vol 4, 2nd edn. Academic Press, New York

Mortimer RK, Hawthorne D (1970) Yeast genetics. In: Rose AH, Harrison JS (eds) The yeasts, vol 1, 1st edn. Academic Press, New York

Phaff HJ, Miller MW, Mark EM (1979) The life of yeasts. Harvard University Press, Harvard University, Cambridge

Roman H (1957) Studies of gene mutation in *Saccharomyces*. Cold Spring Harbor Symp Quant Biol 21:175–183

Simchen G, Piñon R, Salts Y (1972) Sporulation in *Saccharomyces cerevisiae*: premeiotic DNA synthesis, readiness and commitment. Exp Cell Res 75:207–215

Vezinhet F, Blondin B, Hallet JN (1990) Chromosomal DNA patterns and mitochondrial DNA polymorphism as tools for identification of enological strains of *Saccharomyces cerevisiae*. Appl Microbiol Biotechnol 32:568–571

Wheals AE (1985) Biology of the cell cycle in yeasts. In: Rose AH, Harrison JS (eds) The yeasts, vol 1, 2nd edn. Academic Press, New York, p 283

Winge O, Lautsten O (1937) On two types of spore germination and on genetic segregations in *Saccharomyces*, demonstrated through single-spore cultures. CR Trav Lab Carlsberg Ser Physiol 22:99–116

Winge O, Roberts C (1952) The relation between the polymeric genes for maltose, raffinose and sucrose fermentation in yeasts. CR Trav Lab Carlsberg Ser Physiol 25:141–171

Winge O, Roberts C (1954) Causes of deviations from 2:2 segregations in the tetrads from monohybrid yeast. CR Trav Lab Carlsberg Ser Physiol 25:283–329

See also *Notes Added in Proof* on p. 353.

CHAPTER 9

Chemical Warfare Among the Yeasts: the "Killer" Phenomenon, Genetics and Applications

V. Vondrejs and Z. Palková

1 The Killer Phenomenon in Yeasts

Killer yeast strains secrete a toxin (killer toxin, killer factor, zymocin) which kills sensitive strains, but not the killers. They are ubiquitous among laboratory yeasts and isolates from nature (Table 9.1), and the killer character can be transmitted from strain to strain by hybridization, fusion with protoplasts from whole cells, or miniprotoplasts without nuclei. Transformation of protoplasts with DNA or RNA coding for killer toxin is possible. Strains lacking genes encoding killer toxins are often sensitive to one or more toxins.

Killer toxins may have wide action spectra against various yeast species. Some are active against yeasts of other genera. *Hansenula saturnus* CCY38-4-2 kills *Zygosaccharomyces bailii* (a contaminant of wine), *Saccharomyces cerevisiae* some strains of *Candida albicans*, *Candida krusei*, *Kluyveromyces phaffii*, *K. marxianus*, *K. fragilis* and *lactis*, and *Hansenula* species. In others, the effect of killer toxin is more specific.

Yeast strains differ in their spectra of activity against sensitive yeast strains, in their cross-reactivity and other features of the toxins, and in the genetic determinants of killer character and immunity (Young 1989). There may be other killer types as yet not found (Mitchell and Bevan 1983; Tipper and Bostian 1984; Wickner 1986; Young 1987; Tipper and Schmitt 1991; Vondrejs et al. 1991).

The killer phenotype is conferred by cytoplasmic, virus-like particles containing dsRNA, in *Saccharomyces cerevisiae* (Bevan and Makower 1963; Bevan et al. 1973) and *Ustilago* sp., and by a linear dsDNA plasmid in *Kluyveromyces* sp. The genetic determinant of killer character in most other killer strains is not known. The toxin may sometimes be encoded in chromosomal or mitochondrial DNA (in *Saccharomyces paradoxus*).

The majority of zymocins kill sensitive cells only during growth, usually by disruption of the membrane. This group includes toxins from *S. cerevisiae* (K1, K2, or K3) *Pichia kluyveri* and *Ustilago maydis*. The K28 toxin from *S. cerevisiae* causes rapid, reversible inhibition of DNA synthesis, then slow cell death (Schmitt and Tipper 1992). Toxin from *Hansenula mrakii* inhibits cell wall synthesis, and the heterotrimeric toxin from *Kluyveromyces lactis* arrests growth of sensitive yeast strains as unbudded cells, probably blocking completion of the G1 phase of cell division. It does not inhibit adenylate cyclase.

J.F.T. Spencer/D.M. Spencer (eds)
Yeasts in Natural and Artificial Habitats
© Springer-Verlag Berlin Heidelberg 1997

Table 9.1. Occurrence of killers among yeasts and yeast-like organisms

Species	Basic references
Candida sp.	Philliskirk and Young (1975)
Cryptoccocus sp.	Stumm et al. (1977)
Debaryomyces sp.	Philliskirk and Young (1975)
Hansenula sp.	Philliskirk and Young (1975)
Kluyveromyces sp.	Philliskirk and Young (1975)
Pichia sp.	Philliskirk and Young (1975)
Rhodotorula sp.	Golubev (1989)
Saccharomyces sp.	Bevan and Makower (1963)
Sporidiobolus sp.	Golubev et al. (1988)
Torulopsis sp.	Bussey and Skipper (1975)
Ustilago sp.	Puhalla (1968)

The killer systems K1, K2, and K28 of *S. cerevisiae* (Tipper and Schmitt 1991) are determined by two species of dsRNA (L and M) found in the cytoplasm as latent mycovirus particles. M dsRNAs are killer strain-specific (M1, M2, and M28), encode the killer toxin, and determine immunity of the host to the toxin. L dsRNAs are also present in many nonkiller strains, and specify the capsids of mycovirus particles and viral RNA-dependent RNA polymerases. Maintenance of M dsRNA depends on L dsRNA-determined functions. Many chromosomal genes are involved in maintenance and expression of the killer phenotype. The killer types K1, K2, and K28 are distinguished by absence of cross-immunity and by the properties of the toxins. These toxins are all produced from glycosylated precursors, which requires the KEX2 function. The K1 and K2 protoxins (Zhu et al. 1993) are N-glycosylated, but the K28 precursor lacks N-glycosylation sites. The mature K28 is O-mannosylated. The secreted dimeric toxin K1 (20.7 kDa) contains two polypeptide chains (α, β). Its precursor is composed of four domains: (1) an N-terminal leader sequence (delta); (2) an α-domain responsible for killing and binding to the cell wall receptor, (3) an N-glycosylated γ-region of uncertain function, and (4) a β-domain required for binding to the cell wall receptor. The K28 toxin is more stable than K1 and K2. The optimum pH for activity of the K1 and K2 toxins is 4.3–4.7, at temperatures below 23 °C. K28 is active between pH 5.0 and 5.8 at temperatures below 35 °C. Strain S6 of *S. cerevisiae* is killed only by K1 or K2 toxin, while *S. cerevisiae* strain 381 is sensitive to K28 toxin. M28 dsRNA excludes both M1 and M2 from the same cytoplasm in hybrids. Stable coexpression of the K28 phenotype from M28 dsRNA and K1 phenotype from an M1-cDNA were obtained (Schmitt and Tipper 1992). Exclusion probably acts at the level of replication of dsRNA. Likewise, M1 excludes M2 (Wickner 1986). These results may show a hierarchy (M28 > M1 > M2) of some functional interaction related to replication of these RNAs.

The first step in the action of K1 and K2 toxins is rapid binding to the cell wall receptor, where (1–6)-β-D-glucan is essential (Hutchins and Bussey 1983). The

K28 toxin binds to the α-1,3-linked mannose residues of a mannoprotein cell wall receptor. The second step is energy- and membrane-receptor-dependent, but the receptors are different for the three toxins. In sensitive cells, the action of the K1 toxin begins with inhibition of proton-amino acid cotransport and proton extrusion, followed by a marked drop in intracellular pH, inhibition of cell metabolism, efflux of potassium ions, and cell death, with damage to the plasma membrane. The toxin interacts with the cytoplasmic membrane of sensitive cells (Vondrejs et al. 1982) and forms ion-permeable channels (Hianik et al. 1984). The pores are large enough to allow lethal leakage of ions and dissipation of membrane potential.

Genetic manipulation of yeasts is done by the classical methods of selection, mutagenesis, mating, protoplast fusion, or the use of recombinant DNA technology. Clones of manipulated strains are selected for resistance to inhibitors, complementation of auxotrophic requirements and/or detection of particular products or morphological forms.

We have developed a selection procedures using killer toxin K1. The techniques are useful for selection of hybrid and cybrid clones of industrial yeasts and enrichment of auxotrophic mutants. Choice of technique depends on the susceptibility of the parental strain to be eliminated, so we developed rapid methods for semiquantitative estimation of the killer toxin activity. These determinations are an integral part of the selection procedure.

2 Methods for Estimating Killer Toxin Activity and/or Susceptibility to the Toxin

Killing ability of Kil+ strains and sensitivity to killers (R^+) is determined by the formation of clear zones around colonies of killer strains (Fig. 9.1). A well test, using toxin solutions in agar wells, is similar (Fig. 9.2). Changes in the fraction of colony-forming cells or in optical density in liquid cultures after treatment with killer toxin are also used. A faster method uses rhodamine B or bromocresol purple staining of killed cells (Špaček and Vondrejs 1986; Kurzweilová and Sigler 1993). The assays give good resolution at low concentration and allow counting with a flow cytometer. Several toxin samples can be tested in a few hours, and the concentration, LC_{50}, killing 50% of the cells, determined (Fig. 9.3). The same assay can be used to estimate the sensitivity of different yeast strains (Fig. 9.4).

Another rapid assay uses the reorientation of yeast cells in liquid media by intense alternating electric fields (5 MHz). The direction of orientation depends on several properties, including the permeability of the cytoplasmic membrane. At the correct conductivity and field frequency, permeabilized cells undergo the dielectrophoretic transition, parallel $\rightarrow$ perpendicular to the field lines. Factors influencing the dielectrophoretic transition are shown in Table 9.2. When the cells are permeabilized by killer toxin K1, heat shock, nystatin, etc., the cells reorient perpendicularly at 5 MHz. Valinomycin and K. lactis killer toxin do not affect the cytoplasmic membrane and do not induce reorientation. The full transitional

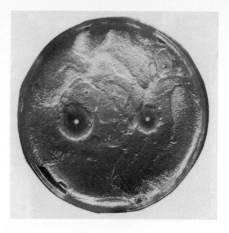

Fig. 9.1. Dot assay for detecting the killer toxin production. *S. cerevisiae* S6/1 forms the background lawn. Killer strains: *S. cerevisiae* GOSMH (*dot on the left*), *S. cerevisiae* GMH (*right*)

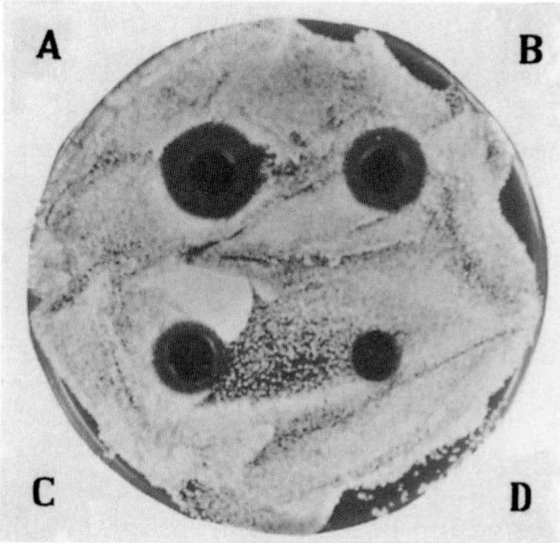

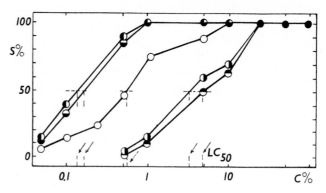

Fig. 9.2. Well assay for detecting the killer toxin activity. *S. cerevisiae* S6/1 forms the background lawn. Solution of killer toxin produced by *K. lactis* IFO1267 (50 μl/well). *A* Nondiluted; *B* dilution 1:5; *C* dilution 1:10; *D* dilution 1:50

Fig. 9.3. The relative killer toxin K1 activity in cell-free media estimated by modified rhodamine B assay. (Špaček and Vondrejs 1986), *S%* is the fraction of stained cells; *C%* is the relative amount of killer toxin medium in the sample. C = 100% when the undiluted cell-free medium J (Vondrejs et al. 1991) containing killer toxin after 24h of aerobic cultivation of killer strains: *S. cerevisiae* GOSMH (-◑-), GOSMC (-◒-), GMH (-◐-), GMC -◓- or IP3 -○- is used

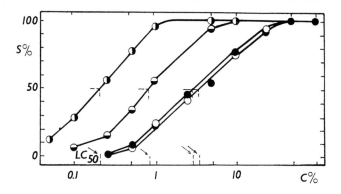

Fig. 9.4. The relative susceptibility of various sensitive strains towards the killer toxin K1 produced by *S. cerevisiae* GOSMC. Sensitive strains *S. cerevisiae* S6/1 (-◑-), *S. cerevisiae* var. *uvarum* P9 (-◒-), P95/1 (-○-) and P96 (-●-) were analyzed in this experiment by modified rhodamine B assay. *Abbreviations* as in Fig. 9.13

Table 9.2. The effect of various factors on dielectrophoretic behavior of *S. cerevisiae* S6/1 exponential-phase cells

Treatment	Time	Orientation of cells to the field lines at:	
		0.5 MHz	5 MHz
Untreated		Parallel	Parallel
Nystatin 10 μg/ml	2 h	Parallel	Perpendicular
Valinomycin 10 μg/ml	4 h	Parallel	Parallel
Heat shock 80 °C	5 min	Parallel	Perpendicular
Heat shock 100 °C	5 min	Parallel	Perpendicular
Killer toxin K1 from *S. cerevisiae* Undiluted cell-free medium (20 °C)	2 h	Parallel	Perpendicular
Killer toxin from *K. lactis* IFO1267 Undiluted cell-free medium (30 °C)	2 h	Parallel	Parallel

effect develops more slowly than the effect detected by rhodamine B (Fig. 9.5). The concentration of killer toxin which causes the transition effect in 50% of the cells (TC$_{50}$) is a measure of killer toxin activity (Fig. 9.6).

These rapid assays cannot be used to determine the effect of toxins which do not permeabilize the cytoplasmic membrane (*Kluyveromyces lactis* toxin). These toxins can be assayed by rhodamine B-staining exponentially grown cells of *S. cerevisiae*, treated with nystatin. If they have been previously exposed to *Kluyveromyces lactis* killer toxin (Palková and Cvrčková 1988), the cells do not stain. The fraction of unstained cells increases with the concentration of the toxin in the medium, since the toxin suppresses the growth of exponential phase cells. An improved nystatin-rhodamine B assay (Fig. 9.7) depends on killing of the cells

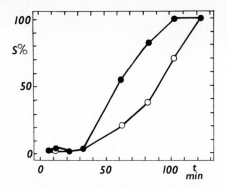

Fig. 9.5. The time course of development of dielectrophoretic transition behavior and the effect detected by rhodamine B in exponential cells of *S. cerevisiae* S6/1 since the addition of killer toxin K1. *S%* is the fraction of cells rhodamine B-stained (-●-) or perpendicularly oriented to the field lines of alternating (5 MHz) electric field (-○-) after t min of incubation at 23 °C with undiluted cell-free medium J containing killer toxin K1 produced during 24 h of aerobic cultivation of *S. cerevisiae* GOSMH at 20 °C. The conditions of killer toxin treatment were as in Fig. 9.3

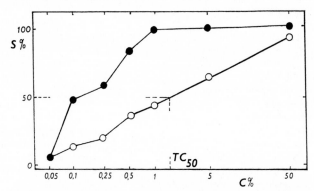

Fig. 9.6. The activity of killer toxin K1 produced by *S. cerevisiae* GOSMH estimated either by rhodamine B (-●-) or dielectrophoretic transition assay (-○-). *S. cerevisiae* S6/1 was used as sensitive strain. The conditions of killer toxin treatment were as in Fig. 9.1. Abbreviations as in Figs. 9.3 and 9.5

by nystatin after 3 h growth of a diluted stationary phase culture at 28 °C in the absence of killer toxin. In the presence of toxin, the cells are protected. The concentration of killer toxin which protects 50% of the cells against killing by nystatin (PC50) can be used to calculate the relative activity of killer toxin preparations.

3 Killer Selection Techniques

The general principles of the method (Vondrejs et al. 1991) are shown in Table 9.3. A brewing yeast (*Saccharomyces cerevisiae* var. *uvarum* P9) and a haploid, K1

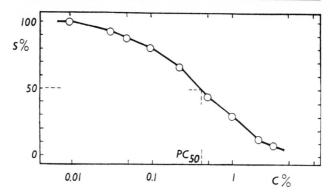

Fig. 9.7. The killer toxin activity produced by *K. lactis* IFO1267 estimated by modified nystatin-rhodamine B assay (Palková and Cvrčková 1988). *S. cerevisiae* S6/1 was used as sensitive strain. Cell-free medium YEG (Vondrejs et al. 1991) containing killer toxin produced during 16 h of aerobic cultivation of *K. lactis* IFO1267 at 28 °C was appropriately diluted and mixed with stationary culture of sensitive cells. After 3 h at 28 °C, small samples of cell suspension were withdrawn, centrifuged, washed, and resuspended in 1 ml of 4 μM nystatin-100μM rhodamine B. After 1 h at 28 °C (in dark), the fractions of stained cells were estimated. Abbreviations as in Fig. 9.3

Table 9.3. Principles of killer selection techniques

Parental strains	*S. cerevisiae* T158C Haploid his⁻K1⁺R1⁺			*S. uvarum* P9 Triploid his⁺K1⁻R1⁻		
Main products of fusion		T158C×T158C his⁻ K1⁺R1⁺	T158C×P9 his⁻ K1⁺R1⁺ Cybrid 1	T158C×P9 his⁺ K1⁺R1⁺ Hybrid	T158C×P9 his⁺ K1⁻R1⁻ Cybrid 2	P9×P9 his⁺
Eliminated in minimal medium without histidin	T158C his⁻ K1⁺R1⁺ Parent	T158C×T158C his⁻ K1⁺R1⁺	T158C×P9 his⁻ K1⁺R1⁺ Cybrid 1			
Eliminated by killer toxin K1 treatment					P9×P9 his⁺ K1⁻R1⁻	P9 his⁺ K1⁻R1⁻ Parent
Selected in minimal medium with killer toxin K1				T158C×P9 his⁺ K1⁺R1⁺ Hybrid	T158C×P9 his⁺ K1⁺R1⁺ Cybrid 2	

An example: The brewing strain *S. cerevisiae* var. *uvarum* P9 is hybridized with the strain *S. cerevisiae* T158C by induced protoplast fusion. K1⁺... production of killer toxin K1; R1⁺... immunity to killer toxin K1; his⁻... dependency on histidine.

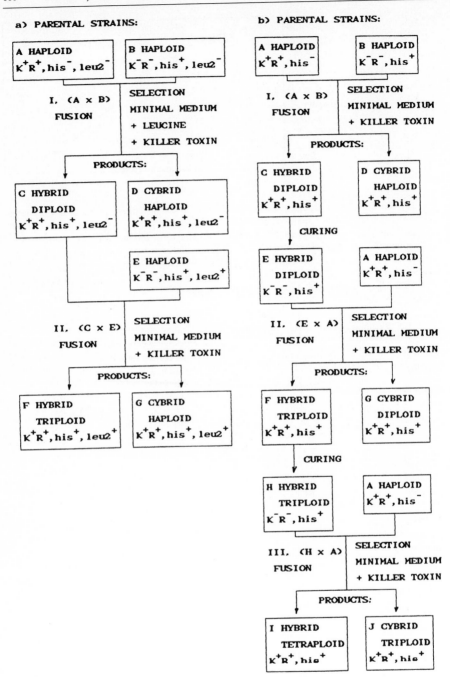

Fig. 9.8a,b. The procedure used in a cascade fusion of protoplasts **a** without or **b** with curing of killer character

killer, *his-* strain, *S. cerevisiae* T158C, were used as parental strains for constructing hybrids or cybrids. The brewing yeast is sensitive to killer toxin produced by the other. The hybrid or cybrid strains obtained by protoplast fusion, acquire resistance to the killer toxin from the killer parent. Cells requiring histidine will not regenerate on minimal medium without histidine. Strains sensitive to killer toxin are killed, while those which are prototrophic for histidine and resistant to killer can regenerate a wall and form colonies. The ploidy of the clones must be determined, to distinguish which are hybrids, cybrids, and other fusion products. If the haploid strain carries additional genetic markers, or if the cells can be separated

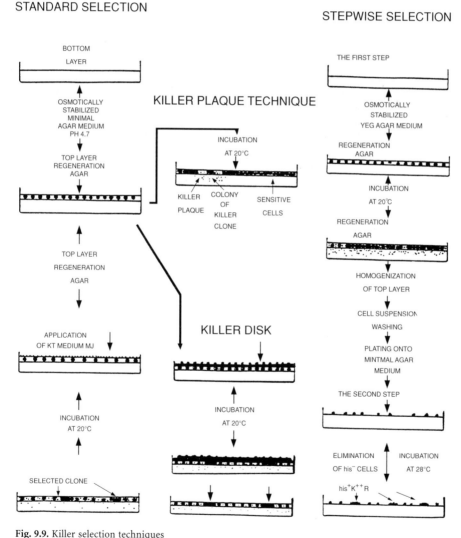

Fig. 9.9. Killer selection techniques

according to ploidy by density gradient centrifugation, isolation of true hybrids is simpler.

Sometimes, killer hybrids can be used immediately, in the killer selection procedure in the next fusion step (Fig. 9.8a). The killer phenotype can then be eliminated if desired (Fink and Styles 1972). The system can be used in multiple cascade fusion of protoplasts (Fig. 9.8b) and the strains "cured" afterwards. True hybrids, not cybrids, must be selected for the next step in this procedure.

We developed a "standard selection technique" (Fig. 9.9) in model systems for protoplast fusion using killer strain *S. cerevisiae* T158C with various sensitive strains. Approximately 5×10^7 protoplasts of each parental strain were fused under

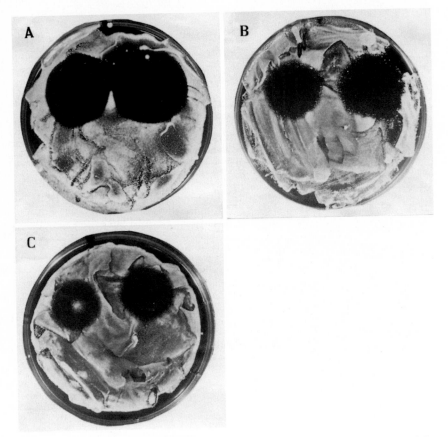

Fig. 9.10A–C. The killer disk technique. Killer strain of *S. cerevisiae* GMH (*left*), supersecreting killer strain GOSMH (*right*) or sensitive strain (*below*) was spread onto the nitrocellulose filter Sympor 6 (porosity 0.45 µm, diameter 2.6 cm, 10^7 cells/filter). Filters were incubated for 24 h at 23 °C on the surface of plate (pilsener wort agar). Then the strains of various sensitivity (cf. Fig. 9.4) were spread onto the surface of agar medium and cultivated for 48 h at 23 °C. **A** *S. cerevisiae* S6/1; **B** *S. cerevisiae* var. *uvarum* P9; **C** *S. cerevisiae* var. *uvarum* P96

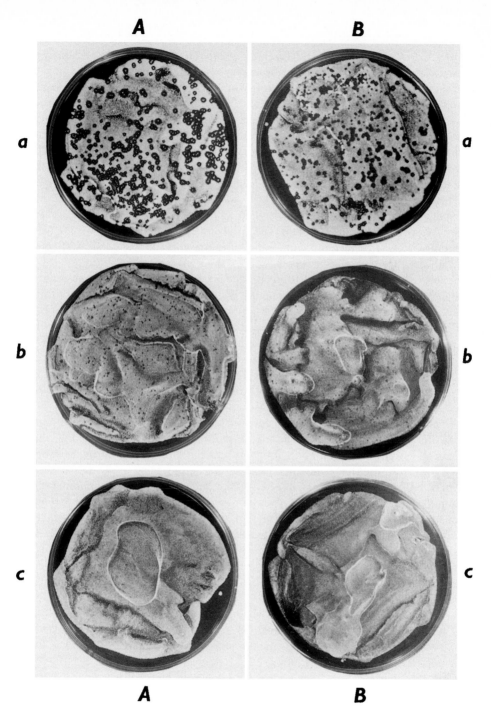

Fig. 9.11A,B. The killer plaque technique. Killer plaques around colonies of *S. cerevisiae* GMH (killer) (**A**) or GOSMH (supersecreting killer) (**B**) on the background of the sensitive strains: **a** *S. cerevisiae* S6/1; **b** *S. cerevisiae* var. *uvarum* P9; **c** *S. cerevisiae* var. *uvarum* P96 (cf. Fig. 9.2); 10^7 cells of sensitive strain were mixed with about 300 killer cells per plate (pilsener wort agar)

known conditions (Maráz and Ferenczy 1979). The protoplast suspension was diluted and suspended in osmotically stabilized agar, and layered over the surface of osmotically stabilized minimal agar (pH 4.7) lacking histidine. These plates were incubated at 28 °C for 26 h, 0.4 ml of cell-free medium containing killer toxin K1 was spread over the surface of each plate, and the Petri dishes were immediately transferred to a 20 °C incubator. Large colonies of hybrids and cybrids appeared along with many very small colonies of strains having different phenotypes. The frequency of appearance of selected hybrid and cybrid clones depends upon the mating types of the fused protoplasts. In general, low or negligible frequencies of cybrid clones in comparison with hybrids were observed when parental strains of opposite mating types were used. When like mating types were fused, cybrid clones predominated over hybrids. The disadvantage to this procedure is the need to prepare cell-free medium containing killer toxin of suitable concentration in advance.

The killer disk (Figs. 9.9, 9.10) technique is very similar, but does not require preparation of cell-free medium containing killer toxin. It avoids some disadvantages of the killer bed technique. The simplest version of the killer disk was a nitrocellulose membrane filter (porosity 0.45 μ). The filters, spread with killer cells, are placed on the top layer of the regeneration agar, carrying the protoplasts. The filter separates the killer cells from the surrounding medium, but the toxin can pass through and kill the sensitive cells. The technique is very efficient and can be used even when the sensitive parental strain is not very susceptible to the killer toxin.

The stepwise selection technique does not require the preparation of killer toxin or a prototrophic killer helper. The auxotrophic parental strain produces the killer toxin in complex medium and eliminates sensitive protoplasts during selection and regeneration. After incubation at 20 °C, the cells growing in the top layer are recovered from the agar, centrifuged, washed, and resuspended in minimal medium or spread on the surface of a minimal agar medium (Fig. 9.9). If the parental killer strain is a stable auxotrophic mutant, it is completely eliminated during the second step of selection at 28 °C. The hybrid or cybrid clones selected in the second step are not independent isolates of those clones arising in the first step; thus, the frequency of hybridization cannot be estimated. This technique is very efficient and should be used when the sensitive parental cells are not sufficiently susceptible to the toxin.

The killer plaque method is used when toxin production from a single colony of hybrids or cybrids is great enough to make a small zone (plaque) in lawns of regenerated sensitive strain in minimal medium (Figs. 9.9, 9.11). The parental auxotrophic killer strain cannot grow. The method was tested using strains producing killer toxin K1 or killer strains of *Kluyvermyces lactis*. Use of this procedure requires high susceptibility of the sensitive cells to the toxin, a high level of killer toxin production by the hybrids and cybrids, and a negligible reversion frequency of the auxotrophic markers in the parental killer strain.

4 Advantages and Limitations of the Method: a Summary

The techniques depend on elimination of one parental strain by its natural sensitivity to the corresponding killer toxin. They are superior to methods based on auxotrophy, respiration deficiency, and/or sensitivity to antibiotics. Parental strains, especially of industrial yeasts, for the latter methods carry damaged or altered genomes or mitochondria. This makes the use of killer selection techniques for the improvement of industrial yeast strains very attractive.

The killer parental strains confer upon the hybrids and cybrids both the ability to produce killer toxins and immunity to the toxin. Resistance to toxin and toxin production is utilized in selection by the methods described previously. Both properties (K^+, R^+) can be tested independently when selected clones are analyzed. The strains can easily be cured of the killer characters K1, K2, and K3 when required. Selection may be used repeatedly, along with curing, in multiple "cascade" fusion of protoplasts (Fig. 9.8). Cells of higher ploidy may be constructed using this procedure. The killer disk and stepwise techniques are suitable for elimination of parental strains which are not very susceptible to killer toxins.

Cybrids as well as hybrids can be isolated by means of killer selection techniques, by methods similar to those used with mitochondrial markers (Spencer and Spencer 1983). Hybrid and cybrid clones can easily be isolated using an auxotrophic, sensitive strain. It selection is done in minimal medium containing a low concentration of an auxotrophic requirement of the sensitive strain, auxotrophic killer cybrids may form smaller plaques around the colonies than prototrophic hybrids.

Many yeasts, including industrial strains, are sensitive to killer toxins, and killers were found in many yeast-genera (Table 9.1), so the methods are widely applicable for selecting different hybrid strains.

The selection procedures used after mating or rare-mating of parental cells can be simpler, since regeneration of protoplasts is not required. Modifications to the techniques are: (1) A cell-free medium containing killer toxin can be added to the cell sample immediately after hybridization is complete. (2) The killer disks are preincubated on top of the minimal medium for about 48 h at room temperature before mating cells are spread on the agar plate if a modified killer disk technique is used. (3) An excess of killer parental cells is added to the sample if a stepwise procedure is used. (4) An excess of sensitive cells is recommended if the killer plaque technique is used.

Otherwise, the protocol of the killer-selection procedures may be essentially the same as for protoplast fusion, when genetic determinants of killer phenotype are transmitted by fusion of miniprotoplasts from donor killer strains and protoplasts of sensitive recipient strains or when they are introduced into protoplasts by transformation (El-Sherbeni and Bostian 1987).

Selection of killer hybrids and cybrids was investigated using the stable killer toxin K1 from *S. cerevisiae* strain T158C. Toxins are labile proteins, sensitive to proteases, with narrow ranges of optimal conditions for specific killing actions. Killer toxins encoded by cytoplasmic determinants, chromosomal DNA, and DNA plasmids may differ in their spectrum of selected products of fusion or mating. The use of some killer systems may require modifications of the procedures used to reach a satisfactory selection effeciency.

[The killer plasmids of killer strains of *Kluyveromyces lactis* are not transferable to respiratory-competent strains of *S. cerevisiae*, since the killer plasmids are incompatible with *Saccharomyces* mitochondria. The plasmids can be transferred to the RD (petite) mutants of *S. cerevisiae*, which lack functional mitochondria].

Killer toxin has been combined with other selection pressures to eliminates the killer strain. Hybrids carrying two killer strains of different classes, giving double selection pressure from two types of killer toxin, might be used for selecting hybrids and/or cybrids. Incompatibility of some killer systems (Vondrejs et al. 1983) might limit the usefulness of such double selection. Exclusion of M dsRNA can be circumvented by replacing M1 dsRNA with M1-cDNA (Schmitt and Tipper 1992).

Killer selection techniques are principally limited by the low efficiency of killing of strains weakly sensitive to the toxin. Killer strains producing high yields of active killer toxins may be used in these methods. The standard and killer disk techniques require prototrophic helper killer strains, overproducing toxin, and auxotrophic superkiller strains (dominant overproducers of toxin) are the best parents for use in the stepwise and killer plaque techniques. The ski1 to ski4 mutants, recessive mutants of the superkiller phenotype, are unsuitable. *S. cerevisiae* strain GRF18/pJDBαCHE, which supersecretes recombinant chymosin, yields suitable dominant supersecretion mutants.

5 Possible Applications of Killer Toxins in Industry, Medicine, and Genetics: a Survey

Industrial fermentations and industrial yeast cultures can be invaded by wild yeast strains, including killers, which are more dangerous than other wild yeasts. They compete for substrate and actively kill the strain used in fermentations. The resulting abnormal fermentations and poor product quality are problems which can be at least partly solved by use of killer toxins and/or industrial strains which are immune to the toxin produced by contaminating yeasts.

Crude toxin preparations will protect beer from the effects of wild yeast toxin. A more sophisticated solution of the problem is to confer killer character on nonkiller strains of industrial yeasts. The resulting killer strains can protect fermentations and products against contamination by sensitive yeasts or wild killer strains.

Killer strains, or strains immune to killer toxins, have been constructed from sake yeasts, wine yeasts, yeasts for ethanol production, and lager beer and ale yeasts (Vondrejs 1987), using conventional hybridization techniques, rare-mating, or protoplast fusion, for transfer of killer character into nonkiller industrial strains. The fermentative behavior of these strains was the same as that of the nonkiller parental strains. The taste of the products was indistinguishable from that of the originals. The toxins are specific to the sensitive yeast strains, and are unstable above pH 5 and inactive below pH 3. They are nontoxic to humans. Killer yeasts carrying the *kar1* mutation, which prevents nuclear fusion (Conde and Fink 1976), are very useful donors of the killer character. The mutation prevents fusion of the nucleus of the donor strain with that of the industrial strain, so no genes or a few only (single-chromosome transfer) are transferred to the nucleus of the industrial strain, which avoids any possible undesirable effects.

We constructed a lager yeast strain from *S. cerevisiae* var. *uvarum* P9-LK-12/1, producing killer toxin K1 (Vondrejs et al. 1991) by standard killer selection methods. Killer yeasts were used for constructing killer strains of brewing yeasts (*S. cerevisiae* var. *uvarum* P95 and P96) by the killer disk and stepwise techniques, and the killer strain P9-LK-12/1 was used to construct brewing strains producing killer toxin K1 and an extracellular glucoamylase as well.

Killer toxins are unsuitable for use as antifungal antibiotics in medicine, because they are proteins and usually have low pH and temperature stability. The use of killer toxins against infections of the oral or vaginal cavities and skin by pathogenic yeasts might be feasible. Some killer yeasts produce more stable toxins (Young and Yagui 1978) and some of the toxins are active against pathogenic yeasts. A further search for killer toxins with possible medical applications may be desirable. Killer toxins might be useful for treatment of patients with AIDS, because they often suffer from infection by pathogenic yeasts.

The killer selection techniques summarized here may serve as tools in construction of new yeast strains by protoplast fusion and other procedures. The killer plaque and stepwise selection techniques were used also for constructing strains supersecreting killer toxin K1. *S. cerevisiae* GRF18 (α his- leu- K1- R1-), transformed with recombinant plasmid pJDBαCHE (Fig. 9.12) was used for selecting faster-growing strains which oversecreted chymosin. One of the strains, *S. cerevisiae* GOS1 (α his-leu- K1- R1-), cured of the plasmid, was derived from a chymosin overproducing strains. Hybrids GOSMH (α K1^{2+} R1$^+$) and cybrids GOSMC (α his- leu- K1^{2+} R1$^+$), overproducing killer toxin K1 (K1^{2+}), were constructed by protoplast fusion of *S. cerevisiae* GOS1 with *S. cerevisiae* M3 (α ade- ilv- trp- K1$^+$ R1$^+$). A 30-fold increase in production of killer toxin K1 activity was demonstrated in these strains compared with the same type of cybrid (GMC) and hybrid (GMH) strains obtained by protoplast fusion of *S. cerevisiae* M3 with *S. cerevisiae* GRF18 (Fig. 9.3). The use of these strains is advantageous in the standard, killer disk, killer plaque and stepwise killer selection techniques (Figs. 9.9, 9.10, and 9.11).

The killer toxin K1 permits a selection for resistant mutants with glucan defects (Brown et al. 1993). Therefore, the system can be used to investigate the pathways of biosynthesis and secretion of some cell wall components which serve as cell wall receptors for toxin. Probably, the K28 toxin can be used to select mannan(mnn) mutants which affect the structure of the cell wall mannoprotein of *S. cerevisiae* (Schmitt and Radler 1990). The use of killer systems in genetic analysis of protein

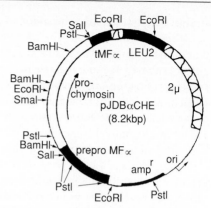

Fig. 9.12. The plasmid pJDBαCHE. 2μ-ARS from 2μ plasmid of *S. cerevisiae*; *ori* replication origin from pBR322; *LEU2* gene from *S. cerevisiae*; *amp^r* gene for ampicillin resistance from pBR322; *preproMFα* promoter + leader secretion sequence from MFα gene of *S. cerevisiae*; *tMFα* MFα terminator; *prochymosin* calf prochymosin cDNA

secretion pathways was reviewed by Bussey (1988). Killer selection techniques will probably be useful in the isolation of mutants defective in genes involved in these processes.

Applications of killer toxins, similar in principle to the penicillin method for concentration of auxotrophic mutants, have been introduced. Unfortunately, this method is less efficient than the commonly used nystatin method devised by Snow (1966).

Further Reading

Bevan EA, Makower (1963) The physiological basis of the killer character in yeast. In: Geerts SJ (ed) Genetics today, XIth Int Congr Cenet vol 1. Pergamon Press, Oxford, pp 202–203

Bevan EA, Herring AJ, Mitchell DJ (1973) Preliminary characterization of two species of dsRNA in yeast and their relationship to the killer character. Nature (Lond) 245:81–86

Brown JL, Kossaczka Z, Jiang B, Bussey H (1993) A mutational analysis of killer toxin resistance in *Saccharomyces cerevisiae* identifies new genes involved in cell wall (1 → 6)-beta-glucan synthesis. Genetics 133:837–849

Bussey H, Skipper N (1975) Membrane mediated killing of *Saccharomyces cerevisiae* by glycoproteins from *Torulopsis glabrata*. J Bacteriol 124:476–483

Bussey H (1988) Proteases and the processing of precursors to secreted proteins in yeast. Yeast 4:17–26

Conde J, Fink GR (1976) A mutant of *Saccharomyces cerevisiae* defective for nuclear fusion. Proc Natl Acad Sci USA 73:3651–3655

El-Sherbeni M, Bostian KA (1987) Viruses in fungi: infection of yeast with K1 and K2 killer viruses. Proc Natl Acad Sci USA 84:4293–4297

Fink GR, Styles CA (1972) Curing of a killer factor in *Saccharomyces cerevisiae*. Proc Natl Acad Sci USA 69:2846–2849

Golubev WI, Tsiomenko AB, Tichomirova LP (1988) Plasmid-free killer strains of the yeast *Sporidiobolus paravosevs*. Mikrobiologia 57:805–809

Golubev WI (1989) The action spectrum of killer toxins produced by *Rhodotorula glutinis* and its taronomic significance. Mikrobiologia 58:99–103

Hianik T, Laputková G, Vondrejs V (1984) Current response of bilayer lipid membrane to killer factor from *Saccharomyces cerevisiae* T158C. Gen Physiol Biophys 3:93–95

Hutchins K, Bussey H (1983) Cell wall receptor for yeast killer toxin: involvement of (1 → 6)-β-D-glucan. J Bacteriol 154:161–169

Imamura T, Kawamoto M, Takaoka Y (1974) Isolation and characterization of (killer)-resistant mutants of sake yeast. J Ferment Technol 52:300–305

Kurzweilová H, Sigler K (1993) Fluorescent staining with bromocresol purple: a rapid method for determining yeast cell dead count developed as an assay of killer toxin activity. Yeast 9:1207–1211

Maráz A, Ferenczy L (1979) Mating-type-independent protoplast fusion in *Saccharomyces cerevisiae*. In: Peberdy JF (ed) Protoplast Application in Microbial Genetics. University of Nottingham, Nottingham, 35pp

Mitchell DJ, Bevan EA (1983) ScV "killer" viruses in yeast. In: Spencer JFT, Spencer DM, Smith ARW (eds) Yeast genetics: fundamental and applied aspects. Springer, Berlin Heidelberg New York, 371pp

Palková Z, Cvrčková F (1988) Method for estimating activity of killer toxin from *Kluyveromyces lactis*. Folia Biol 34:277–281

Peña P, Barros F, Gascon S, Lazo PS, Ramos S (1981) The effect of yeast killer toxin on sensitive cells of *Saccharomyces cerevisiae*. J Biol Chem 256:10420–10425

Philliskirk G, Young TW (1975) The occurrence of killer character in yeasts of various genera. Antonie Leeuwenhoek J Microbiol Serol 41:147–151

Puhalla JE (1968) Compatibility reaction on solid medium and interstrain inhibition in *Ustilago maydis*. Genetika 60:461–474

Schmitt MJ, Radler F (1990) Blockage of cell wall receptors for the yeast killer toxin KT28 with antimamcoproteins antibodies. Antimicrob Agents Chemother 34:1615–1618

Schmitt MJ, Tipper DJ (1992) Genetic analysis of maintenance and expression of L and M dsRNA from yeast killer virus K28. Yeast 8:373–384

Snow R (1966) An enrichment method for auxotrophic yeast mutants using the antibiotic "nystatin". Nature 211:206–207

Špaček R, Vondrejs V (1986) Rapid method for estimating of killer toxin activity in yeasts. Biotechnol Lett 8:701–706

Spencer JFT, Spencer DM (1983) Genetic improvement of industrial yeasts. Annu Rev Microbiol 37:121–142

Stumm C, Middlebeek EJ, de Bries GJML, Croes AF (1977) Killer-sensitive relationships in yeast from natural habitats. Antonie Leeuwenhoek J Microbiol Serol 43:125–128

Sturley S, Bostian KA (1990) Molecular mechanisms underlying the expression and maintenance of the type 1 killer system of *Saccharomyces cerevisiae*. In: Walton EF, Yarranton GT (eds) Molecular and cellular biology of yeasts. Blackie, London, Van Nostrand Reinhold, New York, pp 246–279

Tipper DJ, Bostian KA (1984) Double-stranded ribonucleic acid killer systems in yeasts. Microbiol Rev 48:125–156

Tipper DJ, Schmitt MJ (1991) Yeast dsRNA viruses: replication and killer phenotypes. Mol Microbiol 5:2331–2338

Vondrejs V (1987) A killer system in yeasts: applications to genetics and industry. Microbiol Sci 4:313–316

Vondrejs V, Gášková D, Plášek J, Prosser V (1982) N-phenyl-1-naphtyl-amine as a fluorescent probe for early event in the action of yeast killer factor. Gen Physiol Biophys 1:435–445

Vondrejs V, Pšenička I, Kupcová L, Dostálová R, Janderová B, Bendová O (1983) The use of killer factor in the selection of hybrid yeast strains. Folia Biol 29:372–384

Vondrejs V, Palková Z, Sulo P (1991) Application of killer systems in yeasts to selection techniques, vol 1. In: Cheremisinoff PN, Ferrante LM (eds) Biotechnology – current progress. Technomic Publishing, Lancaster, 227pp

Wickner RB (1986) Double-stranded RNA replication in yeast: the killer system. Annu Rev Biochem 55:373–395

Young TW (1984) Killer yeasts. In: Rose AH, Harrison JS (eds) The yeasts, 2nd edn. Academic Press, New York, vol 2, pp 131–164

Young TW, Yagui M (1978) A comparison of the killer character in different yeasts and its classification. Antonie Leeuwenhoek Microbiol Serol 44:59–77

Zhu Y-S, Kane J, Zhang X-Y, Zhang M, Tipper DJ (1993) Role of the γ component of preprotoxin in expression of the yeast K_1 killer phenotype. Yeast 9:251–266

Inside the Inside: Part I: Yeasts and Molecular Biology, a Recipe for Alphabet Soup. Chromosome Structure, Replication, Transcription, and Translation

J.F.T. Spencer and D.M. Spencer

1 Introduction

The literature on molecular biology is filled with strings of letters; TGACTC, TGGCCGGGGTTTACGGACGATGA, hapl, TACATCA, TATA etc., all part of the complex language. This chapter will attempt to interpret some of this language, that is the shorthand of molecular biology.

The study of cytology and ultrastructure deals with the morphology of structures inside the yeast cell, which can be seen with the light or electron microscope. Molecular biology deals with the behavior of chromosomes (in particular), membranes, the vacuole, the nucleolus, the cytoskeleton, Golgi bodies, vesicles, and other intracellular organelles, and the behavior of the nucleus during the cell cycle. Bud emergence, bud development and, finally, cytokinesis and bud separation, courtship, conjugation, zygote formation, and diploidization of haploid cells may now be studied at the molecular level.

What happens inside the cell or zygote when molecules meet face to face? What happens in the microenvironment of the chromosome? What determines the replication of a strand of DNA? What determines whether a gene will be transcribed (switched on), or remain forever silent? This is the province of molecular biology.

1.1 Early History of the Molecular Biology of Yeasts

The knowledge of the molecular biology of yeasts developed late. Transformation in bacteria had been demonstrated by Avery and coworkers in 1944. Yeasts and other fungi possessed a rigid and almost impervious cell wall. The first attempts to induce transformation in yeast was undertaken by Oppenoorth (1960), who extracted DNA from *Saccharomyces cerevisiae* and attempted to transform a maltose-negative recipient strain to maltose-positive. The results were not convincing. Several investigators attempted, unsuccessfully, to duplicate Oppenoorth's results. No further attempts were reported until 1978.

Hinnen, Hicks, and coworkers in Fink's laboratory in the United States, and J.D. Beggs in England successfully transformed protoplasts of a leu- strain of *S. cerevisiae* with a plasmid containing the LEU2 yeast gene, and obtained prototrophic cells. Foreign genes could now be transformed into yeasts, apparently,

J.F.T. Spencer/D.M. Spencer (eds)
Yeasts in Natural and Artificial Habitats
© Springer-Verlag Berlin Heidelberg 1997

but yeasts proved more complex than bacteria. Beggs attempted to introduce the gene encoding rabbit globulins into yeast and have it expressed and the protein synthesized. However, synthesis did not begin at the point expected, and only part of the protein was formed.

The understanding and use of yeasts in molecular biology now developed rapidly. Yeasts have some advantages over bacteria as hosts for production of heterologous proteins. They do not produce the endotoxins formed by *Escherichia coli*, they are eukaryotes, and contain the same mechanisms for processing proteins as are found in animal cells. The physical size of the yeast cell is considerably larger than that of any bacterium and allows the packing of the interior with correspondingly more DNA, i.e., more genes.

The major developments in the molecular biology of yeasts, after methods transforming yeast protoplasts and intact cells have been:

1. Development of shuttle vectors carrying genes which are expressed in yeast and bacteria (usually *E. coli*).
2. Development of promoters and other controlling elements in production of homologous and heterologous proteins in yeast.
3. Increasing knowledge of the secretory processes for proteins.
4. Knowledge of the factors influencing maintenance of yeast plasmids and integration of genes carried on the plasmids into the yeast genome.
5. Construction of artificial chromosomes having increased stability in yeast and carrying much larger sequences of DNA.
6. Development of techniques for electrophoretic separation, visualization, and isolation of yeast chromosomes. From the first discovery to highly refined techniques has taken only about 5 years.

Progress in yeast molecular biology was greatly facilitated by the discovery of the restriction enzymes, which cut the DNA strand at specific sequences, permitting the isolation and modification of genes and construction of plasmids.

The discovery of the methyltrophic yeasts is also of great actual and potential importance. *Hansenula polymorpha*, (*Pichia angusta, Pichia pastoris, Candida boidinii*, and possibly *Candida sonorensis* can utilize methanol for growth and undergo radical morphological and metabolic changes when the carbon source is switched from glucose to methanol. The cytoplasm of the cell becomes packed with microbodies (peroxisomes) which contain the enzyme, alcohol oxidase. If the yeast is modified genetically using recombinant DNA techniques, disrupting the gene encoding alcohol oxidase and substituting the gene encoding a heterologous protein, the microbodies contain the heterologous protein when the carbon source is switched to methanol instead of alcohol oxidase. Somatostatin, tumor necrosis factor, and hepatitis B surface antigen have been produced (Cregg et al. 1987).

1.1.1 The Molecular Biology of Yeasts

The chromosomes are composed of double-stranded DNA and carry the genetic code which determines the reproduction, maintenance, and metabolic behavior of the yeast cell. The replication, transcription, and translation of the genetic code is regulated by many mechanisms, all of which are ultimately accomplished by the binding of regulatory proteins, synthesized in accordance with the code carried by the chromosomes, to their binding sites in the regulatory sequences of the chromosomal DNA.

The chromosomes are replicated during the multiplication of the yeast cell itself. In *Saccharomyces cerevisiae*, the mechanism is similar to that observed in the other 500 and more known species of yeasts, so it is a useful example.

The yeast genome is not large compared to the genomes of other eukaryotes. The size of the genome in the haploid cell of *S. cerevisiae* is about 1.4×10^4 base pairs. Other yeast species usually have a genome size about equal to that of *S. cerevisiae*. However, the latter yeast has the largest number of chromosomes, approximately 16, though there are strain differences, especially in industrial yeasts. The mitochondria also have their genomes, amounting to about a tenth that of the nuclear DNA. The human genome, which determines the nature of that much-loved primate, is about 500 times as large.

The yeast genome, unlike the naked genome of the prokaryotes, is firmly associated with proteins, to form chromatin. The chromatin is organized into nucleosomes, sequences of DNA, about 200 base pairs long, wrapped in a coil of two turns around a protein octomer of two molecules each of four different basic histones, H2a, H2b, H3, and H4. A fifth histone is loosely associated with the DNA which connects the nucleosomes like a string, accounting for about 60 base pairs of the total DNA of the nucleosome. This histone may be responsible for the higher-order coiling of the chromatin, in which the nucleosomes themselves are arranged in a supercoil to form a 30-nm fiber, the chromosome of higher eukaryotes. In yeast, most of the chromatin remains in an extended form, consisting of a string of DNA, connecting beads, the nucleosomes. The greater part of the DNA, about 140 base pairs, is tightly bound to the histone octomer. The histones are important in regulation of gene expression.

While the nuclear genome accounts for the greater part of the DNA in the yeast cell, mitochondrial DNA (mtDNA) accounts for another 15% of the total, in about 35 copies per cell, of 75 kbp each. There are 60–100 copies of the well-known 2-μM circle (6.3 kbp), accounting for 3–5% of the total yeast DNA.

Other elements composed of nucleic acids are the killer particles, which may be plasmids or virus-like particles (VLPs). *S. cerevisiae* killer particles are double-stranded RNA and have a protein coat. Those of *Kluyveromyces lactis* are of DNA and may be naked. Other killer species, having a very broad spectrum of action, are usually associated with the nucleus. There is also the elusive *psi* factor, which may increase the level of suppression of ocher and frameshift mutations, and segregates 4:0 in meiosis, and the [URE3] element, which affects the ability of cells to utilize ureidosuccinic acid. There may be others in different yeast species. Plasmids similar to the 2-μm circle or the linear DNA plasmids of *K. lactis* have been found in only very few yeast species; *S. cerevisiae*, *K. lactis*, three or four species of *Zygosaccharomyces*, *Schizosaccharomyces pombe*, and possibly *Candida glabrata*.

The chromosomal DNA contains numerous repeated sequences. These may account for 10–15% of the total nuclear genome and include transposable elements (Ty, tau, and sigma), telomeres, a special case, and important repeated sequences, encoding rRNA and tRNA. DNA sequences encoding rRNA (18S and 25S rRNA) account for about 100 copies per haploid genome, are encoded on a 9 kb fragment which also contains the sequences for 5S and 5.8S rRNA, and amount to about 6% of the total nuclear DNA. The genes are located on chromosome XII, are tandemly repeated, and each array is at least 20 repeats long. They may even be arranged in a single tandem array. Some yeast strains carry about five copies of the rRNA repeats as extrachromosomal, circular DNA.

There are many more copies of the genes encoding tRNA. There are approximately 360 such genes, about eight copies for each species of tRNA, which are widely dispersed throughout the genome (unlike the rRNA genes, which are grouped on chromosome XII), and they account for about 0.2% of the nuclear DNA.

The Ty elements (Ty and Ty2) consist of central units about 5.2 kb long, flanked by delta sequences each about 330 kb long. There are also about 100 delta elements which are not associated with Tys. The two elements have large regions of nonhomology in the central regions, but the proteins they encode show considerable amino acid homology. The Ty elements are transposable at low frequency through an RNA intermediate, and belong to the large group of eukaryotic retrotransposons using a retroviral-like transposition mechanism. There are about 30 to 35 Ty elements per haploid genome, constituting about 1.5% of the genome.

Sigma elements are about 314 bp in length, and have 5-bp repeats at the ends. Evidence for transposition is indirect. There are about 30 sigma units in the genome and they are all associated with the tRNA genes. They constitute about 0.08% of the genome.

Tau elements are approximately 371 bp long, have 5-bp repeats at the ends, and a copy number of 15 to 25 per cell. They constitute 0.04 to 0.07% of the genome. There is no visible sequence homology between the elements. However, there are short blocks of homologous sequences through these elements. Not much is known about these elements and their transposition.

Many genes encoding abundant proteins are present in multiple copies. There are two copies of each of the four genes encoding histones and three for glyceraldehyde dehydrogenase. Different strains of yeast have different numbers of copies of the SUC genes encoding invertase, and MAL genes encoding genes for maltose fermentation. Multiple genes are not required absolutely; many yeast strains have one of any given gene, while others have as many as five or six. There are three STA genes and the SSG gene in diastatic strains of yeast, all homologous. There are more than 12 MEL (α-galactosidase) genes, mostly homologous. Some of the genes encoding ribosomal proteins are duplicated, and three genes, including iso-1-cytochrome c found on chromosome X, are duplicated and repeated on chromosome V by transposition. Finally, there are three copies of mating-type information on chromosome III, though only one of these is expressed at a time.

1.1.2 Chromosome Structure and Size

The size of the individual chromosomes of *S. cerevisiae* and other species of yeasts can now be determined directly, by pulsed-field gel electrophoresis, and ranges from 260 kb (chromosome I) to 2000 kb (chromosomes XII and XIV). There is one (double) strand of DNA, passing through the centromere, and there are no protein or RNA linkers in the chromosomal DNA. The DNA is associated with four types of histones, being organized into nucleosomes. Folded chromosomes have been observed which consist of negatively coiled DNA.

1.1.2.1 Centromeres

The centromeres are definite sequences of the chromosomes, one per chromosome, where the spindle fibers are attached. The presence of one centromere increases the stability of plasmids as well as chromosomes. In plasmids, a centromere reduces the copy number to one. The centromeric DNAs are not homologous, but do contain conserved sequence elements. There is a central element, CDEII, a 77–86-bp region which contains more than 90% A + T, which is essential for centromere functiion. The sequence varies in different centromeres.

The centromere chromatin has a nuclear core which is about 220–250 bp in length, and includes the CDEIII-centered region. The centromeres are essential for normal mitotic segregation of the chromosomes.

Centromere-like DNA sequences isolated from *Schizosaccharomyces pombe* do not confer stability on plasmids. They may not be centromere fragments.

1.1.2.2 Telomeres

These are definite structures of great importance, not merely blunt ends of the chromosomal DNA. The general structure of the telomeres is as follows (Fig. 10.1): (1) Several copies of the sequence $(C_{1-3}A)_n$, (2) the Y' sequence, which consists of one to four copies of a 6.7 bp repeated sequence, (3) another tract of $(C_{1-3}A)_n$. This is followed by the X sequence, 2 kb in length, which is less conserved. Both X and Y' sequences contain ARS sequences. Recombination is common between telomeres, between chromosomal telomeres, and between plasmid and chromosomal telomeres, generating new telomeric repeats. Telomeric structures are rather variable. The mechanism of replication of telomeres is not completely known. The structure is thought to be a self-pairing hairpin.

The telomeres are an essential part of the chromosome, and prevent degradation of the chromosomal DNA. They are involved in regulatory functions as well; genes located close to the telomeres may be either delayed in transcription or are completely silent. The sequences at the HML and HMR sites are normally silent, and are only copied and expressed during mating-type switching, in a haploid, mother cell. The mating-type sequences are stored in locations near the telomeres,

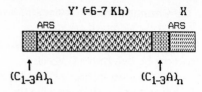

Fig. 10.1. Structure of a yeast telomere. The telomeres have a hairpin form which loops back on itself and contains cross-linked DNA which is resistant to degradation and maintains the integrity of the chromosome. All telomeres have a terminal tract of a few hundred bp of tandem repeats of the $C_{1-3}A$ sequence, usually bordered with one copy of the Y' sequence, but sometimes, with up to four of these. There is also an X sequence which is less well conserved. There is another tract of the $C_{1-3}A$ sequence between the Y' and X sequences as shown. Both X and Y' sequences contain ARSs

where their expression is silenced by the products of the SIR genes. During mating-type switching, they are copied, and a copy is inserted at the mating-type locus, near the centromere where they can be expressed. The small "tails" (the ends of the histone molecules) protrude from the histone core particle, and may play a part in repressing genes found near the telomeres. As the knowledge of the structure and function of the telomeres increases, their importance becomes more apparent.

Other *cis*-acting elements in the chromosomes include small linear plasmids (unstable) and small circular plasmids (stable).

Yeast chromosomes are stable, and replicate once during the mitotic cell division cycle in vegetative growth, daughter chromatids segregating to opposite poles of the mitotic spindle. Chromosomes can, rarely, be lost at the rate of 1×10^{-3} per cell division.

Yeast artificial chromosomes are less stable than the naturally occurring ones, especially when the length is less than 100 kbp. Stability increases as the length increases.

1.1.3 Replication of Chromosomal DNA

Replication takes place in the S phase of the cell division cycle, a stage which may occupy 25–50% of the cell cycle if glucose is the carbon source; less, if other compounds are used. If a less readily metabolized C source is used, the cell cycle may be longer, but this does not affect the S phase or the G2 phase. The G1 phase, when the cell is in the unbudded state, is longer. During starvation, growth is arrested in the G1 phase, and does not recommence until nutrients are again available and the cell cycle can begin again.

Daughter cells which have not yet produced a bud remain in the G1 phase longer, to reach full size before they enter the S phase and DNA replication begins. The S phase is longer in cells undergoing meiosis, than in diploids in mitosis. Starvation for nitrogen, however, can increase the length of the S phase.

Bud emergence frequently coincides with the beginning of the S phase. It may occur at the beginning, before the beginning, or as late as halfway through the S phase. Bud emergence and DNA replication are on independent pathways.

What controls the initiation of replication? Where does replication originate? Prokaryotes have a single origin of replication (and only one chromosome). However, yeast chromosomes have numerous origins of replication. Replication, which begins as a fork or a bubble, takes place bidirectionally. There are about 400 origins of replication in *S. cerevisiae*, spaced about 36 kb apart, plus another 100 in the tandemly expressed rDNA sequences. The forks may move at about 2.1 to 6.3 kb/min in both mitosis and meiosis. Under nitrogen limitation, the rate of fork movement may be slower, possibly accounting for the extension of the S phase. The replicon size is about 90 kb.

The time for DNA replication is not known exactly. Each replicon replicates only once during any one cell cycle, and most of them probably initiate replication early in the S phase. The rate of DNA synthesis is constant throughout the S phase, about 100 min, and each replicon should complete replication in about one-seventh of this time. Treatment with NTG induced ten times as many mutations during the S phase as during G1. If NTG mutagenizes replicating DNA preferentially, replication takes place throughout the S phase. It appears that replicons in yeasts, like those in higher eukaryotes, fire in bursts. This explains the apparent activation of groups of replicons at the beginning of the S phase, and the apparently continuous synthesis of DNA throughtout the S phase. It is not known how a given sequence is prevented from replicating more than once during the cell cycle.

1.1.4 The Molecular Biology of Replication

In prokarayotes, the initiation of DNA replication is regulated by the interaction of a regulatory protein with an origin of replication. In yeast, autonomously replicating sequences (ARSs) are similar to origins of replication. Some plasmids have DNA sequences which permit extrachromosomal replication of the plasmid DNA. These ARSs may be replication origins. There are certain similarities: they are spaced 32–40 kb apart; ARS-containing plasmids are found in the nucleus, where the plasmid DNA is organized into nucleosomes similar to all yeast chromatin; the plasmids replicate under cell-cycle control, in the S-phase but not in G1, and require the products of the genes CDC28, CDC4, CDC7, and CDC8, like the chromosomal DNA. The plasmids carrying ARSs replicate only once in the cell cycle, like chromosomal DNA but unlike mtDNA or the dsRNA of the killer VLPs; they act only in *cis*. One rDNA repeat has an ARS in the nontranscribed spacer regions, in a region where replication bubbles have been observed. Replication initiates preferentially near an ARS element, in some in vitro systems.

Therefore, ARSs are probably replication origins. Nevertheless , some yeasts, i.e., *Schizosaccharomyces pombe,* do not seem to require specific ARS sequences for initiation of replication. Any unspecified DNA sequence will serve.

1.1.5 Structure of ARS Sequences

There are three regions to the ARS sequences. On the left-hand end is a flanking sequence, 200–300 bp long (domain C), which is not essential for function, though it enhances it. Central to the sequence is a core consensus sequence, (A or T)TTTATA(or T)TTTA(or G) (domain A: 11 base pairs). To the right (3′) is another flanking sequence (domain B), necessary for function. There is a conserved sequence, not always present, though some sequence is required (Fig. 10.2). The conserved sequence is 11 bp long, and is located about 55–88 bp 3′ from the

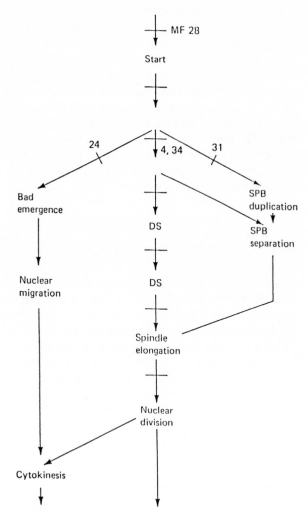

Fig. 10.2. A functional sequence map. *MF* Mating factor step; *cdc* genes are given by numbers; *iDS* initiation of DNA synthesis; *DS* propagation of DNA synthesis; *SPB* spindle pole body

core consensus. The smallest fragment tested which had ARS activity, was a 43-bp sequence from the HO ARS.

ARS sequences are nearly always in noncoding regions, often adjacent to the 3′ end of coding regions. Transcription through an ARS may inactivate its function. There is no absolute role known for an ARS in gene regulation.

ARS sequences in *Schizosaccharomyces pombe* differ slightly from those of *S. cerevisiae*. The sequences determined have been as follows: AA(or TT)TTTATTTAA(or G). Furthermore, ARS sequences from *S. cerevisiae* do not function in *Schiz. pombe*. Some DNA sequences from several other species do show ARS function in *S. cerevisiae*. These sequences have not been tested for ARS function in *Schiz. pombe*.

All replication origins may not be active in *S. cerevisiae*. There are at least ten ARSs on chromosome III, but all of these may not be used during any given round of replication. Four appear to remain silent.

1.1.6 Proteins Required for Replication of DNA in *S. cerevisiae*

DNA replication requires the participation of numerous proteins. There are at least 50 cdc genes, each encoding a different protein taking part in the replication of DNA. There are seven genes whose products are required *before* the initiation of DNA synthesis, at the boundary between G1 and the S phases. These include CDC28, encoding a protein kinase, essential for initiation of replication. During DNA synthesis, eight more gene products are required (CDC21 which encodes thymidylate synthetase, converting dUMP to dTMP, and CDC8, encoding thymidylate kinase, which phosphorylates dTMP to dTTP), then one for joining the different replicons (CDC9, encoding DNA ligase); five more for medial nuclear division, three for late nuclear division, and two for cytokinesis. The cell cycle with its associated genes is illustrated in Fig. 8.1.

These account for 27 genes, barely more than half of the known complementation groups. There are many partly characterized proteins without known DNA replication mutants. Two of these, TOP1 and TOP2, encode topoisomerase I and topoisomerase II, which uncoil the strands during replication. There are also three DNA polymerases, two of which are located in the nucleus and one in the mitochondria, a DNA primase associated with POL1, a number of single-stranded nucleic acid binding proteins (SSBs). Three DNA-dependent ATPases, only one of which has helicase activity, have been identified in yeast. Ribonuclease H enzymes have been observed, which may remove RNA primers. None of these proteins appears to participate in DNA replication.

1.1.7 Replication of the DNA of the 2-μm Circle

The 2-μm DNA plasmid, found in *S. cerevisiae* and in very few other yeast species, is located in the nucleus, but segregates 4:0 in meiosis, and exists in copy numbers of 50 to 100 per cell. Only about 80% of the plasmids are transmitted to meiotic spores, but the copy number does not seem to increase. If the copy number is below normal, even though it is as low as one per cell, the plasmid is amplified until the normal copy number is reached.

Replication is under the control of the same genes as the replication of the chromosomes. It requires a functional CDC8 gene product, and does not replicate at the restrictive temperature in the cdc mutants, cdc28, cdc4, and cdc7, which arrest in G1 at this temperature. It is also inhibited in cells which are arrested in G1 by the action of mating hormones. It replicates once during each cell division cycle, normally in the early S phase, and has two replication origins, one of them near one of the inverted repeats, and one in the middle of the large unique region. The single ARS region contains the same consensus sequence as other ARSs. It behaves like a normal, full-sized chromosome.

The 2-μm plasmid consists of four domains, of which two are inverted repeats, each 599 bp long, plus a single large unique domain (2774 bp) and one smaller one (2346 bp). There are four open reading frames (ORFs); REP1 and REP2 (open reading frames B and C) act in *trans*. REP3 (or STB) is required for segregation of the 2-μm circle. The ARS locus acts in *cis*. REP3 is found in the large unique region, several hundred base pairs from the ARS sequence. The open reading frame D is located between REP3 and REP1 on the larger unique region.

Segregation of the 2-μm circle (Fig. 10.3) is controlled by four loci REP1, REP2, REP3, and the ARS sequence. Only the requirement for the ARS sequence is absolute. Reading frame A (FLP) catalyzes the inversion of the molecule, via a site-specific recombination with the inverted repeat sequence. Copy number control is probably distinct from the plasmid replication system. REP1 and REP2 bind to the 5′ end of the FLP transcript. Expression of the FlP gene and plasmid amplification may be controlled through the repression of FLP by the binding of REP1 and REP2 in this region. Plasmid partitioning is encoded by reading frames B and D. ORF D is involved in plasmid maintenance.

The structures of the REP3 (STB) locus and other elements of the partitioning system are known. The MAP1 gene affects ARS- and contromere-based plasmids and the 2-μm plasmids; "nibbled" colonies (NIBN1) have enhanced numbers of plasmids per cell; some mutants are defective in maintenance and 80–90% of cells lack plasmids. The remaining ones have the normal number (these may be host-encoded maintenance mutants). A spontaneous [cir°] mutant could not be transformed with 2-μm-based plasmids. When crossed with a [cir⁺] strain the plasmid instability segregated 2:2.

The 2-μm plasmid has been found in *Saccharomyces cerevisiae*, *Zygosaccharomyces bailii*, *Z. bisporus*, *Z. rouxii*, and *Schizosaccharomyces pombe*. There is no sequence homology between the *Zygosaccharomyces* plasmids, the *Schizosaccharomyces* plasmids, and those of *S. cerevisiae*; there is a striking similarity in the *organization* of the three types of plasmid, including the presence of the inverted repeats, and the location and number of the open reading frames.

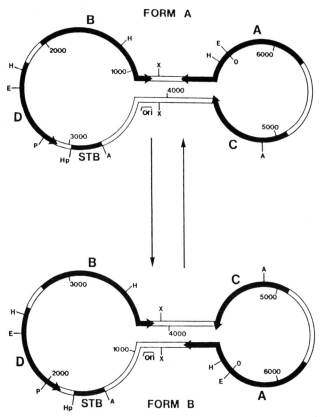

Fig. 10.3. Structure of the 2-μm circle. The diagram shows the two isomeric molecular forms, A and B, of the 6.3-kb 2-μm plasmid, with the four open reading frames *A(FLP)*, *B(REP1)*, *C(REP2)*, and *D*, and the two *cis*-acting loci *ORI* and *STB(REP3)*. The molecules are drawn as *dumbbells* with the *bars* showing the two copies of the inverted repeat sequences. The locations of the cleavage sites for restriction endonucleases *AVAI(A)*, *EcoRI(E)*, *HindIII(H)*, *HpaI(Hp)*, *PstI(P)*, and *XbaI(X)* are as shown

Nevertheless, though of unknown function, the 2-μm plasmid is a very useful sequence of DNA, and has many uses in applied molecular biology.

1.1.8 The Replication of Mitochondrial DNA

Replication of mitochondrial DNA is semiconservative, not dispersive. The mtDNA is visible in preparations stained with DAPI (4,6-diamidino-2-phenylindole) and viewed under fluorescent light, when it appears as a "necklace" of brilliant spots like a chain of beads around the periphery of the cell, just inside the wall. There are other bits of mtDNA deeper in the cytoplasm, which are less noticeable. The location of the mtDNA within the mitochondria is uncertain, since the form and size of the mitochondria are variable. However, functional mitochondria are a feature of every eukaryotic cell, though *Saccharomyces cerevisiae* is one of the few eukaryotic organisms which can survive without functioning mitochondria. The mitochondria are where nonfermentable materials, as energy sources, are metabolized, and ATP is synthesized during respiration. They are the powerhouses of the cell.

The mtDNA, the other major component of the cellular DNA. It encodes part of the cytochromes, and a few other proteins such as the *var1* protein. Many of the proteins involved in mitochondrial function are encoded in the nucleus, and transported to the mitochondrion for assembly and functioning.

The mtDNA is a rather fragile circle, about 75 kb long (25 μm). When the functional mtDNA is lost from the cell, the cell becomes a petite colonie or petite mutant, and no longer respires, all energy being obtained through the glycolytic cycle.

Synthesis of mtDNA continues throughout the cell cycle, in cells growing under "normal" conditions, cells arrested in G1 by mating-type hormones and in cells treated with cycloheximide. Synthesis of mtDNA continues at the restrictive temperature, in mutant strains (cdc4, cdc7, and cdc28, that arrest in G1); in cdc2, defective in synthesis of chromosomal DNA, and cdc14 and cdc23, arresting in G2, at the restrictive temperature. Replication of mtDNA and cell growth are probably linked. The ratio of mtDNA to nuclear DNA can increase three to six times if the synthesis of nuclear DNA is blocked but cell growth continues.

Petite colonie mutants formed spontaneously have a full complement of nonfunctional mtDNA. A short sequence of the DNA is excised and replicated to reform a circle the size of the original. This DNA carries only a few of the genes of normal mtDNA, and cannot carry out its functions. Petites induced by mutagenic agents (acriflavine, ethidium bromide, manganese, or similar compounds) may have nonfunctional mtDNA, but it is much more randomized. Nonfunctional mtDNA in petites may retain cryptic genes. If it recombines with mtDNA from respiratory-competent yeast strains, any cryptic genes may be expressed. Other petites, induced by prolonged treatment with ethidium bromide, may have no mtDNA at all.

rho⁰ petites, lacking any mtDNA, are mated with respiratory-competent haploid strains, the character segregates 4:0 and the progeny are respiratory-compentent. If the petite cell retains nonfunctional mtDNA, it may be a suppressive, and some of the progeny retain the petite phenotype. In hypersuppressive petite strains, up to 95% of the progeny may be petites. Three regions of the mitochondrial genome give rise to hypersuppressive mutants. These regions share a common 300-bp base sequence, and are designated rep1, rep2, and rep3. They may correspond to replication origins. There is a set of five regions which gives rise to hypersuppressive strains (ori1 to ori7). Three of these regions correspond to the *rep* regions just mentioned.

Nuclear or segregational petites are encoded in the nucleus and segregate 2:2 during meiosis.

2 Transcription and Promoters in Yeast

The chromosomes are at the center of the activities of the yeast cell, with the mechanisms for maintainance and replication. A few other smaller units of cellular DNA, the mitochondrial genome and the 2-μm circle, have been described. There are also the mechanisms for transcription of the genetic message into messenger RNA (mRNA), which carries the message to the point of synthesis of protein. Protein synthesis is a complex process. The ribosomes must read the message and translate it to protein. Before the message is read, there must be mechanisms by which promoters and repressors switch on or switch off the enzymes – RNA polymerases – that read the message from the chromosomal DNA and translate it to protein on the ribosomes – the machinery which does the work of the cell.

The genome consists of structural genes which encode the proteins, and regulatory sequences that enable the structural genes to be switched on or off. Other structural genes encode DNA-binding proteins, which bind to the regulatory sequences and decide which is on and which is off. There are also activating sequences which stimulate transcription of the gene.

The yeast genome is a typical eukaryotic genome, more complex than the prokaryotic bacterial genome. (There are border-line orgainsms, apparently halfway between prokaryotes and eukaryotes). The yeast genome contains about 10 000 kbp in 16 chromosomes, in *Saccharomyces cerevisiae* as determined by pulsed-field gel electrophoresis. There are approximately 5000 structural genes (Although *S. cerevisiae* has 16 chromosomes, most yeasts have four to ten. *Schizosaccharomyces pombe* has the same quantity of DNA, organized into three huge chromosomes.) Normally, the genes are not grouped into operons, except for the mating-type locus, and the GAL1, GAL7, GAL10 cluster. Most of these genes are transcribed at low levels, with one or two molecules of mRNA per cell. The genes most studied are nearly all highly expressed and tightly regulated.

2.1 Transcription

Transcription is accomplished by three RNA polymerases, RNA polymerase I, RNA polymerase II, and RNA polymerase III. RNA polymerase I transcribes the clustered genes encoding ribosomal RNA (rRNA), which amounts to 70% of the cellular RNA. RNA polymerase III transcribes the genes encoding transfer RNA (tRNA), which is 30% of the total RNA, and RNA polymerase II transcribes messenger RNA (mRNA), which is transcribed from the approximately 5000 structural genes which encode the cellular proteins, and is about 1% of the total RNA of the cell. We know most about RNA polymerase II and the genes transcribed by it.

2.2 Promoter Regions

These are regulatory regions (Fig. 10.4), upstream from a protein-encoding region of a gene, and consist of four elements: (1) upstream repressor sequences, (2) upstream activator sequences, (3) the TATA element, and (4) the initiator (I) element. The distance upstream of the regulatory region and that separating the elements of the promoter region may vary considerably.

The upstream activator sequences (UASs) confer promoter specificity, are exchangeable, and are defined by multiple short sequences of 17–30 base pairs. For example, the CYC1 promoter has subsites responding to heme and carbon sources. In the absence of heme or oxygen, transcription does not occur. In medium containing glucose, with adequate amounts of oxygen, UAS1 is ten times as active as UAS2 in activating transcription. In medium containing lactate (i.e., gluconeogenic), UAS1 is derepressed ten times, but UAS2 is derepressed 100 times. Transcription of CYC1 is controlled by UAS1 in medium containing glucose, but by UAS2 in lactate-containing medium.

HAP1, HAP2, and HAP3 also activate transcription of CYC1. HAP1 activates UAS1, CYC7, and cytochrome b_2. HAP2 and HAP3 activate UAS2 jointly.

UAS1 has two subsites, both of which are required for the activity of UAS1. Site B has sequences binding to the HAP1 gene products and factor RC2, and formation of both of these proteins requires adequate concentrations of heme. HAP1 also associates with site A when it is complexed with another site-A binding factor.

Maintaining protein synthesis in yeast requires coordinated transcription of the family of genes encoding the ribosomal proteins. There are 25–35 proteins in the

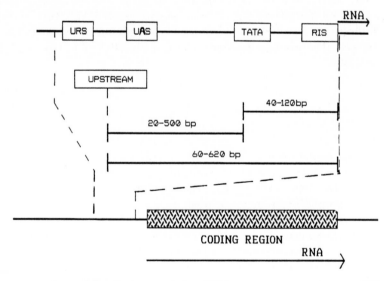

Fig. 10.4. Arrangement of control elements in transcription. *URS* Upstream, repressor sequences; *UAS* upstream activator sequences; *TATA* TAT box; *RIS* RNA initiation site

40S subunit and 35–45 proteins, a total of 60–80 proteins, in a given ribosome. Random synthesis of these proteins would probably stop the metabolic processes and kill the cell. Fifteen genes encoding ribosomal proteins, out of 20 examined, contained three short consensus sequences; HOMOL1 (consensus AACATTC/TG/ATA/GCA), RPG (consensus ACCCATACATT/CT/A), and a T-rich region. All three elements are essential for expression of this gene. The tripartite element formed from them has UAS activity and can reactivate the heterologous CYC1 promoter as well. HOMOL1 and RPG are variants of a consensus, UAS, found in numerous genes involved in the synthesis of yeast proteins.

Other yeast genes (CUP1, GCN1, GAL1, GAL10, GAL7, GAL2, GAL4, GAL80) may have tandemly duplicated upstream activator elements. The GAL80 and GAL4 genes are transcribed divergently and share a bidirectional UAS which is halfway between the genes. The UAS_G contains four 17-bp dyad symmetrical sequences where GAL4 binds, to activate transcription. A single copy of this 17-bp sequence will activate defective GAL4 or CYC1 promoters in the presence of galactose.

The activator sequences may be regulated by negative elements. This may be in two ways: action at a distance by the regulatory elements, or by antagonizing the binding of transcription factors at the upstream activation site. Control of mating type is an example of repression by negative elements at a distance, and control of the expression of the enolase genes, ENO1 and ENO2, by the substrate is another. These two genes are expressed to different degrees, depending on whether the carbon source is glycolytic or gluconeogenic, but ENO1 is regulated by a negative-acting, downstream element which antagonizes UAS function. Another example is heme, which activates several aerobic genes, and inhibits the expression of the anaerobic ANB1 gene under aerobic conditions.

There are negative elements in other yeast promoters. These reduce but do not abolish the expression of the gene or genes. Sometimes, deletion of a negative element increases expression of the gene (CYC7 and TRP1 are examples) by two to five times.

2.3 Catabolite Repression

The reduction or abolition of the utilization of another sugar in the presence of glucose in the medium has been known for some time. It is explained by the *reduction in the level of transcription* of such genes as SUC2, GAL1–10, ADH2, and CYC1. The sequences mediating catabolite repression are upstream of the TATA box. This type of repression in yeast is mediated by negative control. Gene fusions between regions containing the GAL1–10 UAS, including the region required for catabolite repression, and others containing the HIS3 promoter, showed that UAS_G was activated by galactose, and the basal level of expression of the HIS3 promoter was increased. Repression by glucose reduced expression below the normal basal level. The negative site was upstream of the HIS3 regulatory sequences, showing that action at a distance had occurred.

Oxygen also participates in regulating many genes, especially those involved in respiratory functions, and interacts intimately with heme in this role. Glucose represses a number of the same genes, and is probably intimately related to regulation of the pertinent genes by oxygen and heme. Some genes regulated by oxygen are heme-dependent and some heme-independent.

2.4 Genes Activated by Heme (14 Genes)

These genes fall into two classes: those encoding proteins involved in respiratory functions (subunits of the different cytochromes), and those encoding proteins for damage repair (catalase and manganese superoxide dismutase). A second set (7 genes) of hypoxic genes is repressed by heme. These are related to utilization of oxygen in electron transport (COX5B), synthesis of components of membranes OLE1, ERG3, ERG11, HMG2), or biosynthesis of heme itself (HEM13), so that they are induced at low oxygen tensions to compensate for this substrate limitation.

Many genes occur in pairs, each pair having identical functions even though the genes are unlinked, and one of each pair being activated and one repressed by heme. These include genes encoding subunit V of cytochrome c oxidase (COX5A and COX5B), those encoding 3-OH-3-methylglutaryl coenzyme A reductase (HMG1 and HMG2), translational initiation factor eIF5A (TIF51A and ANB1), the gene pair CYC1 (activated by heme) and CYC7 (activated and repressed by heme), encoding cytochrome c, as well as a number of other paired genes. One gene of the pair probably functions more effectively at low oxygen tensions, and the other at higher ones.

Transcription of several genes is activated by the HAP1 protein, the CYC1, CYC7, CTT1, and HEM13 (if, for HEM13, heme is not present). A different group of genes is activated by the HAP2/3/4 complex (CYC1, COR2, CYT1, and CYB2 are activated by both HAP1 and the HAP2/3/4 complex). The activator proteins bind to the UASs in the normal manner. The HAP2/3/4 complex responds to both heme and nonfermentable carbon sources (lactate). Expression is increased by an order of magnitude by heme, and by a further ten times by lactate. In the complex, the HAP2 protein contains the DNA-binding region, and the HAP4 protein the transcriptional activation domain.

2.4.1 Genes Repressed by Heme

These genes are usually regulated through the action of the ROX1 repressor protein. Transcription of the ROX1 gene itself is activated by heme, and is partly dependent on the HAP1 protein. There is a HAP2/3/4 consensus sequence which may be involved in activation by heme of ROX1. Expression of ROX1 is *not* subject to catabolite repression, unlike other genes regulated by the HAP2/3/4 complex. The function of the ROX1 protein is not heme-dependent, even though the expression of the gene is. Repression of genes regulated by the ROX1 gene is by activation by heme of transcription of the repressor gene (ROX1). The ANB1 gene, which encodes the initiator of transcription, eIF5A, is repressed by heme, and activated by anaerobiosis. Induction of its mRNA reaches normal levels within 90 min. This occurs because of loss of expression of the ROX1 gene and a subsequent reduction in the levels of the ROX1 protein. The protein is relatively unstable and is degraded rapidly. The system can respond rapidly to changes in oxygen tension, so that hypoxia is a stress response. Repression of ROX1 is mediated through binding to the ANB1 operator site.

The genes regulated by both HAP1 and ROX1 include CYC7, ERG11, and HEM13. The regulatory elements are antagonist in CYC7 and probably in ERG11, so that aerobic expression is at a low level. HEM13 is unusual, since HAP1 activates transcription in the absence of heme. There is also at least one alternate repression system involving heme.

Other factors affecting heme regulation include mutations in the CYC8 and CYC9 genes (SSN6 and TUP1). In the absence of CYC1, the levels of expression of CYC7 are too low for the cell to grow on lactate. Cell growth on lactate under these conditions allows selection of mutants showing increased expression of the CYC7 gene and the SSN6 and TUP1 genes. Mutations in these two latter genes had pleiotropic effects, including increased expression of the CYC7 gene, release from catabolite repression, flocculation, a-sterility, poor sporulation in homozygous strains recessive for these genes, and partial repression of the ANB1 gene. Mutants in the TUP1 gene show increased plasmid stability and can take up TMP from the medium. The ssn6 and tup1 mutants can function independently of each other.

In catabolite repression, the TUP1 and SSN6 proteins are part of a pathway related to the SNF1 protein kinase, since mutations in the ssn6 and tup1 genes suppress the inability of the snf1 mutations to induce synthesis of invertase in the absence of glucose. Expression of the CYC1 and COX6 genes is also affected. Apparently, this does not relate the catabolite repression pathway to other phenotypes caused by ssn6 and tup1 mutations.

The TUP1 function is required for activation of the ROX1 repressor, though DNA binding does not require it. Activation of ROX1 by TUP1 protein does not require heme, since ROX1 can function in heme-deficient cells. There may be two alternative functions for the TUP1-SSN6 complex, whose functions are still unknown.

There are numerous genes and enzymes in yeast which are regulated by oxygen, but are independent of heme. Two systems well known to be independent of regulation by heme are the mitochondrial translation factor PET494 and the anaerobic genes, ANB2 to ANB15.

2.4.2 Mitochondrial Genes

COXI, COXII, and COXIII encode the three cytochrome c oxidase subunits. The COXIII gene, which encodes subunit III, requires three specific factors for its translation, the products of the nuclear genes, PET494, PET 122, and PET54. The factors recognize the 5′ leader sequence of coxIII mRNA. PET494, unlike a number of other nuclear genes which encode proteins related to mitochondrial function, is regulated by oxygen tension at the translational level, independently of heme. It is not known whether the other two factors are also regulated at this level. Oxygen regulates translation of the *coxI* and *coxII* mRNAs which each require a different set of factors encoded in the nucleus. The protein encoded by the PET111 gene is required for translation of the *coxII* mRNA. This is translated from an mRNA having a remarkably long 5′ leader sequence. Like the PET494 protein, it may be under translational control. Translation of the mitochondrial *cob* gene, which encodes a subunit of the cytochrome bc_1 complex, is similar. Probably, translation of each of the mitochondrial cytochrome subunits is regulated independently. The initial transcripts apparently requre extensive processing, mediated by intron-specific mitochondrial and nuclear genes, also regulated by oxygen.

2.5 Anaerobic Genes

These genes are expressed exclusively in the absence of oxygen, after the onset of anaerobiosis, some early (4 genes; 1.5h) and some later (11 genes; 6h). Genes ANB13, ANB14, and ANB15 are expressed early, independently of induction by heme or ROX1. They are not expressed under aerobiosis in heme-deficient or *rox1* mutants. Four of the "late" genes, ANB2, ANB3, ANB4, and ANB5, required 3h of induction after anaerobiosis before being expressed. Expression of the ANB2 gene was independent of heme or *rox1*. The system is complex.

There are constitutive promoter elements in yeast, and increases in mRNA levels are obtained by regulation at an upstream activator sequence, which is a simple poly(dAT) sequence. The efficiency of translation apparently depends on the length of the sequence. The DED1 gene has a 34-bp region containing 28 thymidine residues. It produces five times as much RNA as the HIS3 or PET56 genes, which have a common 17-bp tract containing 15 thymidine residues. When the ADR2 promoter region was increased from 20-bp of dA-dT, to 54–55 bp, the activity was greatly increased. This phenomenon has not been explained.

2.6 Heat Shock Elements: Proteins in Many Cells

When cells of prokaryotes and eukaryotes are exposed to high temperatures, a family of specific proteins is synthesized. Other proteins are affected, particularly those participating in ribosome biosynthesis, whose levels are reduced. Some, but not all, of the genes encoding heat shock proteins are expressed at a low level without heat shock. Some genes (encoding phosphoglycerate kinase) are not directly involved in heat shock, but show increased expression on being shocked.

The heat shock element, which has the consensus element $C_{nn}GAA_{nn}TTC_{nn}G$, appears at least once in the promoters of genes affected by heat shock. The heat shock reaction is complex and not well understood, the heat shock element (HSE) sometimes functioning as a conventional upstream activator sequence, and sometimes simply augmenting gene expression. Different types of HSEs show different goodness of fit to the consensus sequence, which may affect the strength of the expression of the different heat shock genes. The heat shock proteins may be involved in the development of thermotolerance in yeast.

2.7 Transcriptional Control at Upstream Elements

Although the upstream elements of a number of genes are localized in small regions, elements for different genes do not, apparently, share common sequences.

2.8 Coregulated Genes

If these genes are subject to the same general pattern of regulation they may have common short squences of bases, upstream from the start site for RNA transcription. Members of the same family of genes may not be repressed or induced to the same extent. This occurs in the genes in synthesis of amino acids, sharing the core sequence TGACTC, where the GCN4 protein shows different affinities for this sequence in different promoters. The affinity may be affected by the flanking sequences.

The genes encoding acid phosphatases (PHO5 and PHO7) are regulated by the phosphate content of the medium and promoters induced by galactose (GAL80, GAL2, GAL7, MEL1, and GAL1-10). These share a 23-bp dyad consensus sequence. GAL4 binds to the UAS sequence, whether in induced or uninduced conditions, and GAL80 binds to a region in the GAL4 protein, preventing activation. The number of binding sites varies from 1 to 4, being correlated with the degree of repression by the GAL80 protein.

The HO promoter is regulated according to the number of copies of the consensus sequence PyCACGAAAA, regulation being tighter as the number of copies is increased. Repression of haploid-specific genes occurs in the control of mating-type genes. The HAP1-heme complex binds to UAS1 in the CYC1 promoter and to the UAS in the CYC7 promoter, but the sequences in the binding sites are different. Coregulated sites do not necessarily share related control sites, even though the same activator protein mediates induction.

2.9 Regulation of Transcription by Bidirectional Upstream Elements

The nucleotide sequences of some upstream promoter elements are not symmetrical. These are unidirectionally oriented to the structural gene. In the GAL1-10 genes, the four core sequences are arraged asymmetrically between the structural genes, so transcription is probably controlled by one element acting bi-directionally. Where a promoter should function bidirectionally, and does not, the element may be blocked by sequences which are normally located upstream from it. The promoter may function bidirectionally if these sequences are removed. Terminators may be located upstream of the UAS, and, when the promoter is reversed, can block the transfer of an activating signal from the UAS to the RNA initiation region. While promoters which function in one direction participate only in transcription, others are bidirectional and can be read in either direction. When a promoter appears to function unidirectionally, removal of blocking sequences may allow the promoter to function bidirectionally.

2.10 Position of the Upstream Element

These elements may be located several hundred base pairs from the TATA element and the RNA initiation site, the distance being variable. The efficiency of gene expression is not determined by the upstream sequence, but at other regions (the TATA box or the I site).

2.11 Do Other Sequences Also Determine Transcriptional Efficiency?

Usually, the upstream sequences have the greatest influence on the efficiency of transcription, as for the DED1 upstream sequences. In some cases, downstream sequences may also affect transcriptional efficiency. A region in the coding part of the PGK gene apparently enhances the rate of transcription of the gene, and is called the downstream activator sequence. The influence of downstream activator sequences and other downstream sequences on transcription has not yet been determined.

2.12 Complex Upstream Elements, Signals, and Coeffectors

These elements have different effects when acting alone or in combination. A silencer element, upstream from the HMR locus, represses transcription of mating-type information. It contains three short sequences; the E and B sequences activate transcription from a heterologous promoter, when either one is acting alone. When the two are acting together, or with the third sequence (A, which contains an ARS consensus sequence), they repress transcription. They are also involved in DNA replication and segregation. The essential genes, RAP1 and SBF B, bind to the E and B sequences, and may be concerned with repression and activation of the genes.

Galactose is a coeffector for GAL80 activity and association with the GAL4 carboxy terminus, allowing interaction with unknown transcription factors. Heme levels are also global signals and coeffectors. With the ROX1 gene, heme is a positive and negative effector for aerobic and anaerobic promoters. With the HAP1 gene product, the complex regulates expression of the cytochromes. Heme levels are a measure of the capacity of the cell to carry out oxidative phosphorylation, depending on whether cell growth occurs under aerobic or anaerobic conditions. In this way, regulation of many genes can be done using a single signal molecule and a small number of regulatory proteins.

Expression of the GCN4 gene is regulated in a different way. Levels of the mRNA transcribed from GCN4 remain relatively constant, even under conditions of nitrogen (amino acid) starvation. Levels of the GCN4 protein are 50 times higher under these conditions. Transcription of the many genes for amino acid biosynthesis is regulated by changes in the level of the GCN4 *protein*. This behavior can be explained by the structure of the mRNA transcribed from the GCN4 gene. This mRNA has an extremely long 5'-untranslated region, where four AUGs followed by in-frame termination codons precede the initiating AUG for the structural gene. This should prevent the synthesis of any GCN4 protein at all, and this is the case under normal nutritional conditions. Under conditions of amino acid starvation, the GCN4 gene is expressed, GCN4 protein is made in quantities, and the genes encoding mRNAs for synthesis of many amino acids are expressed.

Phosphorylation, accomplished by protein kinases, is also a signal for regulatory action on expression of genes associated with catabolite repression and other regulatory circuits in yeasts.

2.13 The Mysterious TATA Box

This is a short A/T-rich sequence found in all promoters, prokaryotic and eukaryotic. It has a definite location in prokaryotes, 10 bp upstream from the RNA initiation site. In "higher" eukaryotes, it is located less precisely, 25 to 30 bp from the 5' end of the corresponding mRNA. In yeasts, the TATA box is difficult to define or locate, since yeast promoters are very A/T-rich, and any one of several sequences could be *the* TATA box.

The upstream (5') end of the TATA box was defined reasonably readily, and a second critical promoter element was detected, by deletion of a series of short sequences until a further drop in RNA levels after the deletion of the first TATA element was observed. The 3' (downstream) end was located in the HIS3 promoter, which contained the whole TATA element in a 20-bp region, from −32 to −52 bp upstream of the mRNA initiation site. Two separate regions of the DNA are required to constitute a promoter, neither being sufficient alone.

2.14 Distance, Direction, and Distinctiveness of TATA Elements

The functional TATA box is often found at about 60 bp from the RNA initiation site. This is not fixed, as some are found 100 or more base pairs upstream from the

I site. Some yeast promoters may contain more than one sequence, homologous to the TATA consensus site.

The TATA element can be oriented in only one direction relative to the structural gene, unlike many upstream elements. Experiments with the TATA elements in the HIS3 and HIS4 promoters have shown that the TATA element is unidirectional. There may also be more than one TATA element in the same promoter. The HIS3 promoter contains one constitutive and one regulated upstream element, each of which may interact with a specific TATA element. The TRP1 promoter also has two TATA elements, each of which can set a different pattern of initiation sites. The reason is not known.

2.15 Initiation of Transcription; the Site

2.15.1 Preferred Sequences, the Initiation Window, and the Mechanism of Start Site Selection.

Yeast seems to have an element that determines the initiation site. There are shared sequences around the RNA initation sites. Some highly expressed genes in yeast have a CT-rich block about 20 bp long, and after that, 9 to 12 bp downstream, the sequence GAAG. Many yeast promoters have a sequence at the initiation site, RRYRR, where R is a purine and Y is a pyrimidine, or TC (G/A)A, especially when this site was more that 50 bp downstream from the TATA element. There is also a sequence of 7 bp in the TRP1 promoter, CACGTGA, and a factor which binds to it. This sequence is apparently required for initiation of RNA synthesis a little way downstream from it.

The initiation "window" is apparently a stretch of DNA, 40–120 bp downstream from the TATA box, where functional initiation sites are found. The CYC1 promoter has at least four potential TATA elements upstream from the initiation sites. There of these may be involved in initiation of transcription, at different but overlapping sites. The TATA box at −106 promotes initiation at +1, +10, and +16; the TATA box at −52 initiates at +16, +25, and +34, and the −22 element at −34 and −43. Each TATA element appears to have a "window" of initiation, 55–110 bp downstream of the element itself. In the case of the HIS4 promoter, if the TATA box is moved upstream, the I site is always at least 60 bp downstream from the TATA box. Yeast seems to require both the TATA binding and I site binding proteins, so transcription initiation in yeast is likely to be complex.

2.16 Regulatory Proteins

A basic assumption concerning regulatory systems in yeast is that the regulatory proteins exercise control by binding to specific DNA sequences. Very few of these proteins have been isolated, purified, and tested for binding to specific DNA sequences. Some exceptions are: (1) the positive regulatory protein GCN4, (2) the

GAL4 activator protein, and regulation by GAL80, (3) the HAP1 protein, which binds to two different sequences, and (4) the activator protein TUF.

2.16.1 The GCN4 Regulatory Protein

This protein binds to the promoter regions of genes under general control; HIS3, HIS4, TRP5, ARG4, ILV1, and ILV2. It does not bind to promoters of TRP1, DED1, URA3, and GAL1–10, which are not subject to general control. The transcriptional activation region of the GCN4 protein is separable from the DNA-binding region in the 40 amino acids at the carboxy terminal, and the transcriptional activation region in an acidic region (19 amino acids) in the center of the protein. Thus, there is a specific DNA binding region in this protein.

2.16.2 Binding by the GAL4 Activator Protein and Regulation by GAL80

Regulation of the genes involved in galactose metabolism depends, first, on binding of the GAL4 protein to a sequence in the upstream region. The region of the protein directing binding to the upstream activator sequence is at the amino terminal, while that for activating transcription is found near the carboxy end. The action of the GAL80 gene product is to inhibit transcriptional activation by the GAL4 protein, an effect which is relieved by galactose. Catabolite repression, produced by glucose, is independent of the GAL80 gene and probably causes repression by preventing the binding of the GAL4 protein to the UAS_G sequence. The complex dissociates in the presence of galactose and transcription is activated. Probably GAL80 interferes with the interaction of the GAL4 protein with other proteins and stimulation of transcription.

2.16.3 The HAP1 Protein

This protein also has well-separated regions for DNA binding and transcriptional activation, for the CYC1 gene. The UAS1 binding domain coincides with the heme binding domain, and is found at the amino terminus, and the activation region at the carboxy end. The HAP1 protein binds to the UAS1 region of both the CYC1 and CYC7 genes, but the two binding sites are not similar. The HAP1 protein may have more than one binding domain. Assuming this, the HAP1 binding domain must be able to recognize two different sequences in the two genes. There is no consensus sequence at the two binding sites.

2.16.4 TUF, Another Activator Protein

This protein, length about 150 kDa, interacts with two DNA sequences, HOMOL1 and RPG. These are variants of a consensus upstream activator sequence and are features of a number of genes encoding the yeast translational apparatus. TUF appears to have a 50-kDa binding domain, so there are probably distinct domains on the protein.

Transcriptional activation regions of the protein are in the most acidic regions, and DNA binding domains in the basic regions. Mutants having GAL4 protein types that increase activation show increasd acidity in the activating region of the protein. Much of the protein between the transcriptional activation region and the DNA binding region in the GAL4 protein can be deleted without impairing its function.

2.17 Mechanisms for Transcriptional Activation. A Summary

The initiation of transcription requires that RNA polymerase II begins transcription at particular sites on the chromosomal DNA. The site is apparently determined by a complex of factors on the promoter as scaffold, not by a particular sequence of DNA. The DNA binding domains in the regulatory proteins are in the regions where there is a preponderance of basic amino acids, and the activator sequences, in acidic regions. The interactions between elements at a distance have been explained through a "looping-out" mechanism, analogous to the movement of a caterpillar.

The structure of the chromosome is extremely important. In eukaryotes this is not naked DNA, but a complex of DNA and histones forming chromatin, arranged in nucleosomes. The structure of DNA being expressed differs from that in regions which are not. The latter is almost completely insensitive to degradation by nucleases. The difference is due to the effect of the binding of activating factors and other regulatory proteins to the DNA at these prints. The nature of the rearrangements of the DNA-histone complexes comprising the nucleosomes is unknown. The histones are not merely a sort of skeletal material, supporting the DNA, but play an active (probably negative) role in regulation of gene expression.

The most significant point is that the fundamental knowledge necessary for the understanding of the processes of chromosome replication, transcription, and translation of the resulting mRNA into protein, *and the regulatory mechanisms associated with them*, is also necessary for the most profitable use of these systems for the production of the almost unlimited range of valuable proteins by microorganisms of all species and genera, especially yeasts, which has become possible in recent years.

3 Translation of the Genetic Message. Life and Death of a Messenger RNA

3.1 Introduction

The third step leading from the message at the heart of the yeast cell, from the code written in the chromosomes to the formation of the materials of construction of the cell, is the translation of the message carried by the messenger (mRNA) into protein. The messenger must make contact with, first, the small subunit (40S) of the ribosome, initiate translation by finding the correct initiation codon, bind to it, bind to the large subunit of the ribosome, and the complete ribosomal complex must then move along the strand of mRNA, finding the charged transfer RNAs according to the instructions on the messenger at each codon, adding the amino acid carried by the tRNA to the growing peptide chain, and on reaching a termination codon, dissociating. The ribosomal subunits can then associate with another strand of mRNA and begin the synthesis of another peptide chain.

Eventually, its work finished, the strand of mRNA is degraded by the cytoplasmic ribonucleases. All of this is under tight and rigid control during the entire process.

3.2 The Ribosome

The ribosome is the key organelle involved in protein (polypeptide) synthesis. It functions in a way similar to the reading head of a tape player; it detects the message on the tape (the mRNA) and converts it into a form that can be understood by the recipients; in one case by the listeners as sound; in the other by the machinery of the yeast cell as proteins.

The individual ribosome is made up of two subunits; the smaller (40S) containing one molecule of ribosomal RNA (rRNA) (18S) and a number of proteins, between 25 and 35. The larger subunit contains three molecules of rRNA, 25S, 5.8S, and 5S, and from 35 to 45 proteins. Nearly all of these components are present in equimolar amounts. The bacterial ribosome is somewhat smaller than the cytoplasmic ribosomes of yeast, though the mitochondrial ribosomes of the latter are similar in many ways to the bacterial ribosome.

Many of the genes encoding ribosomal proteins have been investigated and have been cloned and sequenced. There are two consensus sequences which may be necessary for transcription of some of the ribosomal genes encoding proteins. There are also regulatory mechanisms which ensure that the ribosomal proteins are synthesized in the correct amounts. Control of the amounts of the proteins is quite rigorous and prevent their synthesis in excess; an example will be described later in this section. There are large numbers of protein factors and cofactors which are involved in protein synthesis in the yeast cell, at different stages of the process.

3.3 Initiation, Elongation, and Termination

3.3.1 Initiation

Initiation of protein synthesis in prokaryotes is relatively sample. The small ribosomal subunit recognizes a binding site on the mRNA, through sequence complementarity with the 3′ end of the 16S ribosomal RNA, and a nearby codon initiates translation.

The situation is more complex in eukaryotes, including yeasts. Ribosomes actively translating mRNA and synthesizing polypeptides are found in the form of polysomes bound to the outer surface of the rough endoplasmic reticulum. As the polypeptides are synthesized, they are passed through the membrane into the lumen of the ER, where they are trimmed to size by a number of proteases.

Before this happens, however, the message on the mRNA must be translated at the ribosome and appear as protein. First, a complex containing the small (40S) subunit of the ribosome binds to the 5′ end of the 5′ UTR, and moves along the mRNA strand until it reaches the first AUG codon. It stops at this point and the large (60S) subunit of the ribosome binds to the complex, along with a number of protein factors. Besides these factors, initiation requires the binding of a methionyl tRNA residue to the 40S subunit, plus GTP and ATP.

This relatively simple picture is further complicated by the influence of the sequence environments around the AUG codons. These squences alter the ease with which the 40S subunit can recognize the initiator codon; if it does not recognize a particular AUG codon, the subunit continues scanning along the mRNA strand until it encounters an AUG which it can recognize. The most favorable environmental sequences appear to be ANNAUGGN and GNNAUGR, where N = any base, R = any purine, and Y = any pyrimidine, and the least efficient, GNNAUGY, YNNAUGR, and TNNAUGY. Unrecognizable AUG codons are usually in unfavorable environments and are the ones passed over during scanning by the 40S complex. However, there is no absolute requirement for a particular sequence around the AUG codon.

It may be important, in production of heterologous proteins using yeast as a vector, to keep in mind that some sequences which are 5′ to the initiation may inhibit translation of a heterologous mRNA such as hepatitis B core antigen. The mRNA was expressed much better when ten contiguous G resides, upstream from the initiation codon, were deleted.

The 5′ cap is also necessary to promote translation of most yeast mRNAs. The 5′ cap consists of a methylated guanosine residue (m_7G) linked to the 5′ end by a rare 5′–5′ phosphotriester link. The 5′ cap also protects the mRNA from degradation by exonucleases, which we will describe in further detali in Section 4.2.

Two other points pertinent to initiation of translation of mRNAs in yeast: first, the length of the 5′-UTR region may vary from 30 to nearly 600 nucleotides, but the length has no effect on the efficiency of translation. Second, the number of G residues in the 5′-UTR region *can* affect translation, which decreases according to the number of these residues, especially near the AUG initiation codon.

When all of these conditions have been fulfilled, the process can proceed to the next stage in translation, which is:

3.4 Elongation

Elongation begins when the 60S subunit of the ribosome has bound to the complex, which is positioned over the "Start" codon and completes it. There are four substages in the elongation phase:

1. The necessary tRNAs must be charged with amino acids by aminoacyl-tRNA synthetases. This requires hydrolysis of one ATP molecule per tRNA charged, and must be done very accurately for the correct polypeptide to be synthesized.
2. The correct aminoacyl-tRNA is bound to the A-site on the ribosome. This requires elongation factor EF-1, and the particular aminoacyl-tRNA is specified by the codon in the position at the active site on the mRNA.
3. Formation of a peptide bond, accompanied by transfer of the peptide chain from the tRNA in the P site, to the A site on the ribosome.
4. The elongating complex moves a distance of one codon along the mRNA chain (translocates), and the new peptidyl-tRNA is moved to the P-site from the A-site, a step which requires EF-2 (elongation factor 2). It should be noted at this point that yeast has a third elongation factor, EF-3, whose function is unknown.

Sometimes, this process fails to follow the even course just outlined. Elongation may go through a frame-shift, which occurs within a small region of the Ty mRNA, to bring about synthesis of one of the retrotransposon proteins, p3. The frame-shift is needed for correct synthesis of this and a few other proteins in yeast. There is a low natural level of frame-shifting in yeast, specified by the mRNA sequence, which is used by the Ty element, but the mechanism of control of the frame-shift is not known.

3.4.1 Codon Bias

This is another factor in the elongation process. Out of the 61 possible codons for amino acids, yeast appears to favor only 25 of them. For instance, mRNA transcribed from the PYK1 gene, encoding pyruvate kinase (a strongly expressed enzyme) contains 36 codons for leucine. Of these, 32 are UUG, and only 4 are selected from the five remaining leucine codons; UUA, CUU, CUC, CUA, and CUG. The preferred codons are not the same in all organisms, whether eukaryotes or prokaryotes. In each organism, however, the preferred codons are well correlated with the corresponding tRNAs; in yeast, the major leucine acceptor tRNA has the anticodon CAA, which pairs with the preferred codon for leucine, UUG. Finally, strongly expressed genes show a strong codon bias, while genes with little or no codon bias are expressed at low levels. Codons can act more or less additively as well; a string of four nonpreferred codons may reduce expression of the protein encoded.

There may be pauses in translation, as well, which may be brought about by the occurrence of nonpreferred codons or by the formation of secondary structures in the mRNA. These may allow time for folding of the peptide chain. Folding and the role played by delay in translation and other causes may be significant in the production of heterologous proteins; the codon bias in yeast may be different in the organism which is the source of the heterologous gene. This, in turn, may affect the correct folding of the desired protein.

This process could go on in theory, and if the mRNA chain were long enough, for a tremendous distance. However, the yeast has no requirement for a polypeptide chain of indeterminate but remarkable length, and as the ribosomal complex moves along the actual rather than a theoretical chain, it reaches its predetermined length and encouters the end, which is a stop codon, and faces – termination.

3.5 Termination

The normal termination codons in yeast are UAA, UAG, and UGA. There are no normal tRNAs which correspond to these codons. Mutants where termination of translation is suppressed are known, and the mutant has an altered tRNA which can translate a termination codon.

In normal, wild-type yeast, when the mRNA reaches a termination codon, which is transferred to the A site on the ribosome, there is a release factor (RF) which binds to the ribosome. The ribosomal peptidyl transferase then causes the polypeptide chain to separate from the last tRNA. After this, the RF then dissociates from the ribosome, GTP is hydrolyzed to GDP-P_i, and the ribosome dissociates into its subunits which are then available for a new round of translation. The UGA codon is the terminator in 20% of mRNAs, UAA in 50%, and UAG in 30%. In many yeast mRNAs, a second terminator codon follows the first, which probably ensures efficient termination. Efficient nonsense suppressor mutations are deleterious to growth, so the explanation seems reasonable.

The role of the 3′-untranslated region (3′-UTR) is not well understood. Sometimes, sequences in this region appear to influence translation, but at others, not. Once more, further investigation is required.

4 Regulation of Translation

Translation, like other aspects of yeast growth, reproduction and metabolism, is closely regulated. Global control probably occurs mainly at the initiation level, by competition among mRNAs for a specific initiation factor, which is most likely eIF-2. The mechanism (once again!) is not known.

Translation of some particular mRNAs is individually regulated, often by interaction of upstream ORFs with a variety of *trans*-acting factors. the mRNA transcribed from the CPA1 gene, which encodes one subunit of carbamoyl-phosphate synthetase, involved in the synthesis of arginine. Expression of the gene is regulated at the transcriptional level, but it is also repressed in the presence of arginine at the translational level. This is accomplished through a 5′-UTR of about 250 bases, containing an open reading frame of 25 codons and a number of possible hairpin structures. If this ORF is disrupted, translation of the CPA1 mRNA occurs in the presence of arginine. This and other evidence suggest that the leader amino acid sequence is essential for repression of translation. It has been suggested that the 40S complex has a choice of upstream ORFs. If translation occurs at the upstream ORF, then translation of the leader peptide takes place and this interacts with the product of CPAR (a regulatory gene) and prevents further scanning downstream, by other 40S complexes, to the major ORF. If arginine is not present, the CPA1 gene product does not recognize the leader peptide, and the 40S complex continues scanning to the major ORF, where translation of the sequence for synthesis of carbamoyl phosphate synthetase then begins.

Other genes which are regulated in a similar way include GCN4, which has a very long 5′ UTR (590 bases) which contains four AUG codons, each of which is followed by a termination codon. All four of the AUGs are functional to some degree in the regulation mechanism, though deletion of all four bases, in the 280 base sequence, allows derepression of translation of the GCN4 mRNA. It is interesting to note that the sequence in the neighborhood of the AUG codon for the major reading frame is closer to optimum than for those farther upstream.

More genes whose mRNAs are subject to similar regulation include PET3, PPR1, SUC2, HTS1, LEU4, TRM1, and MOD5, all having multiple AUG codons. In the SUC2 mRNA, there are at least two AUG codons, and the cytoplasmic enzyme is carried out using the downstream "Start" codon, while the upstream codon is used for translation of a leader sequence which targets the polypeptide for secretion. However, in this case, control is at the transcriptional level, as transcription may start at two different points, so that the upstream AUG may be included or omitted from the mRNA.

4.1 Secretion

As such, secretion depends on the leader sequence just mentioned. This sequence is normally an amino-terminal extension, hydrophobic, 20–30 amino acids in length, which interacts with signal recognition proteins (SRP). This binds the polysomes (ribosomes + mRNA) to the outside of the rough endoplasmic reticulum, and as translation continues, the precursor protein is passed through the membrane into the lumen of the ER. This is the first step in secretion of the protein; the leader sequence is cleaved from the polypeptide by a "signal peptidase", resulting in the formation of the mature protein. The protein is transported to the Golgi body and from there, via secretory vesicles, to the cytoplasmic membrane and released into the external environment, which may be the periplasmic space or the culture medium. The process will be discussed further in the section on production of heterologous proteins (Chap. 14).

Another form of regulation at the translational level is that of synthesis of the ribosomal L3 protein, which is encoded by the TCM1 gene. Synthesis is tightly coordinated with that of the other ribosomal proteins. Extra copies of the gene may be transformed into yeast, but the level of the L3 protein in the cell does not change significantly; dosage copmpensation mechanisms come into operation to maintain normal levels of L3 protein. One of these is translation; cells may contain as much as 3.5 times the normal levels of L3 mRNA, but only 1.2 times the normal level of the L3 protein is synthesized.

Finally, when the synthetic and regulatory mechanisms have performed their function and the appropriate levels of the required cellular proteins have been synthesized, there remains only the matter of the disposal of the mRNA, whose function is finished. There follows:

4.2 Death of a Messenger: Degradation of mRNA

There are numerous mRNAs in the yeast cell, of a wide variety of sizes, and also having a wide variety of half-lives, the latter ranging from approximately 1 min to more than 100 min. Once their functions have been completed, they are degraded to greater or less degress to the point where their component parts can be recycled and reused by the cell. The process or processes have been relatively little studied, and no major genetic investigations have been made of the degradation of mRNA. Mutants having lesions in RNA metabolism exist; many of the conditional lethal mutants isolated by Hartwell are of this type, most being of the rna2 type in which RNA splicing is disrupted. Apparently, the *RNA1* gene has nothing to do with splicing, being involved in transport of mRNA from the nucleus to the cytoplasm. The *RNA1* gene has been cloned and this may permit a more detailed analysis of the function of the gene product.

There are three major groups of RNases in yeast; Mg^{2+}-dependent endoribonucleases, Mg^{2+}-independent endoribonucleases, and Mg^{2+} exoribonucleases. More than 50% of the nonspecific nuclease activity is found in the mitochondrion, and is encoded by a single nuclear gene, *NUC1*. It is a Mg^{2+}-dependent endoribonuclease, which also has DNAase activity. However, this enzyme probably has nothing to do with degradation of cytoplasmic mRNAs, and probably the major nuclease involved in degradation of these mRNAs is a pyrimidine-specific, Mg^{2+}-independent endonuclease. Mg^{2+}-dependent exoribonucleases are probably important in degradation of cytoplasmic mRNAs, as is shown in Fig. 10.5.

The structure of the mRNA itself dictates the mode of degradation of mRNA. The molecule has a 5' cap and a 3' poly(A) tail, which act to protect the ends of the molecule from degradation by exonucleases. This dictates a mechanism by which the molecule is cleaved by an endoribonuclease, resulting in two fragments each with an unprotected end, so that the fragments can be further degraded by two types of exonucleases: 5' to 3' and 3' to 5' exonucleases, which split off single units from the molecule. Yeast mRNAs seem to be classifiable into two general groups, stable and unstable, having broadly different half-lives. However, the general mechanism of degradation is the same.

At this point, the chromosomes have been replicated, the message has been transcribed from them to messenger RNA and translated by the ribosomes with subsequent synthesis of protein. The messenger RNA has been degraded, and the

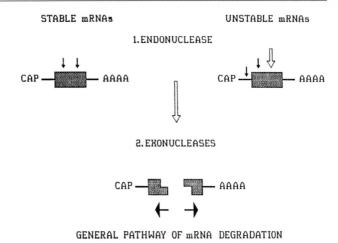

Fig. 10.5. General pathway of mRNA degradation. See text for details

ribosomal subunits have gone on to other things and other proteins. The proteins, newly formed, may be directed to their appointed places in the cell or modified by specific proteases and/or passed by way of the Golgi body through the secretory pathway to the secretory vesicles and into the periplasmic space, and sometimes, into the culture medium surrounding the cell. This phase is extremely important commercially in the production of everything from extracellular enzymes to valuable heterologous proteins of pharamaceutical and therapeutic importance. We will discuss these processes under the heading of the application of molecular biology of yeasts, to practical problems, but first we must investigate the matter of how the cell regulates and recognizes itself.

References

Beggs JD (1978) Transformation of yeast by a replicating hybrid plasmid. Nature 275:104–109

Bowdish KS, Mitchell AP (1993) Bipartite structure of an early meiotic upstream activation sequence from Saccharomyces cerevisiae. Mol Cell Boil 13(4):2172–2181

Bowdish KS, Yuan HE, Mitchell AP (1995) Positive control of yeast meiotic genes by the negative regulator UME6. Mol Cell Biol 15(6):2955–2961

Brearley RD, Kelly DE (1991) Genetic engineering in yeast. In: Wiseman A (ed) Genetically-engineered proteins and enzymes from yeast: Production control. Ellis Horwood, Chichester, pp 75–95

Brown AJP (1989) Messenger RNA translation and degradation in Saccharomyces cerevisiae. In: Walton EF, Yarranto GT (eds) The molecular biology of yeast. Blackie, London, Van Nostrand Reinhold, New York, pp 70–106

Bruhn L, Sprague GF Jr (1994) MCM point mutants deficient in expression of a-specific genes: residues important for interaction with a1. Mol Cell Biol 14(4):2534–2544

Clark KL, Dignard D, Thomas DY, Whiteway M (1993) Interactions among the subunits of the G protein involved in Saccharomyces cerevisiae mating. Mol Cell Biol 13(1):1–8

Cregg JM, Tschopp JF, Stillman C et al. (1987) High level expression and efficient assembly of hepatitis B surface antigen in the methylotrophic yeast Pichia pastoris. Bio/Technology 5:479–484

Drysdale CM, Dueñas E, Jackson BM, Reusser U, Braus GH, Hinnebusch AG (1995) The transcriptional activator GCN4 contains multiple activation domains that are critically dependent on hydrophobic amino acids. Mol Cell Biol 15:1220–1233

Errede B (1993) MCM1 binds to a transcriptional control element in Ty1. Mol Cell Biol 13(1):57–62

Evans IH, McAthey P (1991) Comparative genetics of important yeasts. In: Wiseman A (ed) Genetically-engineered proteins and enzymes from yeast: Production control. Ellis Horwood, Chichester, pp 11–74

Fangman WL, Brewer BJ (1991) Activation of replication origins within yeast chromosomes. Ann Rev Cell Biol 7:375–402

Grandin N, Reed SI (1993) Differential function and expression of *Saccharomyces cerevisiae* B-type cyclins in mitosis and meiosis. Mol Cell Biol 13(4):2113–2125

Hartwell LH (1974) The yeast cell division cycle. Bacteriol Rev 38:164

Herskowitz I (1988) Microbiol Rev 52:536–553

Herskowitz I (1989) A regulatory hierarchy for cell specialization in yeast. Nature 342:749–757

Heslot H, Gaillardin C (1991) Molecular biology and genetic engineering of yeasts. CRC Press, Boca Raton 324pp

Hinnen A, Hicks JB, Fink GR (1978) Transformation of yeast. Proc Natl Acad Sci USA 75:1929

Holm C (1982) Clonal lethality caused by the yeast plasmid 2-μm DNA. Cell 29:585

Kuo M-H, Grayhack E (1994) A library of yeast genomic MCM1 binding sites contains genes involved in cell cycle control. Mol Cell Biol 14(1):348–359

Lee JC, Yeh LCC, Horowitz PM (1991) The initiation codon AUG binds at a hydrophobic site on yeast 40S ribosomal subunits as revealed by fluorescence studies with bis(1,8-anilinonaphthalenesulfonate). Biochimie 73(9):1245–1248

Linder C, Thoma F (1994) Histone H1 expressed in *Saccharomyces cerevisiae* binds to chromatin and affects survival, growth, transcription and plasmid stability but does not change nucleosomal spacing. Mol Cell Biol 14(4):2822–2835

Lucchini R, Sogo JM (1994) Chromatin structure and transcriptional activity around the replication forks arrested at the 3′ end of the yeast rRNA genes. Mol Cell Biol 14(1):318–326

Marsh L, Neiman AM, Herskowitz I (1991) Signal transduction during pheromone response in yeast. Annu Rev Cell Biol 7:699–728

Matsumoto K, Kaibuchi K, Arai K, Nakafuku M, Kaziro Y (1989) Signal transduction by GTP-binding proteins in *Saccharomyces cerevisiae*. In: Walton EF, Yarranton GT (eds) Molecular and cell biology of yeasts. Blackie, London, Van Nostrand Reinhold, New York, pp 201–222

Meacock PA, Brieden KW, Cashmore AM (1989) The two-micron circle model replicon and yeast vector. In: Walton EF, Yarranton GT (eds) Molecular and cell biology of yeasts. Blackie, London, Van Nostrand Reinhold, New York, pp 330–359

Mellor J (1989) The activation and initiation of transcription by the promoters of *Saccharomyces cerevisiae*. In: Walton EF, Yarranton GT (eds) Molecular and cell biology of yeasts. Blackie, London, Van Nostrand Reinhold, New York, pp 1–42

Mendel JE, Korswagen HC, Liu KS, Hadju-Cronin YM, Simon MI, Plasterk RHA, Sternberg PW (1995) Participation of the protein G_0 in multiple aspects of behavior in *C. elegans*. Science 267:1652–1655

Messenguy F, Dubois E (1993) Genetic evidence for a role for MCM1 in the regulation of arginine metabolism in *Saccharomyces cerevisiae*. Mol Cell Biol 13(4):2586–2592

Mizuta K, Warner JR (1994) Continued functioning of the secretory pathway is essential for ribosome synthesis. Mol Cell Biol 14(4):2493–2502

Moore TDE, Edman JC (1993) The a-mating type locus of *Cryptococcus neoformans* contains a peptide pheromone gene. Mol Cell Biol 13(3):1962–1970

Muhlrad D, Decker CJ, Kreck MJ (1995) Turnover mechanisms of the stable yeast *PKG1* mRNA. Mol Cell Biol 14:2145–2156

Newlon C (1989) DNA organization and replication in yeast. In: Rose AH, Harrison JS (eds) The yeasts, vol 4, 2nd edn. Academic Press, New York, 57pp

Oppenoorth WFF (1960) Modification of the hereditary character of yeast by ingestion of cell-free extracts. Eur Brewery Convention. Elsevier, Amsterdam, 180pp

Piper PW, Kirk N (1991) Inducing heterologous gene expression in yeast as fermentations approach maximal biomass. In: Wiseman A (ed) Genetically-engineered proteins and enzymes from yeast: Production control. Ellis Horwood, Chichester, pp 147–184

Rose MD (1991) Nuclear fusion in yeast. Annu Rev Microbiol 45:539–567

Ségalat L, Elkes DA, Kaplan JM (1995) Modulation of serotonin-controlled behaviors by G_0 in *Caenorhabditis elegans*. Science 267:1648–1651

Thompson JS, Johnson LM, Grunstein M (1994) Specific repression of the yeast silent mating locus (HMR) by an adjacent telomere. Mol Cell Biol 14(1):446–455

Wheals AE (1987) Biology of the cell cycle in yeasts. In: Rose AH, Harrison JS (eds) The yeasts, vol 1. Academic Press, New York, pp 283–390

Yun D-F, Sherman F (1995) Initiation of translation can occur only in a restricted region of the *CYC1* mRNA of *Saccharomyces cerevisiae*. Mol Cell Biol 15:1021-1033

Zhou Z, Gartner A, Cade R, Ammerer G, Errede B (1993) Pheromone-induced signal transduction in *Saccharomyces cerevisiae* requires the sequential function of three protein kinases. Mol Cell Biol 13(4):2069–2080

CHAPTER 11

Inside the Inside: Part II: The Regulators.
Cell Specialization: a Regulatory Hierarchy

J.F.T. Spencer and D.M. Spencer

1 Introduction

In outward appearance, the yeast cell is not exciting. One cell looks much like another, and each will give rise to clones having the same biochemical reactions. Nevertheless, appearances are deceiving, and each cell in its time plays many parts. There are, in fact, three cell types in yeast, each with a subtype, so that from some points of view there can be said to be six types in actual fact (Table 11.1). These are, as main types, two haploid cell types, a and α, and one diploid type, a/α, which can be considered a derivative of the other two, since it arises as the product of a mating between them. a cell types can mate with α types, after exposure to α mating hormone (subtype 1); α cell types can mate with a types, after exposure to a mating hormone (subtype 2), and a/α cells will not mate, are insensitive to both mating hormones, and have the potential to sporulate. a/α cells which are induced by starvation will sporulate, and the induced a/α cell constitutes the third subtype. The story of the control system by which all these activities are regulated by the cell is absorbing one, and shows the remarkable complexity of the mechanisms of the hierarchy of DNA binding proteins which govern the reproductive activities of what appears to be a simple yeast cell.

Similar control systems undoubtedly operate in other yeast species, basidiomycetous as well as ascomycetous.

The master control system consists of the MATa and MATα genes. Cells which carry the MATa allele show the a mating type, those carrying the MATα allele are mating type α, and those which have both alleles do not mate at all and have the a/α phenotype. The MAT locus encodes three polypeptides which are components of regulatory proteins; MATa encodes polypeptide a1, so that this peptide is found only in a-mating-type haploid cells; MATα encodes two polypeptides, α1 and α2, which are therefore produced in α mating-type cells; these three gene products form combinations which program the expression of four classes of genes. (The MATa gene encodes an a2 polypeptide which is produced in both a-type and a/α cells, but its function is not known.) Of these, a-specific genes are expressed in a cells, α-specific genes are expressed in α cells, haploid-specific genes are expressed in haploid cells of either mating-type, and sporulation-specific genes, in diploid cells only, and only after starvation.

J.F.T. Spencer/D.M. Spencer (eds)
Yeasts in Natural and Artificial Habitats
© Springer-Verlag Berlin Heidelberg 1997

Table 11.1. Cell types of yeast

Cell type	Properties
α cell	Potential to mate
α cell induced by a-factor	Ability to mate
a cell	Potential to mate
a cell induced by α-factor	Ability to mate
a/α cell	Potential to sporulate
a/α cell induced by nutritional signal (starvation)	Sporulates

1.1 The α1 and α2 Polypeptides

The action of the α1 polypeptide is the most straightforward, as it is a positive regulator of transcription of the α-specific genes, and is necessary for their expression. The α2 polypeptide, however, is a negative regulator of transcription of the a-specific genes, and turns off transcription of these genes when it is found in the same cell with them.

1.2 The a1 Polypeptide

The a1 polypeptide does not act directly to regulate any gene set, but combines with the α2 polypeptide to form a new regulatory protein, a1-α2, which turns off the expression of such haploid-specific genes as STE3, HO, and RME (whose gene product is an early inhibitor of meiosis). It also turns off synthesis of the α1 polypeptide, which in its turn is not able to activate transcription of α-specific genes. Likewise, it switches off expression of haploid-specific genes, a very large group, which are therefore not expressed in normal diploid cells.

The polypeptides encoded by the two alleles of the MAT locus are components of DNA-binding proteins which bind to upstream regulatory regions in the genes under their control.

1.3 Sporulation-Specific Genes

These genes are not regulated directly by products of alleles of the MAT locus. The a1-α2 polypeptide activates them by switching of synthesis of an inhibitor which prevents their expression, not by binding directly to the regulatory regions of the genes themselves. The inhibitor is the product of the RME1 gene, which is an inhibitor of an early step in meiosis.

1.4 The MCM1 Protein

This is the actual activator of all of the a-specific and α-specific genes, and has binding sites in the upstream regulatory regions of all of them. The MCM1 protein was originally named the PRTF or GRM protein, but was found to be the product of the product of the MCM1 gene. However, the MCM1 protein by itself cannot always activate these genes.

1.4.1 In a Cells

The MCM1 protein can activate the a-specific genes in a-mating-type haploid cells and initiate their transcription, without assistance, by binding to its site in the upstream regions, so that these genes are expressed and α-specific genes are not, since the MCM1 protein cannot, by itself, bind to the regulatory regions of α-specific genes. It requires a helper.

1.4.2 In α Cells

The helpers here are the α1 and α2 proteins. First, the α1 protein binds to its target site and, at the same time, enables the MCM1 protein to bind in the same region, next to the α1 binding site. This activates transcription of the α-specific genes.

The a-specific genes are repressed, not transcribed. There are two binding sites for the α2 protein, one on each side of the binding site for MCM1. When the two

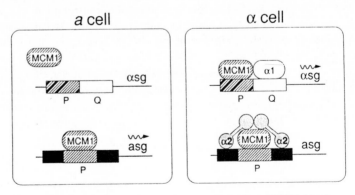

Fig. 11.1. Transcriptional control of α- and a-specific genes by proteins α1, α2, and MCM1 (also known as PRTF, GRM). Expression of these genes is mediated by interactions of α1 and α2 with the MCM1 protein, in the upstream regulatory regions of the genes. Note that the MCM1 protein does not bind to the sites in the a and α cells in the same way. Binding of MCM1 to the regulatory site in a cells stimulates transcription, but binding of both MCM1 and α1 to the corresponding site in α cells is required for activation of the α-specific genes. Binding of two molecules of α2 (a dimer), with MCM1, to the regulatory site in a-specific genes, represses their transcription

sites are occupied by two molecules of $\alpha2$ protein, the MCM1 protein bound to its site is prevented from activating transcription of a-specific genes. The general pattern is illustrated in Fig. 11.1.

The $\alpha2$ protein may mask the MCM1 protein, so that it cannot establish contact with the other elements of the mechanism of transcription. This would indicate that it is similar to the regulation of the GAL4 protein activator, which is masked by GAL80, the negative regulator of GAL4 in yeast. However, this possible mechanism does not explain everything, for instance, how MCM1-$\alpha2$ might repress transcription when it acts in an upstream position to a region for transcriptional activation.

1.5 The a1–2 Protein

This is present only in a/α cells. Not only does this protein repress transcription of a-specific genes, in cooperation with MCM1 protein, but when associated with the a1 protein, it represses the whole range of haploid-specific genes. It recognizes the a1-$\alpha2$ operator sequence; this does not contain an MCM1 binding site. It is interesting to note that the $\alpha2$ polypeptide is also found free in a/α cells, that is, not associated with a1 as the heteromeric form. These two polypeptides recognize different binding sites to turn off both the a- and haploid-specific genes; a1-a2 blocks transcription of the Ty element and the HO gene; these require different transcriptional activators.

The way in which different combination of polypeptides can give rise to different regulatory activities is shown in Table 11.2. The MCM1n protein interacts with $\alpha1$, $\alpha2$, and STE12, but not with a1-$\alpha2$, to bring about cell-specific gene expression.

Table 11.2. Polypeptide subunits and regulatory activities in three different yeast cell types

A) Polypeptide subunits

Cell type	a1	$\alpha1$	$\alpha2$	MCM1	STE12
a	+	−	−	+	+
α	−	+	+	+	+
a/α	+	−	+	+	−

B) Regulatory activities

a	α	a/α
MCM1	MCM1	MCM1
	MCM1-$\alpha1$	
	MCM1-$\alpha2$	
		MCM1-$\alpha2$
MCM1-STE12	MCM1-Ste12	a1-$\alpha2$

Note. The gene products, present in different cell types, interact with eeach other to form the regulatory activities shown in B. STE12 can probably also act independently of MCM1. (After Herskowitz, 1989 with permission)

2 Genes Expressed in Mixtures of Mating Types a and α Cells

2.1 Cell Differentiation

Some genes are only expressed in mixtures of **a** and α cells, cultured together. The FUS1 gene produces a polypeptide required for cell fusion in the mating reaction, and is induced to 500 times its normal level in the presence of cells of the opposite mating type. There are other genes encoding the mating factors, and all of **a**- and α-specific genes so far tested, are induced to levels of three to five times normal, under these conditions. Induction is, in fact, due to the presence of the mating factors.

2.2 The Mating Factors

These are small peptides, of 12 (**a**-factor) and 13 (α-factor) amino acids each. The **a**-factor terminates in a farnesylated cysteine residue which also contains a methyl group. The α-factor is diffusible; part of the **a**-factor may be bound to the cell surface.

The mating factors cause cell cycle arrest in G1, this being the final stage in differentiation of a haploid yeast cell, occurring only when the mating hormone, and in natural conditions, the mating partner, is present (the cell can be be fooled into reacting as if cells of the opposite mating type are in the neighborhood, by adding the appropriate mating hormone. Also, if an **a**-mating-type cell is engineered to have receptors for **a**-mating hormones instead of α-factor as in wild-type **a**-mating cells, it will respond by going into G1 arrest, when exposed to **a**-mating hormone rather than to α-hormone as would normally be expected). The mating factors may inhibit the action of the protein kinase encoded by the CDC28 gene, thereby causing arrest in "Start".

2.3 The Mating Factor Receptors

These are products of the STE2 and STE3 genes, and are members of a family of receptors as diverse as rhodopsin and the β-adrenergic receptor. The yeast mating factor receptors are made up of a central core of seven helical structures have hydrophobic domains which span the membrane from outside to inside; they have a long hydrophilic "tail" at the carboxy terminus. The **a**- and α-receptors have very little sequence similarity to each other, or to various mammalian receptors.

All of these communicate with a G protein having three subunits, α, β, and τ. The G proteins are, in turn, members of a family of GTP-binding proteins, involved in signal transmission, which will be discussed shortly.

The mechanism of transmission of signals from the mating hormones is not yet completely understood. However, the general pathway is as follows: the receptors, at the outer surface of the cytoplasmic membrane, bind the mating pheromone of the opposite type. In the absence of the mating factor, GDP is bound to the G protein, whose subunits are associated. When a mating factor binds to the receptor, GDP dissociates from the G protein, and GTP binds instead. This causes subunit α of the G protein to dissociate, and the liberated $G\beta_\tau$ subunit activates a target, whose nature is unknown, and this, in turn, activates a number of other downstream elements. These include the STE5 gene product, several protein kinases encoded by FUS3 or KSS1, STE7, and STE11, then to the transcriptional activator encoded by STE12, which activates synthesis of a number of genes associated with mating. There is a short sequence of DNA, TGAAACA, the "induction box", or pheromone response element (PRE), which is present in multiple copies (two to nine) in the upstream regions of a number of inducible genes. The final result is the induction of the FUS1 gene and hyperexpression of a-specific genes. These include FUS1, whose gene product acts in cell fusion, FUS3, encoding a protein kinase, FAR1, encoding a product which brings about cell cycle arrest; STE3, and STE3, encoding the α- and a-mating factor receptors, depending on the cell type; MFa1 and MFα, encoding the mating pheromones themselves; SST2, which encodes a factor which allows recovery from hormone-induced cell-cycle arrest; STE6, involved in pheromone secretion; and KAR3, involved in nuclear fusion.

3 Regulation of the Life of the Yeast Cell Continues

3.1 The STE12 Gene Continues to Function

The STE12 gene product sets the basal level of expression of a- and α-specific genes, as well as inducting their expression. (Binding of STE12 gene product to the upstream region of an a-specific gene requires simultaneous binding of the MCM1 protein. Expression of α-specific genes requires the cooperation of the α1 protein as well.)

The practical result of signal transmission initiated by the mating pheromones is first, courtship, in which binding of, say, α-factor to the receptors of an a-mating-type cell, with subsequent induction of the MFa gene(s) which increases the production of higher levels of a-factor and also, more receptors for α-factor. The increased amounts of a-factor trigger a similar response in the other cell, so that in both cells increasing amounts of mating pheromones, of both types, and receptors are produced. At the same time, other genes are activated, which leads to deformation of the cell walls, contact between the cells at that point, fusion of the two cells, the opening of a pore at this point, and eventual karyogamy and zygote formation. (Incidental to this is the switching off of the synthesis of the mating factors and receptors, by the end product of the process; that is, the a/α diploid cell.) The subject of "courtship" in yeasts is discussed in more detail in Chapter 8.

3.2 Who Regulates the Regulators? Control of Mating Type
in the Haploid Yeast Cell

Having seen the process of cell and nuclear fusion, we must now examine a mechanism that ensures that given haploid cells of one mating type, cells of the other mating type are available for mating and reforming the diploid state.

The MAT locus encodes regulatory proteins that determine which set of genes is activated and determines the mating type of the cell. So what controls the MAT genes? To discuss this sensibly, it is necessary to examine, first, the nature of the MAT locus itself. For one thing, there are three, not one, loci where the MAT sequences are found, in chromosome III in *Saccharomyces cerevisiae*. Only one of these is active; the other two, at HMLα and HMRa, are silent. (The names are self-explanatory; the sequence encoding the α-specific genes is to the left of the active MAT locus, and to the left of the centromere, and the sequence encoding the a-specific genes is to the right.) So what brings about the substitution at the active MAT locus, of one of the sequences at the "silent" loci? This is under the control of the HO sequence.

3.3 The Cassette Mechanism

It was discovered a number of years ago that yeast had more than one copy of the MAT genes in each cell, and that while only one locus held an active copy of the gene, the cell held two other copies, one for each mating type, and inserted another copy into the MAT locus as desired, in much the same way as a disk jockey changes a tape; hence the designation "cassette" mechanism. The HO gene encodes a site-specific endonuclease, that causes double-stranded breaks at the ends of the MAT sequence in the active locus. At the same time, the sequence in one of the other loci, conferring the opposite mating-type to the original, is copied and inserted into the active MAT locus, guided by homologous sequences at the ends of the structural gene. Thus, the HO gene is a regulator of the master regulatory locus (see Fig. 11.2).

Later, a regulatory mechanism related to the telomeres was discovered. The two "silent" cassettes are located very near the ends of chromosome III, remote from the centromere and close to the telomeres. They are regulated by the SIR gene products, these genes being located some distance away from the HML and HMR sequences; the mechanism is not known, though some ingenious ones have been suggested. The telomeres apparently play a role which is also not known in inhibiting transcription of the HML and HMR sequences and other genes located near them, which are thus not expressed. When the cell undergoes mating-type switching, the sequence which confers the mating type opposite to that currently present at the actual mating-type locus is copied and the copy is introduced into the MAT locus, thereby changing the mating type of the cell. The active mating-type sequence is thus the one close to the centromere and is transcribed; the sequences at the HML and HMR loci, near the telomeres, remain silent.

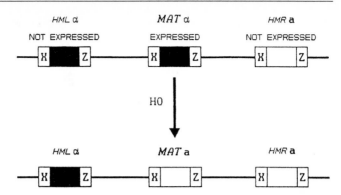

Fig. 11.2. Structure of the mating-type system showing the cassette mechanism. During mating-type switching, an endonuclease encoded by the HO gene causes a double-strand break on each side of the MF gene actually in place at the mating-type locus. This sequence is then replaced by a copy of the MF gene from the other locus

3.4 What Regulates the Regulator of the Regulator?

The sequence of events is complex. The HO gene is subject to three types of constraints: first, it is expressed only in haploid cells (cell-type control). Second, it is expressed only in mother cells, not daughters (mother-daughter or asymmetric control), and third, it is expressed only in the late G1 phase (cell-cycle control). All three of these conditions must be present if HO is to be expressed.

The asymmetric control factor leads to the development of stem-cell lineages in which the original mating-type is maintained (Fig. 11.3). Beginning with an original α-cell, this gives rise to a bud. The bud cannot switch mating types, but the (now) mother cell can. Having switched mating types, the mother cell can bud again, giving rise to a bud of the opposite mating type to the first bud. In the meantime, the first bud can give rise to a bud of its own, which again cannot switch mating types until it, too, has budded. The process can then continue, with each mother cell being able to switch mating types once in any given cell division cycle, and each bud being unable to do so, until it has budded and thereby becoming a mother cell itself.

In this way, a mixed population of **a** and α cells can arise. These can then mate and return the population to the diploid, a/α state. The process of mating-type switching does not exist for the edification of molecular biologists, but occurs as a means of returning to the diplophase, possibly after recombination and rearrangement of the genome, and because there are other advantages to the diploid state.

To accomplish all this, the HO gene must be activated and expressed. This requires the products of 13 genes; the process is under very close control, which demonstrates its importance to the cell. Regulation takes place in two upstream regions of the HO gene, of 1500 base pairs, which is subdivided into two general regions, URS1 and URS2. The first of these is responsible for mother-daughter

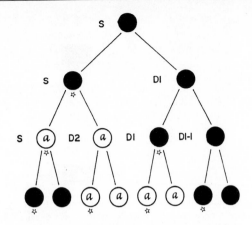

Fig. 11.3. Stem cell lineage during budding and mating-type switching in *Saccharomyces cerevisiae*

control, and the second for cell cycle control. Regulation is by the a1-α2 protein (the product of two of the MAT genes), six SWI genes "Switch", and five SIN genes "Switch-Independent". The SWI genes are positive regulators of transcription of HO. The SIN gene products are probably negative regulators and may be antagonized by some of the SWI products.

3.4.1 Regulation of HO

This is by the a1-α2 protein for negative control. There are ten binding sites for a1-α2, scattered over the URS1–URS2 region, and binding of a1-α2 to these sites presumably prevents activation of events related to activation of HO (see Fig. 11.4).

The URS2 contains ten copies of a "cell-cycle box", having a conserved sequence of CACGAAAA. Complementary to this is a DNA-binding factor which recognizes it, named with great originality, the cell-cycle box factor (CCBF). The SWI4 protein is involved in the process, and SWI6 is required for CCBF activity as well and may be part of the complex. The mechanism for cell-cycle control at the late G1 phase is not known, though the product of the CDC28 gene (protein kinase) may take part. Mother-daughter control may involve the SWI5 and SIN3 gene products; there are many ideas concerning the nature of the actual mechanism, but not many solid facts. The level of the SIN3 product is higher in daughter cells than in mother cells, which may be part of the regulatory mechanism.

Given that the initial requirements are met, that the cells are mother cells, and that they are in the G1 phase of the cell division cycle, the chain of events leading to expression of the HO gene can begin. It begins with the action of SWI5. This protein binds to the sites in URS1 and clears the region of such inhibitory factors as the SIN3 gene product, which has been inhibiting expression of SWI1,2,3, a protein which is the combined product of the genes SWI1, SWI2, and SWI3. This protein probably inhibits the SIN1 gene product, which has been inhibiting access

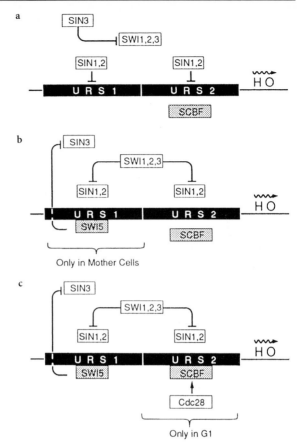

Fig. 11.4a–c. Regulation of expression of the HO gene in *Saccharomyces cerevisiae*. Part of the regulatory region (*URS1*) is responsible for mother-daughter regulation, and the other (*URS2*), for cell-cycle-dependent expression. **a** The upstream regulatory region in a daughter cell. Two regulatory factors, *SIN1* and *SIN3*, prevent expression of the HO gene, by blocking the access of these factors to their target sites. **b** The probable series of events occurring in mother cells. *SWI5* protein inhibits *SIN3* (mother cells only). *SWI1,2,3* then inhibits the SIN1 protein, lifting repression by this factor. Now, the CCBF (cell-cycle box factor can bind to the cell-cycle boxes in URS2. This activates transcription, which occurs only in the G1 phase of the cell cycle, and requires the CDC28 protein kinase, and the HO gene is transcribed. Mating-type switching can now take place

by the CCBF to its binding sites in URS2. On inhibition by SWI1,2,3, the CCBF immediately binds to its sites and with the aid and cooperation of the product of the protein kinase encoded by CDC28, the HO gene is expressed. Its site-specific endonuclease is produced, and the gene at the MAT locus is exchanged for the opposite mating type, and the switch in complete. When the opportunity arises for mating, there are mating partners for each type of haploid cell.

Simplified, this type of rather complex regulation of the expression of the HO gene is explained as the control of the access of an activator protein (the

CCB factor) to its binding site on the DNA of the sequence upstream of the structural gene.

The mechanisms of regulation of the expression of the yeast genes we have given here is of obvious importance for the understanding of yeast metabolism and protein synthesis at the molecular level, and as such, the applications of yeast molecular biology will be discussed in Chapter 13. The other aspect of these investigations, also of paramount importance, is that many of the developmental processes which have been investigated, or are under investigation in yeast, have their parallels in other eukaryotes, up to – or down to – and including man.

3.5 A Microtubule-Dependent Process: Nuclear Fusion

There are at present at least 12 genes closely involved in nuclear fusion in yeast. These include KAR1, KAR2, KAR3, TUB2, BIK1, the KEM genes, CIN1, and the CIK genes (chromosome instability and karyogamy), and the four cell division cycle genes, CDC4, CDC28, CDC34, and CDC37.

The cytology of the nuclear fusion reaction has been described in Chapter 8, but will be reviewed here with special reference to the role and behavior of the SPB and the microtubules. The time course of fusion (Fig. 11.5) begins with the **a-** and α-

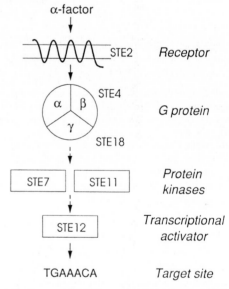

Fig. 11.5. Pathway of signal transduction from membrane to nucleus. The mating factor (**a** or α) binds to its receptor in the membrane. This causes the G protein to dissociate. This activates a chain of other genes and their products, including a series of protein kinases. Eventually the STE12 gene product is activated, and the *FUS1* gene and the members of the **a**-specific gene set are transcribed and hyperexpressed. All this leads to "courtship" between two cells of opposite mating type, and at last, to fusion of the cells, production of a diploid zygote, and a new line of diploid vegetative cells

mating hormones binding to the appropriate receptors in the α- and a-mating type cells respectively. The cells enter G1 arrest (see "Courtship" Chap. 8). The shmoo (copulation tubes or mating processes) develop on the mating partners, with the nuclei being close to the apex of each shmoo, the spindle pole body (SPB) facing the apex, and a short bundle of cytoplasmic microtubules projecting into the shmoo. The tips of the shmoos make contact, then first the cell walls and then the cytoplasmic membranes develop a pore connecting the two cells. The microtubules interleave, connecting the two nuclei, and the nuclei draw closer together. The nuclear membranes fuse along one edge of the SPBs. The nucleus is now diploid, and has an unusual SPB which has a crease along one side, at the site of fusion. Eventually, a diploid zygote, and then a diploid bud, form. Meanwhile, the zygotic nucleus has undergone mitosis and one of the diploid nuclei formed enters the bud. The bud frequently, but not always, forms at the midpoint of the zygote, at the center of the site of fusion of the cells. Normal mitotic division and vegetative cell growth follow.

Three elements are thus involved in nuclear fusion. The cytoplasmic and nuclear microtubules are required at several steps in fusion, and mutations interfering with microtubule formation and/or function are known and block cell fusion. The spindle pole body (SPB) also has a number of functions in fusion, since it is not only the center where the cytoplasmic microtubules assemble, but is the initiating site for fusion as well.

The fusion of the membranes is the final element, though the least is known about it. It may be catalyzed by a membrane-fusing protein, though no such has been isolated to date. It is also not known whether there is one fusion event or two; the edge of the SPB is a point where the inner and outer membranes are continuous, and one such event would be sufficient.

1. The *kar1* mutation was the first of the genes concerned with nuclear fusion to be discovered, in a search for mutants yielding a high frequency of haploid cytoductants, having the nucleus of one strain and the mitochondrial system of another, on mating to normal wild-type cells. In such a unilateral defect in karyogamy, nuclear fusion fails if only one parent is mutant. In bilateral karyogamy defects, both parents must be mutant, for fusion to fail, and fusion is normal when the mutants are mated to wild-type srains. There are numerous theories on the basis for unilateral and bilateral karyogamy defects, but as yet, insufficient facts to decide on one that can explain all the facts unequivocally.

2. The genes required for nuclear fusion include, first, *KAR1*, *KAR3*, *BIK1*, and *KAR2*. The *KAR1* gene product is essential for mitosis, and the mutant phenotype is due to a defect or defects in the structure or function of the microtubules. It is interesting to note that overexpression of the KAR1 protein is lethal and causes cell cycle arrest. The protein is probably part of the SPB complex. The *KAR1* gene product has at least two domains, specifying location; one of 70 residues in the center of the protein, which directs it to the outer face of the SPB, and another, of 40 residues, at the carboxy end, which directs it to the nuclear envelope. The newly formed SPB, containing nonfunctional protein, segregates into the bud, for reasons unknown. The first region is essential for mitotic function of the KAR1

protein, and another, adjacent to the region essential for mitosis, also of about 70 residues, which is essential for karyogamy (nuclear fusion).

3. The *KAR3* gene product also is essential for nuclear fusion, and the mutation has a number of pleiotrophic effects as well. Cultures of mutant strains show a large number of dead cells which are arrested with a large bud, an undivided nucleus, and a short spindle spanning the nucleus. Many of the cells have a DNA content characteristic of the G2 stage of cell division. The mutants have an unusual reaction to benomyl, which apparently reduces the number of dead cells at some concentrations. The *kar3* mutants are also more sensitive to ionizing radiation.

Even more interesting, the gene product has a considerable similarity in sequence to the well-known motor protein, kinesin, though the similarity occurs only in the region involved with movement of the microtubules, and the homologous domains are at opposite ends of the molecules in kinesin and the KAR3 gene product. The two proteins are structurally similar, nevertheless, having long internal coiled-coil domains and small globular domains at the ends opposite the motor domains. The KAR3 gene product is probably a member of the large family of kinesins, which are associated with microtubule-dependent processes such as SPB separation, spindle elongation, meiosis, chromosome segegation, and nuclear fusion.

There is a second microtubule-associated domain in the KAR3 gene product, located near the SPB, which is not associated with the motor domain.

4. The *BIK1* gene likewise causes a 10–100-times decrease in nuclear fusion. The gene product is required for microtubule stability, but is not required for meiosis. The mutant strains show increased chromosome loss, and deletion of the *BIK1* gene leads to increased cold-sensitivity. The *bik1* mutation suppresses the formation of the very long cytoplasmic microtubules found in the *kar1-1* mutants.

The *BIK1* gene encodes a protein of 58kDa, which is associated with the microtubules, and is thought to have three domains; that at the amino terminus is globular, basic, and shows limited sequence similarity to the MAP2 protein and to tau. The central region probably includes an α-helical coiled coil, so that the protein is probably oligomeric, and the carboxy-terminal domain has a small metal-binding motif which is often found in retroviral nucleocapsid proteins.

5. The kar2 mutation produces a unilateral tenfold decrease in nuclear fusion, the block being different from that resulting from mutations in the *KAR1* and *KAR3* genes. It may function in the membrane of SPB events in fusion, which must occur after nuclear association.

The *KAR2* gene product is a member of the HSP70 (heat shock protein) family, and encodes the HSP70 protein which is resident in the lumen of the endoplasmic reticulum (ER) in yeast. It was discovered as a protein that was induced by starvation of the cells for glucose. The *KAR2* gene product is essential for the import of secreted proteins into the ER lumen. The block in nuclear fusion probably does not occur because of a defect in protein translocation. KAR2 protein may be directly required in nuclear fusion, but it is doubtful that protein translocation is involved.

However, it may be necessary for reassembly of the diploid SPB after nuclear fusion.

6. *CIK* genes (chromosomal instability and karyogamy). Four genes, *CIK1*, 2, 3, 4. Mutants show a 10–100-times increase in the number of cytoductants, and a 10–50-times decrease in the fidelity of chromosome segregation. CIK1 is not an essential gene. Hybrid proteins resulting from CIK1-lacZ gene fusions are found in the neighborhood of the SPB.

7. *CIN genes. CIN1, CIN2*, and *CIN4*. Mutations in the *CIN* genes bring about increased chromosome instability and increased sensitivity to benomyl. They are not essential for growth, but the gene product(s) are necessary for growth at low temperature; mutants carrying defects in these genes are cold-sensitive, and arrest at a point which indicates a general defect in stability and function of both nuclear and cytoplasmic microtubules.

Mutations in the *CIN1* gene cause some increase in the frequency of occurrence of cytoductants, there is very little increase in the failure of nuclear fusion in the zygotes, amounting to only about 5%. The role of the *CIN1* gene product in stabilizing microtubules may be primarily limited to vegetative cells.

8. The *KEM* genes (*KAR*-enhancing mutation; *KEM1, KEM2, KEM3*) reduce the residual nuclear fusion which was unaffected by the *kar1-1* mutation, by 10 to 100 times, whether the mutation is in the wild-type parent or the one carrying the *kar1-1* mutation. If neither parent contains the *kar1-1* mutation, the presence of a *kem* mutation has no effect on nuclear fusion. Mutations in two of the *KEM* genes, *KEM3* and *KEM1*, have pleiotrophic effects; the *kem3-1* mutation results in temperature sensitivity, and mutations in the *KEM1* gene may cause increased sensitivity to benomyl and also increased chromosome instability in diploids, homozygous for the mutation. It is possible that *kem1* mutations bring about general defects in microtubule and SPB function, though it is also possible that the KEM1 gene product may primarily have a regulatory function and affect the microtubules and SPB only indirectly.

The *KEM1* gene probably encodes a protein of molecular weight 175 kDa.

3.5.1 The *CDC4, CDC28, CDC34*, and *CDC37* Genes

CDC4. Mutations in the *CDC4* gene block mitotic cell division after SPB duplication but before the two SPBs separate and DNA synthesis begins. This usually produces cells having multiple abnormal buds when the cultures are shifted to the restrictive temperatures. Furthermore, if the cells are mated at the restrictive temperatures, nuclear fusion fails in 80–90% of the zygotes.

The karyogamy defect in the *cdc4* mutation is noteworthy for a number of reasons. The *CDC4* gene apparently acts in mitosis after the point where the cells can mate; *cdc4* cells that have arrested cannot mate. However, the karyogamy defect indicates that the gene must act somewhere at an earlier point in the cell division cycle if normal mating is to take place. One possibility is that the gene product is required for assembly of the SPB. In the zygotes, only preexisting SPBs fuse; the gene product would be required to act before cell fusion for karyogamy to take place. Alternatively, the *CDC4* gene may have different functions in karyogamy and mitosis, so could be expected to act at different points in the cell cycle.

The *CDC4* gene product has considerable sequence similarity and structural homology to the β-subunit of a trimeric G protein. This being so, it may indicate that the protein is more than a structural component of the SPB, but may also be part of a signaling pathway allowing the SPB to respond to cell cycle and binding of mating pheromones. The corresponding α- and τ-subunits, if any, are not known.

3.5.2 *CDC28, CDC34, CDC37*

CDC28. The *CDC28* gene encodes a protein kinase that is required for the cell to exit from G1 arrest and enter the S phase. The kinase is activated by four proteins, the B-type cyclins, Clbl, Clb2, Clb3, and Clb4, which act during both mitosis (Clb2) and meiosis (Clb1, Clb3, and Clb4). Arrest is at the same point as occurs on exposure to mating hormones. Cells carrying the *cdc28* mutation remain capable of mating, the gene is not required for pheromone response or cell fusion. It may be required for some step leading to nuclear fusion. At present, it is uncertain whether the gene acts before or after cell fusion.

Mutations in the *CDC28*, *CDC34*, and *CDC37* bring about a weak unilateral defect in nuclear fusion. However, there is reason to believe, on the basis of microscopic examination of bilateral zygotes obtained at the nonpermissive temperature, that mutations in the *CDC28* and *CDC37* genes may be responsible for a strong bilateral defect. The genes block the cell division cycle in G!.

CDC34 Gene. This gene encodes a ubiquitin-conjugating enzyme, but at present its further function is not known.

CDC37 Gene. The *CDC37* gene encodes a protein which is predicted to have a molecular weight of 51 kDa. The protein may be a component of the yeast cytoskeleton, and has a considerable component of α-helix structures, but it has no sequence similarity to any other proteins. Its function, like that of many other proteins associated with nuclear fusion in yeasts, is not known at present.

To recapitulate, a model for nuclear fusion must consider all of the elements so far described, including the SPB component, a G-protein subunit (G-proteins are discussed later in this chapter), the cytoplasmic microtubules, several proteins associated with the microtubules or stabilizing them, a microtubule motor protein, without which the various components cannot move relative to each other, and a component of the nuclear envelope-ER complex.

The process, which requires the presence of haploid cells of both mating types, begins with the binding of a mating pheromone to its receptor on a cell of the opposite mating type. This triggers a response which begins with the release from a G-protein of its β-subunit and the activation of a chain of signals, through the STE18 gene, STE7, STE11, the transcriptional activator STE12, to the target site TGAAACA (the pheromone response element, and eventually to the FUS1 gene, which is activated to 500 times its basal level. The effects are manifested in shmoo formation, contact between the cells at the ends of the conjugation tubes, breakdown of the walls and cytoplasmic membranes between the cells and reformation of a continuous wall at the junction, and the commencement of nuclear fusion. Among the genes required for this and induced by the binding of mating hormone is *KAR3*, whose product is required, and this protein is shifted from the nucleus to the cytoplasm at this time. About this time, a short bundle of microtubules is formed, leading from each SPB to the tip of the corresponding shmoo. When the cell wall and cytoplasmic membranes break down to form a pore between the two mating partners, the microtubules form a bridge between the two nuclei.

The KAR3-1 protein is associated with the bridge, the bridge being a bundle of antiparallel microtubules. In both dominant and null *kar3* mutants, the nuclei do not fuse. In the null mutant, the protein is not formed and does not connect the microtubules. In the dominant mutant, the protein is present and binds to the microtubules, but has no motor activity to move the tubules, and the nuclei do not come together and fuse.

The orientation of the nuclear motor (the KAR3 protein) has not been determined. However, it is thought to be a plus-end-oriented motor, and this view will probably be held until evidence showing that it may be minus-end-oriented is found.

Once the nuclei have been brought together, the SBPs and the nuclear envelopes fuse. The mechanism for this is not so far known. Mutations in the *KAR2* gene can interfere with fusion. One indication of the requirement for a specific mechanism for fusion is that proximity alone is not enough for nuclear fusion. This being so, the fusion process must be independent of mating type of the original partners, since diploids can be formed artificially by fusion of protoplasts, and behave normally except for retaining their original mating type, **aa** or $\alpha\alpha$.

There are obviously other genes involved in control of nuclear fusion, but these have not yet been identified.

3.6 The GTP-Binding Proteins

The characteristics and functions of this group of proteins warrant special mention, for these are part of the signal transmission mechanisms which convey signals of a wide variety of types from the original receptors to whatever their final destination may be. They are of interest to the molecular biology of yeasts because, as noted earlier, they constitute one link in the chain leading from the a- and α-receptors on the cell surface to their ultimate target in the cell nucleus (see Fig. 11.5).

They are all tripartite proteins, in other words, having three subunits. When a signal is received, the protein separates into its subunits with concomitant hydrolysis of bound GTP to GDP.

Their functions are diverse, though they have in common that of transduction of signals from a receptor, through an amplifier to the cellular target. Some, Gs and Gi, participate in hormonal regulation of adenylate cyclase activity, one activating the enzyme when β-adrenergic stimuli are received, and the other inhibiting the enzyme. The Gt1 protein is found in retinal rods and regulates the activity of cyclic GMP-phosphodiesterase and transduces visual signals – in other words, sight. Gt2 seems to perform the same function in cones. G0 may be involved in neuronal responses. Nothing is known about its precise function. In yeast, as previously noted, they are an integral part of the signal transmission system between the mating pheromone receptors and the genes encoding the proteins directing the mating response.

Given the wide distribution of these proteins, and their high degree of homology in G proteins and the *ras* and *ras*-related families of proteins, which are closely similar, their importance in the functioning of the cell is obviously considerable.

3.6.1 GTP Binding Proteins in Yeast

1. The G protein family. Two of the G proteins found in yeast are GPA1 and GPA2, which encode the proteins GP1α and GP2α, having 472 and 449 amino acids, respectively. Their mammalian equivalents are about half the size. GP1α is part of the signaling series for transduction of the messages induced by binding of a mating factor to the receptors on the cell surface, to the amplifiers which regulate the response of the cell at the nuclear level. These proteins show remarkable homology with mammalian and other eukaryotic G proteins, which also have similar functions; that is, the transduction of signals from receptors to am amplifier protein for further transmission. Homologies range from near 50 to 90% (in some *ras* proteins).

Three regions in which the homologies are especially pronounced are:

1. The GTP hydrolysis (P) site. This site shows GTP activity, and is conserved in all G proteins and related families.
2. The site interacting with the guanine ring (G site). It is also conserved in G proteins and other GTP binding proteins. There is a consensus sequence, Asn/Lys/Xaa/Asp, found in all of these families of proteins.
3. The G' site is unique to G proteins, and is again strongly conserved, being found at amino acid residues 319–335 of GP1α, 296–312 of GP2α, and 201–217 of Giα. There is a sequence of 12 contiguous amino acid residues shared by yeast an mammalian G proteins.

There are other regions of homology among a number of mammalian G proteins. The sites for ADP-ribosylation by cholera toxin, of the Gsα, and Gtα proteins, and the Giα, Goα, and Gtα proteins show a high degree of homology. In addition, the sequence around Arg-201 of the Gsα protein is homologous to that of transducin, a protein which is ribosylated by cholera toxin, but also to the Gi1α and Goα, which are not.

2. The *ras* family of genes were first identified as transforming genes of sarcoma viruses, that is, as the Ha-, Ki, and N-ras genes, and, later, were found in genomes of a number of eukaryotes, including yeast. RAS1 and RAS2 have been isolated from yeast, using the viral *ras* gene as a probe. They encode proteins of 309 and 322 amino acids, respectively, which is about twice the size of mammalian *ras* proteins, which have about 188 or 189 amino acids. The structure of the yeast *ras* genes is interesting; they show 90% homology to mammalian *ras* genes in the first 90 amino acids, and then about 50% homology over the next 80. There follows a variable region, and then they terminate in a short conserved sequence (Cys-A-X, where A is any aliphatic amino acid).

Disruptions in either of the RAS1 or RAS2 genes are not lethal, but if both genes are disrupted the cell dies. The RAS1 gene inhibits inositol phospholipid turnover, and RAS2 stimulates adenylate cyclase activity.

Recently, a *RAS* gene has been found in the nematode *Caenorhabditis elegans*, which forms part of a signaling pathway beginning with a tyrosine kinase. The RAS protein acts downstream of the tyrosine kinase and an activator and an inactivator, and upstream from its target. Unfortunately, at the moment, no one knows the message or its target. In *C. elegans*, a gene encoding tyrosine kinase and a *RAS* gene have been found among genes controlling development of sex organs. Phosphorylation of proteins by kinases is known to occur in a number of regulatory systems, including those in yeasts, but whether a tyrosine kinase is to be found among them has yet to be determined. It is also as yet unknown whether a GTP-binding protein is part of the signaling pathway, or what the signal may be, or whether, indeed, a similar pathway is to be found in yeasts at all.

3. The ras-related family includes *RHO1*, *RHO2*, and *SEC4*. *RHO1* and *RHO2* are a conserved family having significant homology to the *ras* genes, and molecular weights of 23 000 ns 21 000 Da. These two genes are 53% homologous to each other, and 70 and 57% homologous to the *rho* gene of *Aplysia* (a marine snail).

RHO1 is essential for yeast growth; *RHO2* is not. These proteins may link a GTP-dependent amplifier to receptors on the cell surface, as to other G-family proteins.

SEC4 complements the *sec15* growth defect. It is essential for transport of secretory proteins from the Golgi body to the cell surface. It is a protein of molecular weight 23 500 daltons, and which has 32% homology to the Ha-*ras* gene. The highest homology with the regions involved in binding and hydrolysis of GTP. The gene and its product are essential for growth; the *SEC4* gene is probably a GTP-binding protein which is required for control of a late stage of the secretory pathway.

YPT1 is a gene which is located between the actin and the β-tubulin structural genes. It is essential for cell growth, it binds and hydrolyzes GTP, and may be involved in organization and function of the microtubules. This would explain the requirement for this gene in cell growth.

3.7 Role of Yeast GTP-Binding Proteins in Signal Transduction

We will review the mode of signal transduction in yeast briefly. Yeast has two such systems, glucose being a positive extracellular regulator of adenylate cyclase and inositol phosphate turnover in early G1. The mating pheromones are negative regulators and act to cause arrest in late G1.

GP1α, for instance, is involved in mating factor-mediated signal transduction in haploid cells, as shown by the following data:

1. The GPA1 transcript is found only in haploid cells, but not **a**/α diploids.
2. Disruption of the GAP1 gene is lethal in haploids but not in diploid cells. It is a haploid-specific gene, essential for cell growth.
3. The mutants *sgp*1 and *sgp*2, which suppress the lethal effect of the *gp*α1 mutation, have a sterile phenotype which is not cell-type-specific.
4. If the GPA gene is placed under the control of the GAL1 promoter, which is then induced by galactose, loss of GP1α also brings about cell-cycle arrest in G1.
5. If the GP1α function is lost through mutation, this can relieve the sterility brought about by a mutation in the STE2 gene.

The pathway of signal transduction is dependent on the fact that the protein GP1α has different conformations, depending on whether it has been bound to GDP or GTP. It is inactive when bound to GDP and active (i.e., can transmit a signal) when bound to GTP. When the bound GTP is hydrolyzed, the signal is switched off.

The pathway of action is as follows:

1. When there is no mating factor bound to the α-factor (or **a**-factor receptor, in α-mating type cells), GDP binds to GP1α and inhibits the amplifier protein from generating a cell cycle arrest signal.
2. When the mating factor is present and bound to the receptor, it brings about the exchange of bound GDP for GTP.
3. The bound form of GP1α activates the amplifier, which transmits a signal which results in cell cycle arrest in late G1.
4. When the mating factor is released from the receptor, the GTP is hydrolyzed to GDP + P$_i$ and the signal is no longer transmitted. Either the cell has mated with one of the opposite mating type, or in the absence of a mating partner, the cell has resumed its progress through the cell cycle and will eventually bud and give rise to another cell.

3.8 Control of the Cell Division Cycle by Nutrient Limitation

3.8.1 Participation of Cyclic AMP

Under conditions of nutrient limitation (glucose), mutants (cyr1) arrest in early G1 if cyclic AMP is not present. If cAMP is present, it binds to gene products which are subunits of regulatory proteins (protein kinases), which frees active catalytic subunits. These are products of the BCY1 gene, and the bcy1 mutants have decreased levels of the regulatory subunits, so that the protein kinase can function in the absence of cAMP.

There are two G proteins, Gs and Gi, involved in the signal transduction system for adenylate cyclase in mammalian systems. The system itself has at least two protein components, the catalytic and regulatory subunits. It is regulated by guanine nucleotides plus Mg^{2+} ions. In yeast, the adenylate cyclase system is regulated by the RAS proteins (also G proteins, as described earlier). The RAS protein probably stimulates adenylate cyclase in much the same way as the Gs protein stimulates the enzyme in mammalian cells. The G protein of yeast, GP2α, also participates in the regulation of the level of cAMP. GP2α probably acts in the same way, by binding to a receptor molecule in the cytoplasmic membrane to transmit a signal to an amplifier molecule, as most or all G proteins do.

In mammals, inositol phospholipids are hydrolyzed by phospholipase C to diacylglycerol and inositol triphosphate InsP$_3$. These act as second messengers for activation of protein kinase C and Ca^{2+} respectively. They may have roles, along with cAMP, in the transition from the G0/G1 phase of growth in mammalian cells.

In yeast, these roles may be played by glucose itself. It stimulates phospholipase C to hydrolyze inositol phospholipids and influx and efflux of Ca^{2+}. Glucose probably activates adenylate cyclase as well, and its metabolites activate phospholipase C. Protein kinase C has not been demonstrated in yeast. Further investigation of the role of the RAS1 and RAS2 genes has shown that the RAS1 gene product inhibits inositol phospholipid turnover, though the RAS2 gene product was less effective.

There are numerous theories and suppositions concerning the role of the RAS genes and other factors involved in regulation of inositol phospholipid turnover, and concerning the mechanism of signal transduction in general, but so far, not enough hard facts. Nevertheless, one idea is well supported by factual evidence; that when a signal, ranging in nature from a burst of photons of light to an impulse to a mating reaction, is to be received from an external source, whether in yeast, invertebrates, or mammalian tissues and organs, for transmission to an internal target within a cell, the nearest intermediate in the pathway for transduction of the signal and speeding it on its way will be a membrane-bound G protein.

References

Beggs JD (1978) Transformation of yeast by a replicating hybrid plasmid. Nature 275:104–109

Bowdish KS, Mitchell AP (1993) Bipartite structure of an early meiotic upstream activation sequence from Saccharomyces cerevisiae. Mol Cell Boil 13(4):2172–2181

Bowdish KS, Yuan HE, Mitchell AP (1995) Positive control of yeast meiotic genes by the negative regulator UME6. Mol Cell Biol 15(6):2955–2961

Brearley RD, Kelly DE (1991) Genetic engineering in yeast. In: Wiseman A (ed) Genetically-engineered proteins and enzymes from yeast: Production control. Ellis Horwood, Chichester, pp 75–95

Brown AJP (1989) Messenger RNA translation and degradation in Saccharomyces cerevisiae. In: Walton EF, Yarranton GT (eds) The molecular biology of yeast. Blackie, London, Van Nostrand Reinhold, New York, pp 70–106

Bruhn L, Sprague GF Jr (1994) MCM point mutants deficient in expression of *a*-specific genes: residues important for interaction with *a*1. Mol Cell Biol 14(4):2534–2544

Clark KL, Dignard D, Thomas DY, Whiteway M (1993) Interactions among the subunits of the G protein involved in *Saccharomyces cerevisiae* mating. Mol Cell Biol 13(1):1–8

Cregg JM, Tschopp JF, Stillman C et al. (1987) High level expression and efficient assembly of hepatitis B surface antigen in the methylotrophic yeast *Pichia pastoris*. Bio/Technology 5:479–484

Drysdale CM, Dueñas E, Jackson BM, Reusser U, Braus GH, Hinnebusch AG (1995) The transcriptional activator GCN4 contains multiple activation domains that are critically dependent on hydrophobic amino acids. Mol Cell Biol 15:1220–1233

Errede B (1993) MCM1 binds to a transcriptional control element in Ty1. Mol Cell Biol 13(1):57–62

Evans IH, McAthey P (1991) Comparative genetics of important yeasts. In: Wiseman A (ed) Genetically-engineered proteins and enzymes from yeast: Production control. Ellis Horwood, Chichester, pp 11–74

Fangman WL, Brewer BJ (1991) Activation of replication origins within yeast chromosomes. Ann Rev Cell Biol 7:375–402

Grandin N, Reed SI (1993) Differential function and expression of *Saccharomyces cerevisiae* B-type cyclins in mitosis and meiosis. Mol Cell Biol 13(4):2113–2125

Hartwell LH (1974) The yeast cell division cycle. Bacteriol Rev 38:164

Herskowitz I (1988) Microbiol Rev 52:536–553

Herskowitz I (1989) A regulatory hierarchy for cell specialization in yeast. Nature 342:749–757

Heslot H, Gaillardin C (1991) Molecular biology and genetic engineering of yeasts. CRC Press, Boca Raton 324pp

Hinnen A, Hicks JB, Fink GR (1978) Transformation of yeast. Proc Natl Acad Sci USA 75:1929

Holm C (1982) Clonal lethality caused by the yeast plasmid 2-μm DNA. Cell 29:585

Kuo M-H, Grayhack E (1994) A library of yeast genomic MCM1 binding sites contains genes involved in cell cycle control. Mol Cell Biol 14(1):348–359

Lee JC, Yeh LCC, Horowitz PM (1991) The initiation codon AUG binds at a hydrophobic site on yeast 40S ribosomal subunits as revealed by fluorescence studies with bis(1,8-anilinonaphthalenesulfonate). Biochimie 73(9):1245–1248

Linder C, Thoma F (1994) Histone H1 expressed in *Saccharomyces cerevisiae* binds to chromatin and affects survival, growth, transcription and plasmid stability but does not change nucleosomal spacing. Mol Cell Biol 14(4):2822–2835

Lucchini R, Sogo JM (1994) Chromatin structure and transcriptional activity around the replication forks arrested at the 3′ end of the yeast rRNA genes. Mol Cell Biol 14(1):318–326

Marsh L, Neiman AM, Herskowitz I (1991) Signal transduction during pheromone response in yeast. Annu Rev Cell Biol 7:699–728

Matsumoto K, Kaibuchi K, Arai K, Nakafuku M, Kaziro Y (1989) Signal transduction by GTP-binding proteins in *Saccharomyces cerevisiae*. In: Walton EF, Yarranton GT (eds) Molecular and cell biology of yeasts. Blackie, London, Van Nostrand Reinhold, New York, pp 201–222

Meacock PA, Brieden KW, Cashmore AM (1989) The two-micron circle model replicon and yeast vector. In: Walton EF, Yarranton GT (eds) Molecular and cell biology of yeasts. Blackie, London, Van Nostrand Reinhold, New York, pp 330–359

Mellor J (1989) The activation and initiation of transcription by the promoters of *Saccharomyces cerevisiae*. In: Walton EF, Yarranton GT (eds) Molecular and cell biology of yeasts. Blackie, London, Van Nostrand Reinhold, New York, pp 1–42

Mendel JE, Korswagen HC, Liu KS, Hadju-Cronin YM, Simon MI, Plasterk RHA, Sternberg PW (1995) Participation of the protein G_0 in multiple aspects of behavior in *C. elegans*. Science 267:1652–1655

Messenguy F, Dubois E (1993) Genetic evidence for a role for MCM1 in the regulation of arginine metabolism in *Saccharomyces cerevisiae*. Mol Cell Biol 13(4):2586–2592

Mizuta K, Warner JR (1994) Continued functioning of the secretory pathway is essential for ribosome synthesis. Mol Cell Biol 14(4):2493–2502

Moore TDE, Edman JC (1993) The α-mating type locus of *Cryptococcus neoformans* contains a peptide pheromone gene. Mol Cell Biol 13(3):1962–1970

Muhlrad D, Decker CJ, Kreck MJ (1995) Turnover mechanisms of the stable yeast *PKG1* mRNA. Mol Cell Biol 14:2145–2156

Newlon C (1989) DNA organization and replication in yeast. In: Rose AH, Harrison JS (eds) The yeasts, vol 4, 2nd edn. Academic Press, New York, 57pp

Oppenoorth WFF (1960) Modification of the hereditary character of yeast by ingestion of cell-free extracts. Eur Brewery Convention. Elsevier, Amsterdam, 180pp

Piper PW, Kirk N (1991) Inducing heterologous gene expression in yeast as fermentations approach maximal biomass. In: Wiseman A (ed) Genetically-engineered proteins and enzymes from yeast: Production control. Ellis Horwood, Chichester, pp 147–184

Rose MD (1991) Nuclear fusion in yeast. Annu Rev Microbiol 45:539–567

Ségalat L, Elkes DA, Kaplan JM (1995) Modulation of serotonin-controlled behaviors by G_0 in *Caenorhabditis elegans*. Science 267:1648–1651

Thompson JS, Johnson LM, Grunstein M (1994) Specific repression of the yeast silent mating locus (HMR) by an adjacent telomere. Mol Cell Biol 14(1):446–455

Wheals AE (1987) Biology of the cell cycle in yeasts. In: Rose AH, Harrison JS (eds) The yeasts, vol 1. Academic Press, New York, pp 283–390

Yun D-F, Sherman F (1995) Initiation of translation can occur only in a restricted region of the *CYC1* mRNA of *Saccharomyces cerevisiae*. Mol Cell Biol 15:1021–1033

Zhou Z, Gartner A, Cade R, Ammerer G, Errede B (1993) Pheromone-induced signal transduction in *Saccharomyces cerevisiae* requires the sequential function of three protein kinases. Mol Cell Biol 13(4):2069–2080

Yeasts and the Life of Man: Part I: Helpers and Hinderers. "Traditional" Yeast-Based Industries; Spoilage Yeasts

J.F.T. Spencer and D.M. Spencer

1 Introduction

Yeasts are newcomers to the economic life of man, and among his oldest associates. As newcomers, they are used as vehicles for production of heterologous proteins of many types. As old associates, yeasts have been used in the oldest of the yeast industries, baking, brewing, and winemaking, from the earliest days of recorded history. We must also consider the role of yeasts as spoilage agents. Yeasts have not always been a pure friend to mankind. They will raise bread, convert grapes and barley and other grains into beverages, and improve the flavor and nutritive value of his foodstuffs. They will also compete for his daily bread, should the "wrong" species of yeast invade it, and convert other foods into undesirable products which are inedible or toxic. They may invade human tissues with serious or fatal results (see Chap. 4).

Some yeasts obtain their growth requirements by degrading industrial waste products, which are usually converted into food yeast. *Candida utilis* is grown on wastes from the sulfite process for wood pulp manfacture, *Yarrowia lipolytica* and other species, on alkane fractions.

Products of the "traditional" yeast processes may be the metabolic products of the yeast, or the cell mass itself (baker's yeast). Yeasts used in the pharmaceutical industries for production of heterologous proteins (hormones, antigens, and other pharmaceuticals) may excrete the end product into the medium or retain it within the yeast cell. This affects the nature of the recovery method.

2 "Traditional" Yeast Industries

2.1 Brewing and Brewing Yeasts

The history of the brewing and winemaking industries is well known. Almost every sweet fruit or starchy material known has been used to make wine or beer. These include African and Indonesian palm wines and numerous maize beers. "Beer" in the Americas and Europe usually means a light lager, drinkable in hot climates. The heavy ales and stouts of the British Isles are less suitable for hot weather.

Improvements in brewing have consisted of attempts to reduce costs without changing the nature of the beer. Improvements in the process have been the use of

J.F.T. Spencer/D.M. Spencer (eds)
Yeasts in Natural and Artificial Habitats
© Springer-Verlag Berlin Heidelberg 1997

starch adjuncts to increase the amount of alcohol produced without increasing the amount of barley malt required; use of hop extracts to save shipping costs of crude hops, development of continuous brewing processes, which have increased the possible maximum production, but have often not produced acceptable beer.

The tower fermentor was originally developed for the brewing industry, but did not produce saleable beer. However, it has been used successfully in the production of industrial ethanol in Brazil. Mash is fed into the bottom of the tower, fermented by a flocculent (to aid settling) yeast; and fermented mash dead and other nonflocculating cells are drawn off at the top. The characteristics required by the yeast, other than flocculence, are not yet known.

Brewing yeast strains have been constructed by genetic techniques (Hinchliffe and Vakeria 1989; Johnston 1990). Investigators at the Carlsberg Laboratory have constructed strains carrying mutations in the *ilv* genes (isoleucine and valine) and tested them in the brewery. In Finland, research programs have been aimed at reducing the diacetyl content of beer, to reduce the "lagering" time necessary to reduce the diacetyl content.

"Engineered" brewing yeasts have been constructed containing the gene coding for α-acetolactate dehydrogenase. These have diacetyl levels low enough that the beer need be held in lager only a few days (Kielland-Brandt et al. 1983).

The most important requirements for a good brewing yeast are:

1. Rapid fermentation rate with minimum yeast growth.
2. Good conversion of sugars, especially maltose and maltotriose, to ethanol.
3. Resistance to high concentrations of ethanol and osmotic tension. This is especially important in yeast for high-gravity brewing.
4. Consistent production of acceptable levels of flavor and aroma-producing compounds.
5. Flocculation capacity, suited to the process in use.
6. Genetic stability and retention of viability during storage.

The genetic control of characteristics of brewing yeasts is under investigation. Electrophoretic karyotyping of yeasts (by OFAGE, FIGE, CHEF, TAFE) allows detailed mapping of the technologically important genes in industrial strains of *Saccharomyces* species. This will increase the efficiency of processes using these strains.

Development of improved brewing yeast strains has proceeded through mutation, meiotic hybridization, hybridization by rare-mating, cytoduction and single-chromosome transfer, protoplast fusion, and transformation with hybrid plasmids carrying the desired genes, promoters, secretion vectors, terminators, and other genetic elements, and finally by transformation with artificial chromosomes. These have a centromere, telomeres at each end, and selected DNA sequences. Much larger sequences of DNA can be cloned into artificial chromosomes than into plasmids.

Recent work has been concerned with improved substrate utilization, especially starches and β-glucans, reduction of vicinal diketone levels (diacetyl and pentanedione), production of protease to reduce "chill haze", improved ethanol

tolerance, control of flocculence, and production of appropriate levels of organoleptic compounds.

2.2 Wine Yeasts

The technology of winemaking has probably changed least of all of the yeast
industries. The grapes are harvested, pressed, and the juice fermented. When the
wine is ready for consumption, it is bottled. The process is slightly more complex
for production of sherries, other fortified wines, and champagne (Martini and
Vaughan Martini 1990). (Champagne was originally manufactured as a way of
making a poor-quality wine into a saleable product.)

The development of active dry yeast for starters to replace the natural yeast
microflora in fermentation has been a major application of biotechnology to wine
yeasts. Most major winemakers now use commercial starters.

Construction of strains which will carry out the malolactic fermentation at an
adequate rate to reduce the acidity of the wines when necessary is of major importance to the wine industry. Successful attempts to introduce the necessary genes
from *Schizosaccharomyces pombe* have been made in Uruguay to construct such
strains using protoplast fusion, between wine yeasts and *S. pombe*.

Snow (1983), in California, investigated the genetic improvement of wine
yeasts. Vezinhet et al. (1990) used detailed karyotyping of wine yeasts by pulsed-
field gel electrophoresis (TAFE) to classify strains of wine yeasts. They showed that
few strains have identical karyotype patterns. The chromosomal patterns often
segregated according to Mendelian laws. Karyotypes of desirable strains may be
used for identification and characterization of wine yeast strains. They may also
aid in answering the question: "What is it that makes a good wine yeast good?"

Sherries are fermented with a flor (film-forming) yeast, in oak barrels partly
filled with a wine containing about 15% alcohol. The yeast reduces the acid content
and increases the content of acetaldehyde and other aldehydes, esters, ketones,
and higher alcohols of the wine. When the fermentation is complete, brandy is
added to raise the alcohol content to 17–19% by volume.

Sake (*rice wine*) probably originated in Japan. It is manufactured from steamed
rice, fermented using a starter culture (koji) of *Aspergillus oryzae* grown on
steamed rice (Onishi 1990). It contains a mixture of fungal amylases and proteases.
A sake yeast (special strain of *Saccharomyces cerevisiae*) is added. When this
starter culture (moto) is fermenting well, it is added to the main mash of steamed
rice, additional koji, and water. The main fermentation takes about a month, at 10–
15 °C. Lactic acid bacteria may be present. The alcohol content should reach 17%,
in normal fermentation. The ethanol tolerance of the yeast is increased by a
proteolipid, consisting of a phospholipid, 58.5 and 26.7% protein, produced by the
Aspergillus oryzae. The phospholipid contains 78.3% phosphatidylcholine, 18.5%
sphingolipid, and 3.2% lysophosphatidylcholine. The predominant fatty acid is
linoleic acid. A mixture of Tween 80 or mono-olein, ergosterol, and albumin was
also effective. Sake is warmed to about 37 °C before serving.

Sake yeast has been improved by isolation of foamless mutants and alcohol-resistant mutants. Strains carrying the killer character were constructed by protoplast fusion or by classical mating using a killer *kar1-1* mutant. The sake yeast carrying the killer character was resistant to invasion by wild strains of *Saccharomyces cerevisiae* (Onishi 1990).

Other fermented beverages include mead (fermented honey) and Tibi, which is a Swiss beverage made by fermentation of a 15% sugar solution containing dried figs, raisins, and lemon juice. The fermenting agent is a mixed culture of a capsulated bacterium, *Betabacterium vermiforme*, and a strain of *S. cerevisiae*, originally named *Saccharomyces intermedius*. In Indonesia, a tea infusion is fermented with a mixture of *Acetobacter xylinum* and *Saccharomycodes ludwigii*. Teekvass, from Russia, is fermented by a mixed culture of *Acetobacter* species and two yeasts. Kefir (fermented milk from the Caucasus), Kumiss (Asia), Leven (Egypt), and Mazun (Armenia) are also drunk locally in these regions.

2.3 Baker's Yeasts

Baker's yeasts differ from brewer's and winery yeasts. The cell mass is the desired product, and it must have definite characteristics. Baker's yeast can be used as a wine yeast or brewing yeast, but a brewing or wine yeast can seldom be used as a baker's yeast. The practice of using spent brewing yeast in the bakery, with the resulting poor results, was the primary cause leading to the development of new processes for production of baker's yeast and the development of new strains of these yeasts.

At present, baker's yeast is supplied either as wet compressed yeast or active dried yeast (ADY). "Instant" active dried yeast (IADY or SPADY) is yeast which can be added to the dough during mixing, without previous rehydration. These yeasts are packed under vacuum or under nitrogen, to reduce the loss of activity occuring on contact with air. They are generally produced by forcing a compressed yeast (70% moisture) through a fine (0.5–3.5 mm mesh) screen to give fine strands of yeast which are broken up and dried in a fluidized bed airlift drier. Improvement in active dried yeast can be either in the process, by using different drying schedules and addition of swelling agents (methyl cellulose, carboxymethyl cellulose or hydroxypropyl cellulose), wetting agents ("emulsifiers", such as sorbitan fatty acid esters, glycerol fatty acid esters, mono- and diglycerides of fatty acids, modified with lower organic acids (e.g., glycerol stearate lactate), propylene glycol fatty acid esters, and mixtures of the above. It is known also that the trehalose content and lipid content, including ergosterol, and the degree of unsaturation of the fatty acids, affect the tolerance of the yeast to drying and rewetting. These factors are influenced by genetic control as well as by the cultural conditions.

The requirements for a good baker's yeast are different from those of a brewing yeast and more stringent (Evans 1990). First, the product is the biomass, not the metabolic products of yeast growth during cell production. Second, the condition of the yeast is the most important role in baking. It forms CO_2, which raises the loaf. Raising modifies the protein (gluten) of the flour and improves the flavor and nutritive value of the bread. The yeast is grown in a highly aerobic environment, which maximizes its respiratory capacity. When used, it is transferred to an anaerobic environment, the uncooked dough. The growth of the yeast in an aerobic environment is essential to its performance under anaerobic conditions. Also, the petite colonie mutation in baker's yeast strains may release them from glucose repression. A strain which is grown under "unfavorable" conditions in the respiratory-competent form and performs poorly as a baker's yeast, may show a much improved performance as a petite (Spencer et al. 1988).

There are other desirable characteristics in a baker's yeast. Evans (1990) lists seven important factors to be considered in a baker's yeast for production on an economic scale:

1. Ability to grow rapidly and produce a good yield of biomass using a medium containing mostly sucrose as carbon source, since molasses is at the moment the major constituent of the medium. Beet molasses contains approximately 1.5% raffinose, which is not normally metabolized by baker's yeasts and is not only lost to the manufacturer, but becomes a disposal problem due to the increased BOD of the effluents. Beet molasses is also deficient in biotin, which must be added.
2. Growth rate on other carbon sources, if necessary.
3. Resistance to drying is important in the manufacture of active dry yeast, which is becoming very widely used.
4. Resistance to freezing, since frozen doughs are also highly popular.
5. Consistency and color are important, being affected by the water content of the yeast and the molasses used.
6. Keeping quality is affected by the presence or absence of oxygen, and the use of antioxidants such as butylated hydroxyanisole. It may also be affected by genetic factors.

In the baking process itself:

1. Good fermentative activity in the dough, especially of maltose, is required. It is strongly influenced by the oxygen supply during production. The yeast should tolerate elevated osmotic tensions and acid, Salts of propionic acid are often added to bread as an inhibitor of mold growth.
2. The yeast should be tolerant of rehydration and "leaching". Yeast may leak a mixture of intracellular materials into the water used at rehydration, or into the dough itself. These may include glutathione, which may cause "slackening" of the dough by chemical reduction of the glutens. Inhibitors (chemical oxidants; iodates or bromates) and emulsifiers (sorbitan monostearate) may be used to minimize these effects.
3. No formation of "grit" (undissolved material) during rehydration. Treatment with cysteine or papain reduces grit formation.
4. Good flavor. Yeast adds many flavor components to bread, but in trace amounts.

The nutritional requirements of a baker's yeast, during growth, are supplied by cane or beet molasses as a carbon source (sucrose), an ammonium salt or urea as N source, and phosphorus as phosphoric acid. K, Mg, and S are supplied as the inorganic salts; and trace elements such as Na, Cu, Fe, and Zn may be supplied adequately in the molasses or water supply. Six vitamins (thiamin, pyridoxine, niacin, inositol, pantothenate, and biotin) are required. Biotin is deficient in beet molasses. Enough oxygen to maintain the R.Q. near 1.0 is necessary, and temperature control to prevent a rise in the culture temperature is necessary.

Other potential carbon sources for producing baker's yeast are: lactose, in cheese wheys, which can be hydrolyzed enzymatically to galactose and glucose. Both sugars can be used by most baker's yeasts. The lactose may also be converted to lactic acid by bacteria, and baker's yeast produced from this carbon source. Finally, it may be possible to engineer baker's yeast to utilize lactose directly. So far, strains able to metabolize lactose have been unstable. Hexoses and pentoses in spent sulfite liquor, starch, and even, perhaps, lignocellulose, may be utilized after suitable pretreatments. Engineered organisms capable of utilizing these compounds may be constructed.

Breeding new baker's yeast strains is relatively simple. Testing the new hybrids is the limiting factor, and is mostly empiricial. The strains are grown under standard conditions, the yield of biomass calculated, and the dough-raising capacity determined. Other tests include determination of the amount of cytochromes. A yeast lacking cytochrome c usually does not perform well in a dough-raising test. Yeast grown under inadequate aeration usually has little cytochrome c.

Genetic improvement of tolerance to drying and rewetting, and to freezing and thawing is under investigation. Chemical analysis of the yeast, to determine trehalose, ergosterol, and other lipids, followed by direct investigation of the yeast strain's response to drying and rehydration, are part of the standard test methods in current use.

The heat shock effect in yeast may be related to the response of yeast to drying. Yeast expresses a series of specific heat shock proteins when heated gently, but their function is not known. Heat treatment, 45–55 °C for 4–20 h, of dried yeast, may improve its activity when it is rehydrated. The heat shock proteins are probably not involved. Permeability of the cell wall may be responsible.

Increasing the osmotic tolerance of baker's yeasts is a major objective in breeding new strains for raising sweet doughs. Normal baker's yeasts may be "acclimatized" to elevated concentrations of NaCl. This is not true genetic change, as the strain reverts almost immediately on transfer to a medium of normal osmotic tension. This increases the difficulty of designing a medium for production of these yeasts on a commercial scale.

Some workers reported mating a baker's yeast with an osmotolerant species such as *Zygosaccharomyces rouxii* to obtain a commercially useful, osmotolerant hybrid. Little information on the construction of the hybrid or its characteristics is available. Osmotolerant hybrids have been obtained by fusing protoplasts of baker's yeast with those of *Z. rouxii*. Some of these showed good performances as baker's yeasts. Legmann and Margalith (1983), in the same way, have obtained strains tolerating elevated sugar concentration in mashes for ethanol production. The method may be useful for obtaining superior strains of baker's yeasts.

In Japan, *Torulaspora delbrueckii* has been used successfully as an osmotolerant baker's yeast. The cells are relatively small, which increases the difficulty in recovery. Cell size may be increased by increasing the ploidy using protoplast fusion, since the cell size increases with increasing ploidy. Useful osmotolerant strains of baker's yeast might be obtained by fusion of *T. delbrueckii* with commercial strains of baker's yeasts.

Baker's yeast strains which will survive in frozen doughs are now commercially important. These convenience foods are becoming increasingly popular. The yeast must survive recovery from the original medium, conversion to active dry yeast, reconstitution in liquid, incorporation into the dough, which is frozen, thawed, and allowed to rise, and finally, baked and eaten. The yeast should probably be osmotolerant also. A formidable task for a small unicellular fungus! Breeding of strains which will survive these conditions is important. Testing of the hybrids is empirical but simple. The yeast is mixed with the other ingredients of the dough, the dough is frozen and stored in the frozen state, and samples thawed at intervals and the efficiency of raising determined (Evans 1990).

Sour-dough bread and salt-rising bread. The organisms used in manufacture of these specialty breads probably have not been selected or otherwise improved. Salt-rising bread is probably not raised by yeasts at all, but by the bacterial flora present in the flour. Sour-dough bread is raised by a mixed culture of a yeast, *Candida milleri*, and a heterofermentative lactic acid bacterium, *Lactobacillus sanfrancisco*. The yeast utilizes glucose but not maltose, and the bacterium maltose but not glucose, producing lactic and acetic acids. The yeast carries out an alcoholic fermentation, raising the dough. Commercial starters are now available for home and industrial use. The bread has a pleasant and distinctive tangy flavor, much sought after.

2.4 Salt-Tolerant Yeasts and Fermented Foods

Fermented foods have been widely used in Japan, Indonesia, and adjacent regions for many years. They are made by the action of (usually) mixed cultures of yeasts, filamentous fungi, or bacteria, on soybeans, fish, and other foodstuffs, to improve the flavor, keeping qualities, and other characteristics of the original material. Soy sauce and miso paste, used in Japan, are made by the action of salt-tolerant yeasts on soybeans.

In manufacture of soy sauce, whole soybeans or defatted soybean flakes are wetted and autoclaved. Whole wheat is roasted and crushed, and mixed with the soybeans and inoculated with koji-mold (*Aspergillus sojae*). The mash is incubated for up to 72h at 30 °C in shallow vessels with perforated bottoms. After the initial hydroysis by mold enzymes, it is then mixed with 120%/vol with brine (23% NaCl) and held in a deep tank for 4–8 months. First, there is a lactic fermentation, then an alcoholic fermentation by yeasts. Last, there is a maturation stage. The liquid is decanted, pasteurized, the proteinaceous precipitate settles out, and the sauce is filtered.

The first-stage fermentation is mediated by the bacterium *Pediococcus halophilus*. The mash is acidified and some of the flavoring components are formed. The principal yeast species in the second stage is a strain of *Zygosaccharomyces rouxii*. *Torulopsis famata*, *Candida polymorpha*, *Pichia farinosa*, and *Trichosporon behrendii* are also present. Later, *Zygosaccharomyces bailii* (*Saccharomyces acidifaciens*), *Torulopsis versatilis*, *T. etchellsii*, *T. halonitratophila*, *T. mannitofaciens*, *T. halophila*, and *T. nodaensis* have been found. The yeasts produce the rich aroma of good soy sauce. A product made using koji-mold and/or koji-mold plus lactic acid bacteria lacks flavor. The salt-tolerant yeasts produce significant amounts of polyhydroxy alcohols (glycerol, erythritol, arabitol, etc.) in response to high salt concentrations. These compounds are probably present in soy sauce and miso (Onishi 1990).

Miso paste is made from soybeans, or soybeans with rice or barley. The different types of miso are made by cultivating koji-mold (*Aspergillus oryzae*) or whole soybeans on steamed rice or barley (not defatted soybean flakes, which produce an inferior grade of miso). The rice or barley koji is then mixed with cooked soybeans in ratios of 0.5 to 3, plus salt solution to adjust the water and salt concentrations. A thick paste results, which is allowed to ferment. An inoculum of *Zygosaccharomyces rouxii* improves the quality of the miso. Sweet white rice-miso is matured mostly by the enzymes in the koji. The product is ready for use in a few weeks. Dark-colored, stronger-flavored misos are fermented for several months, and owe much of the stronger flavor to the fermentation by yeasts and lactic acid bacteria.

The microorganisms are much the same as those in soy sauce fermentation. Strains of *Zygosaccharomyces rouxii* predominate. *T. versatilis*, *T. etchellsii*, and *Hansenula subpelliculosa* are also present (Onishi 1990).

Other Japanese fermented foods include mirin, produced from rice using *Aspergillus oryzae*, shochu, from rice, sweet potato, or barley, using *Aspergillus awamori* and *Saccharomyces cerevisiae*, and natto (*Bacillus natto*).

2.5 Food Yeasts and Yeast Extracts (Bui and Galzy 1990)

Food yeasts are also grown for production of biomass, though the requirements are less strict than for baker's yeast. Food yeasts are used in livestock feeds, for fish (salmonids), poultry, or fur-bearing animals. The physiological state of the yeast at harvest is not important. Food yeasts may be grown on a wider range of substrates than baker's yeasts. These may include a mixture of sugars (mostly xylose) occurring in spent sulfite liquor from wood pulp manufacture. Whey or whey permeate (*Kluyveromyces lactis*), insulin from Jerusalem artichokes (other *Kluyveromyces* spp.), and starch wastes from potato-processing plants (the Symba process, using a mixed culture of *Saccharomycopsis fibuligera* and *Candida utilis*) are also used. Other potential carbon sources are triglycerides from animal and plant sources, fatty acids or ammonium soaps of fatty acids (*Candida tropicalis*), straight-chain alkanes, either pure or as crude petroleum fractions (*Yarrowia lipolytica*), or methanol (*Hansenula polymorpha, Pichia pastoris, Candida boidinii,* and possibly a few other species). These species, grown on such diverse substrates, are used mostly for animal feeds rather than for humans.

Food yeasts or yeast products for human consumption, usually are limited to *Saccharomyces cerevisiae,* and are derived from spent yeast from the brewing industry. "Marmite", a yeast hydrolysate popular in England, is produced in that way. Yeast extracts are widely used as flavorings. They can mimic the flavor of roast beef, cheddar cheese, several types of nuts, and dried bonito. The flavor depends largely on the method of extraction. Yeast cell-wall preparations can be used to replace fats in low-calorie salad dressings and ice creams (Bui and Galzy 1990).

Food yeasts and baker's yeasts can be grown in continuous culture in large-scale production. *Candida utilis* has been grown on spent sulfite liquor in continuous culture. *Yarrowia lipolytica* was produced in continuous culture on n-alkanes. The use of the process has been discontinued because of unfavorable economics.

Recently, continuous culture methods for high cell density fermentations were developed by Phillips Petroleum for production of heterologous proteins using *Pichia pastoris* as a vehicle (Sudbery and Gleeson 1989; Burden and Eveleigh 1990; Evans and McAthey 1991). This species metabolizes methanol as substrate. The process can be used for production of food and fodder yeasts by this yeast and other species, from most substrates. Cell densities can reach 100–200 g/l dry weight. By proper formulation of the medium, it may be possible to send the effluent from the fermentor directly to the recovery stage, without centrifugation and washing, with a consequent significant saving in costs. The process has been little investigated for production of extracellular metabolites, but should be adaptable for this purpose as well.

2.6 Yeasts, Cheese, and Flavor

Bacteria and filamentous fungi contribute to the flavor of many famous cheeses such as Camembert, Roquefort, and others. Yeasts also are involved in cheesemaking and may contribute to cheese flavor. There are not many of these yeasts, but their effects are significant (Devoyod 1990).

Geotrichum candidum is a common organism involved in cheesemaking. It may appear only in the early stages of the process, or may be found during the entire ripening period. It is sometimes regarded as a dangerous contaminant, but in other cases is used as a starter. As a starter, it is either spread over the entire surface of the cheese, or the milk is inoculated with it. *G. candidum* and other species are normal inhabitants of cloths and wooden utensils used in small farmhouse cheesemaking operations, and the milk is inoculated almost immediately.

Only a few yeast species occur in the inner part of cheeses. They include *Kluyveromyces marxianus* (yeasts of this genus include *Kl. lactis, Kl. fragilis, Kl. bulgaricus,* and *Torulopsis sphaericus,* and should be reclassified. Their true relationships are obscure), *Debaryomyces hansenii, Saccharomyces cerevisiae,* and, less often, *Zygosaccharomyces rouxii* and *Candida versatilis.* Yeast species on the outer surfaces of cheeses are usually the same species as previously mentioned. *Yarrowia lipolytica, Candida sake,* and *Candida intermedia* are found on St. Nectaire cheese. Yeasts on goat cheese are mostly *Candida* species.

The actions of yeasts on cheese are: (1) To metabolize lactic acid, increasing the pH and thereby permitting the growth of *Brevibacterium linens,* a major cheese-ripening organism. They also reduce the lactose content of the cheese. (2) Lactose-fermenting yeasts produce alcohol, acetaldehyde and CO_2 directly from lactose. The CO_2 prevents fusion of the curd granules and hence maintains the curd and later the cheese in an "open" form. In blue cheeses, this permits growth of *Penicillium roquefortii* in the fissures formed within the cheese, and gives the cheese its characteristic appearance and flavor. *Leuconostoc* spp. can also improve the blue-veined cheeses in the same way. (3) *Yarrowia lipolytica* improves the flavor of cheddar cheese by its proteolytic activity.

Other yeasts having proteolytic activity might also prove useful in cheese ripening and the development of cheese flavor. However, very few ascomycetous yeasts possess this characteristic. If genes encoding proteases and other enzymes were transferred from desirable species of basidiomycetes to one of the yeasts known to develop in cheese, *Debaryomyces hansenii* or a *Kluyveromyces* species, it might be possible to develop a new set of cheese flavors.

2.7 Spoilage Yeasts

Yeasts not only raise bread, brew beer, ferment wine, and provide alcohol for distilled liquors and fuel alcohol. Yeasts also spoil foods (Fleet 1990). The yeast species and the changes they induce in the foods differs considerably, depending on the nature and composition of the foods. Some of the yeasts, and the changes they produce in spoiled food, are described here.

Yeast species which have been isolated from spoiled foods are shown in Table 12.1. Some of the yeast species are widely distributed, and some food products contain a large number of species. Some foods high in salt or sugar content or of high aciditiy contain only yeasts tolerant of these conditions.

Red meats are seldom spoiled by yeasts. Counts are higher in ground meats. Yeast populations on meats are highest on products stored at low (refrigerator) temperatures, and basidiomycetous species often predominate. Sea foods, especially shellfish, contain higher counts of more spoilage yeasts. Pathogenic yeasts are found in shellfish harvested in estuarine water polluted with domestic sewage. The species of spoilage yeasts in processed meats varies with the preservatives used. Salty meats usually contain salt-tolerant yeast species (*Debaryomyces hansenii*) or lipophilic yeasts (*Yarrowia lipolytica*). Milk and milk products, in-

Table 12.1. Food spoilage by yeasts

Species	RM	SF	PM	MBC	YG	FR	FJ	VG	HS	MS	BK
Aureobasidium pullulans		+									
Candida albicans		+	+	+							
C. famata			+		+						
C. humicola		+				+					
C. krusei							+				
C. lambica							+			+	
C. mogii									+		
C. parapsilosis		+		+							
C. rugosa			+								
C. sake		+				+	+			+	
C. stellata		+								+	
C. tropicalis		+	+	+			+				
C. versatilis						+					
C. zeylanoides	+		+							+	
Cryptococcus albidus						+		+			
Cr. infirmo-miniatus	+										
Cr. laurentii	+			+		+					
Cryptococcus sp.		+									
Debaryomyces hansenii	+	+	+	+				+		+	
Endomycopsis burtonii											+
E. fibuligera											+
Geotrichum candidum						+				+	
Hanseniaspora spp.		+									
H'spora valbyensis							+				
Hansenula anomala							+	+			+
Hansenula spp.		+									
Kluyveromyces fragilis				+	+						
Kl. lactis				+	+						
Kl. marxianus				+							
Kl. bulgaricus							+				
Kloeckera spp.						+					
Pichia fermentans				+							
P. membranaefaciens	+		+	+				+		+	
Pichia spp.		+									
Rhodotorula sp.	+	+		+				+			
Rh. glutinis	+	+									
Saccharomyces cerevisiae					+		+				
S. exiguus										+	
S. kluyveri								+			
S. oleaginus								+			
Schizosaccharomyces octosporus						+			+		
Torulopsis apicola							+		+		
T. bombicola									+		
T. candida					+						
T. haemulonii					+						
T. holmii						+					

Table 12.1 (*Contd.*)

Species	RM	SF	PM	MBC	YG	FR	FJ	VG	HS	MS	BK
T. glabrata			+								
T. inconspicua			+								
T. lactis-condensi									+		
T. sphaerica				+							
Trichosporon cutaneum	+										
Tr. penicillatum				+							
Tr. pullulans	+	+				+					
Yarrowia lipolytica			+	+						+	
Zygosaccharomyces bailii									+	+	
Z. rouxii						+	+		+	+	
Maximum numbers	10^2	10^5	10^6		10^6			10^4	ND	High	

RM = red meats; SF = sea foods; PM = processed meats; MBC = milk, butter and cheese; YG = yoghurts; FR = fruits; FJ = fruit juices; VG = vegetables; HS = foods of high sugar content; MS = mayonnaise, salad dressings and mixed salads; BK = bakery products; ND = not determined.

cluding yogurts, usually contain members of the lactose-utilizing group of *Kluyveromyces* species. Yeasts, including lipophilic yeasts, also develop during cheesemaking, contributing to the flavor. Sweet yogurts are often spoiled by fermentative yeasts.

Ice creams, dried infant foods, and frozen cream pies are all classed as dairy products, and all carry low levels of nonspecific yeast contaminants. Spoilage is not normally a problem.

Fresh fruits often carry a considerable yeast microflora on the skin, and juice escaping through damaged parts may be a yeast nutrient. Undamaged fruits are seldom fermented by yeasts. Insects may transport yeasts to fruits (citrus fruits, for example) during feeding or oviposition, and spoilage may result. Dried fruits may also support surface growth of osmotolerant yeasts such as *Schizosaccharomyces octosporus* and *Zygosaccharomyces rouxii*. Fruit juices are spoiled by fermentative yeasts and such sugar-tolerant species as *Zygosaccharomyces bailii*. This species is osmotolerant and resistant to preservatives (sorbates, benzoates, and acetic acid). It is sometimes controlled with ascorbic acid.

Fresh vegetables are seldom spoiled by yeasts, though they may sometimes carry populations of *Cryptococcus albidus* and other basidiomycetous yeasts. Brined (pickled) vegetables are preserved by lactic acid formed by bacteria is the preservative. They are spoiled by several yeasts; the red yeasts (*Rhodotorula* spp.), which give a pink color, and by *Pichia membranaefaciens*, a fermentative yeast.

Jams and jellies, honey, syrups, and fondant centers of candies may be spoiled by osmotolerant yeasts; (*Zygosaccharomyces* spp.) and by several *Torulopsis* (*Candida*) species, some of which produce glycolipids in artificial culture (Beuchat 1983).

Mayonnaises and salad dressings are also spoiled by species of *Zygosaccharomyces, Yarrowia lipolytyica*, and *Pichia membranaefaciens*. The pH of the dressings is low enough to inhibit many spoilage yeasts, but others are acid-tolerant. Bakery products (uncooked doughs) may be spoiled by *Saccharomyces cerevisiae* itself, and by *Hansenula anomala* and *Endomycopsis burtonii*. Chalky bread may be caused by infections with the starch-degrading yeast, *Saccharomycopsis fibuligera*.

2.8 Prevention of Spoilage

The chemical and physical structure of the food determines what nutrients are available to the yeasts. This is a major factor in the growth and survival of yeasts. The wide range of foodstuffs spoiled by yeasts means that there is a wide range of nutrients available for yeast growth and a wide range of yeast species which spoil foods.

Other factors influencing the development of spoilage yeasts in foods include:

2.8.1 Temperature

Most yeasts normally grow well in the temperature range 20–30 °C. Exceptions are some of the common species of *Kluyveromyces* growing at 45 °C, and numerous yeasts which grow well at refrigerator temperatures and above, at 5–10 °C. The latter species cause spoilage of such foods as poultry meats (*Cryptococcus laurentii* var. *laurentii, Candida zeylanoides*), red meats (*C. laurentii* var. *laurentii, Cryptococcus infirmominiatus, Trichosporon pullulans, Candida zeylanoides. Cryptococcus laurentii* constituted the greater part of the yeast population), fruit juice concentrates (*Zygosaccharomyces rouxii, Hanseniaspora valbyensis*), and mayonnaises, salad dressings and mixed vegetables (*Saccharomyces dairensis, Saccharomyces exiguus*).

Heat treatment (pasteurization) is an effective method of control of most yeast infections, but refrigeration is not. The time-temperature relationships for effective control of yeasts in foods by pasteurization have not been investigated. Even for pasteurization of fruit juices, the conditions vary significantly depending on the type and concentration of the juice, and the presence or absence of antioxidants and preservatives. These substances generally increase the sensitivity of the yeasts to heating. The pH and sugar concentration influence the sensitivity of contaminating yeasts to increased temperature. Refrigeration is ineffective in preventing spoilage of many foods by yeasts. Yeasts are often favored by storage of foods at refrigerator temperatures for long periods.

2.8.2 Water Activity

Water activity is reduced by increasing the concentration of solutes such as sugars or salt. Foods such as preserves and jams or jellies, dried fruits, honey and syrups (sugars), and sausages and bacon, olives, sauerkraut, and dill pickles (salt) are protected against contamination by many yeasts (Beuchat 1983). Nevertheless, osmotolerant species can tolerate low water activities (high solute concentrations; high osmotic tension). *Saccharomyces bisporus* var. *mellis, Torulopsis lactis-condensi,* and *Schizosaccharomyces pombe* are most common in foods of high sugar content. *Debaryomyces hansenii, Pichia ohmeri,* and *Hansenula anomala* var. *anomala* are found in foods having high salt contents. *Zygosaccharomyces rouxii* and *Zygosaccharomyces bailii* grow in either. The effect of reduced water activity on the temperature tolerance of spoilage yeasts, especially osmotolerant species, is to increase the minimum, maximum, and optimum temperatures for growth of osmotolerant yeasts species. Their tolerance to benzoates and ethyl-paraben as preservatives was not much affected.

2.8.3 Preservatives

Sorbic and acetic acids and benzoates have been widely used, where permitted, to control spoilage of foods by yeasts. Propionic, citric, and lactic acids are also used, alone or in combination. In this case, the effects are sometimes synergistic. SO_2 has been used as a preservative for many years, though there is considerable strain variation in sensitivity of spoilage yeasts. It is used in the dried fruit industry, where it probably inhibits growth of sugar-tolerant yeasts on the surface of these products.

Newer preservatives for spoilage control by yeasts in foods include p-coumaric acid, ferulic acid, xylitol, tuberine, and the antioxidants butylated hydroxyanisole, *tert*-butylhydroquinone, and propyl gallate. Most of these, except for xylitol (0.5%), are used in concentrations of 50–500 ppm. Essential oils from onion, garlic, oregano, and thyme inhibit the growth of spoilage yeasts, but most other plant extracts have little effect.

Some yeast species are quite resistant to the effects of preservatives such as benzoates, sorbates and acetic acid, and possibly other inhibitors of yeast growth. These include *Zygosaccharomyces bailii, Schizosaccharomyces pombe, Zygosaccharomyces bisporus,* and *Zygosaccharomyces rouxii. Z. rouxii* and *Torulopsis versatilis* are inhibited by a mixture of lactic and acetic acids in the soy sauce fermentation, and *Kloeckera apiculata,* by 0.1 mM free SO_2. *Z. bailii* and *Saccharomycodes ludwigii* were inhibited only at 2.5–3.0 mM concentrations of SO_2, at pH 3.5. Inhibition was greater at lower pHs (3.0–4.0). Free sugars either inhibited or enhanced yeast growth, depending on their concentration and the species of yeast.

2.9 Biochemical Activities of Food Spoilage Yeasts

Spoilage by yeasts frequently is via alcoholic fermentation accompanied by the production of CO_2, giving the typical gassy type of spoilage. Products such as glycerol, higher alcohols, organic acids, esters, and diacetyl are also formed and may affect the flavor of the food. The amounts of these products varies with the yeast species and conditions of spoilage; aldehydes, ketones, and acids may be produced under more oxidative conditions.

Other yeast species produce pectinases, which may soften fruits and vegetables. Some (*Schwanniomyces* sp.) produce amylases. *Saccharomycopsis fibuligera* infections may cause spoilage of bread. A few yeasts (*Trichosporon* and *Cryptococcus* spp.) have been reported to produce xylanases, and some, cellulases, but these have seldom been responsible for food spoilage. A number of yeast species, *Yarrowia* (*Candida*) *lipolytica*, produce lipases. *Y. lipolytica*, also produces proteinases, which cause spoilage in meat and dairy products. Some yeasts found in cheeses convert phenylalanine into phenethyl alcohol, which gives a distinctive aroma.

Yeasts may also produce foul-smelling organic sulfides and H_2S from sulfur-containing amino acids, and may metabolize nitrates or nitrites added to meat products as preservatives.

2.10 Counting and Identification of Yeasts in Food Products

This can be done by standard methods for yeast culture. A suspension of the product is made, diluted, and the numbers determined by spreading an aliquot on an agar plate of the appropriate medium, or if the numbers are low, by collecting on a membrane filter. On isolation, the cultures are characterized by standard methods, preferably by those described in Kreger-van Rij (1984).

3 Control Systems for Yeast Fermentations

Originally, control of yeast fermentations, either for production of beverage or industrial alcohol, or for baker's, food, or fodder yeasts, was largely empirical. Conditions were established for particular fermentation vessels by trial and error, to obtain reasonable yields of product and acceptable quality. These methods are still in use today in small breweries and production plants. Recently, control systems have become more sophisticated, to obtain the maximum yield per unit of substrate and minimum expenditure of time and energy and good product quality. Yeast and yeast products are produced in fed-batch and, sometimes, continuous culture. The latest development is continuous high cell-density culture. Because of the high density of the culture, one or more stages in the recovery system can often be eliminated, with a corresponding reduction in cost, and the cells sent directly from the fermentor to the final recovery stages.

This requires much closer control than a simple batch or even fed-batch fermentation. Computer control, combined with use of high-quality sensors, is now used extensively, and expert systems, which can "learn" from the experience of previous fermentations, are being developed.

The problem of sensors becomes important in these systems. No one device is capable of measuring all the parameters required for obtaining the data necessary for optimum control. Parameters such as total biomass, dissolved oxygen and carbon dioxide, other dissolved gases, temperature, pH, redox potential if desired, and mass transfer can be measured with standard instruments relatively easily and accurately. Concentrations of substrates and products can be measured more rapidly and easily by the use of enzyme electrodes and other specific analyzers.

Some direct measure of metabolic activity is highly desirable. One such parameter is the rate of heat production, which can be measured by calorimetry. Techniques in microcalorimetry have been developed to the point where they can be used routinely for control systems for yeast fermentations. The instruments respond rapidly and accurately, and are rugged enough to be used in the production plant. It is beyond the scope of this book to describe these systems in detail, but microcalorimetry, in addition to its use in process control, can be used for identification, within limits, of unknown microorganisms, characterization of commercial yeast strains, evaluation of media, fundamental studies of yeast metabolism, and many other purposes. Perry et al. (1990) have reviewed the field thoroughly, and their review is highly recommended to investigators who wish to obtain more detail about the techniques.

References

Beuchat LR (1983) Influence of water activity on growth, metabolic activities and survival of yeasts and moulds. J Food Prot 46:135–141

Bui K, Galzy P (1990) Food yeast. In: Spencer JFT, Spencer DM (eds) Yeast technology. Springer, Berlin Heidelberg New York, pp 241–265

Burden DW, Eveleigh DE (1990) Yeasts – diverse substrates and products. In: Spencer JFT, Spencer DM (eds) Yeast technology. Springer, Berlin Heidelberg New York, pp 199–227

Devoyod J-J (1990) Yeasts in cheese-making. In: Spencer JFT, Spencer DM (eds) Yeast technology. Springer, Berlin Heidelberg New York, pp 228–240

Evans IH (1990) Yeast strains for baking: recent developments. In: Spencer JFT, Spencer DM (eds) Yeast technology. Springer, Berlin Heidelberg New York, pp 13–54

Evans IH, McAthey P (1991) Comparative genetics of important yeasts. In: Wiseman A (ed) Genetically-engineered proteins and enzymes from yeast: production control. Ellis Horwood, Chichester, pp 11–74

Fleet GH (1990) Food spoilage yeasts. In: Spencer JFT, Spencer DM (eds) Yeast technology. Springer, Berlin Heidelberg New York, pp 124–166

Hinchliffe E, Vakeria D (1989) Genetic manipulation of brewing yeasts. In: Walton EF, Yarranton GT (eds) Molecular and cell biology of yeasts. Blackie, London, Van Nostrand Reinhold, New York, pp 280–303

Johnston JR (1990) Brewing and distilling yeasts. In: Spencer JFT, Spencer DM (eds) Yeast technology. Springer, Berlin Heidelberg New York, pp 55–104

Kielland-Brandt MC, Nilsson-Tillgren T, Petersen JGL, Holmberg S, Gjermansen C (1983) Approaches to the genetic analysis and breeding of brewer's yeasts. In: Spencer JFT, Spencer DM, Smith ARW (eds) Yeast genetics: fundamental and applied aspects. Springer, Berlin Heidelberg New York, pp 421–437

Kreger-van Rij NJW (ed) (1984) The yeasts, a taxonomic study. Elsevier, Amsterdam

Legmann R, Margalith D (1983) Ethanol formation by hybrid yeasts. Appl Microbiol Biotechnol 1:320–322

Martini A, Vaughan Martini A (1990) Grape must fermentation: past and present. In: Spencer JFT, Spencer DM (eds) Yeast technology. Springer, Berlin Heidelberg New York, pp 105–123

Onishi H (1990) Yeasts in fermented foods. In: Spencer JFT, Spencer DM (eds) Yeast technology. Springer, Berlin Heidelberg New York, pp 167–198

Perry BF, Miles RJ, Beezer AE (1990) Calorimetry for yeast fermentation monitoring and control. In: Spencer JFT, Spencer DM (eds) Yeast technology. Springer, Berlin Heidelberg New York, pp 276–347

Snow R (1983) Genetic improvement of wine yeasts. In: Spencer JFT, Spencer DM, Smith ARW (eds) Yeast genetics: fundamental and applied aspects. Springer, Berlin Heidelberg New York, pp 439–459

Spencer JFT, Spencer DM, Reynolds N (1988) Effects of changes in the mitochondrial genome on the performance of baker's yeasts. Antonie Leeuwenhoek J Microbiol Serol 55:83–93

Sudbery PE, Gleeson MAG (1989) Genetic manipulation of methylotrophic yeasts. In: Walton EF, Yarranton GT (eds) Molecular and cell biology of yeasts. Blackie, London, Van Nostrand Reinhold, New York, pp 304–329

Vezinhet F, Blondin B, Hallet JN (1990) Chromosomal DNA patterns and mitochondrial DNA polymorphisms as tools for identification of enological strains of *Saccharomyces cerevisiae*. Appl Microbiol Biotechnol 32:568–571

Yeasts and the Life of Man: Part II: Genetics and Molecular Biology of Industrial Yeasts and Processes

J.F.T. Spencer and D.M. Spencer

1 Introduction

The object of genetic improvement of industrial yeast is to modify the genome of the yeast so that the process is more efficient and/or the product is more desirable and useful. The oldest method is random selection of strains of the original culture and testing them for improved performance. No attempt is made to investigate the changes in the genome itself.

Other methods of classical genetics include:

1. Mutagenesis: obtaining mutants by treatment of the strain with mutagens.
2. Classical hybridization by mating of haploid strains.
   a) This procedure includes rare-mating, which is used for introducing auxotrophic and other markers into industrial yeast strains which mate or sporulate poorly.
3. Use of the *kar1* mutant in construction of hybrids of industrial yeasts (single-chromosome transfer).
4. Protoplast (spheroplast) fusion.
   a) Fusion using PEG and related chemical agents.
   b) Electrofusion.
   c) Nucleus-protoplast fusion.
5. Transformation.
   a) With chimeric plasmids.
   b) With yeast artificial chromosomes.
   c) With other linear sequences of DNA; expression cassettes constructed for the purpose or yeast chromosomes isolated by pulsed-field gels from different strains and species of the yeast.

In most cases, the first four methods require only simple equipment and few reagents, and are used for improvement of baker's and brewer's yeasts and other yeasts for industrial processes. Electrofusion is an exception, although the equipment can be constructed in a good electronics laboratory more cheaply than it can be bought.

The fifth group of techniques are primarily used for construction of genetically engineered yeast strains intended for use in production of heterologous proteins. Construction of the necessary plasmids or expression cassettes involves sophisticated genetic engineering.

See also *Notes Added in Proof* on p. 353.

J.F.T. Spencer/D.M. Spencer (eds)
Yeasts in Natural and Artificial Habitats
© Springer-Verlag Berlin Heidelberg 1997

In strain selection, individual cells are isolated by plating or micromanipulation and tested for the presence of the desired characters. The culture may be plated on a selective medium, or grown in an enrichment medium favoring the growth of isolates having these characteristics. This method uses the same system of selection as in high-level genetic engineering. Desirable isolates are separated from the undesirable by a properly designed selective medium. Mechanical sorting of individual cells (flow cytometry) can also give excellent results, but the equipment is expensive.

1.1 Mutagenesis

Mutations to resistance to copper, the herbicide sulfometuron methyl, methotrexate, the antibiotic Geneticin G418 (resembling kanamycin and gentamycin), and antibiotics affecting mitochondrial function (erythromycin, oligomycin and chloramphenicol) have been used for selectable markers in making hybrids. They are also quite widely used as selectable markers in construction of new yeast strains by transformation, the genes being encoded on plasmid vectors. However, mutations to auxotrophy are recessive and difficult to obtain, so the mutations are seldom expressed. Therefore, these mutations are seldom used as selectable markers. However, resistance to copper, sulfometron methyl, etc., is dominant.

Construction of new industrial yeast strains by mutational treatments has been used in yeast-based industries. In baker's yeasts, selection and testing of the numerous strains obtained are difficult, as there are no direct tests for baking performance other than using the yeast in a model bakery. Testing new strains of brewing yeasts is easier. Many of the flavor-producing compounds are volatile enough that they can be determined by GLC. The testing program is still laborious.

The first improvement in penicillin yields were obtained by passing the original strain of *Penicillium* through many cycles of mutation and selection. The product has a much higher unit value than beer or bread, but the methods can still be used for improvement of these yeast strains as well as of *Penicillium* species.

Mutation has been used for improvement of brewing yeasts, especially beer flavor, by control of organoleptic compounds (diacetyl, acetolactate, propanol, and other higher alcohols). Sometimes this can be done by isolating mutants blocked in the biosynthetic pathways for synthesis of the corresponding enzymes, reducing the amounts formed of the flavoring compound(s).

Strains resistant to catabolite repression by glucose have been isolated by selecting mutants resistant to 2-deoxyglucose. These mutants having an increased fermentation rate. Von Borstel (1990) has summarized the principles and strategies of mutagenesis as applied to yeast.

1.2 Hybridization

Meiotic. Mass mating or spore-spore and spore-cell mating. This is a standard method for study of genetic phenomena in laboratory yeasts. It is more difficult to apply in industrial yeasts, because of poor sporulation in many industrial yeasts and the very low viability of the spores formed. Third, the strains are frequently homothallic. When viable spores are isolated, the haploid strains self-diploidize soon after germination of the spores and no longer mate.

Sporulation may sometimes be improved by growing the strain on a presporulation medium containing acetate as a carbon source, at 20 °C, and sporulating the culture at this temperature. This sometimes increases the number of three- and four-spored asci, and the viability of the spores. The proportion of ascospores (random) can be increased by lysing the vegetative cells enzymatically, or by killing the vegetative cells by heat or ether. If the viability of the spores is low, the probability of more than one spore per ascus being viable is also low. Isolation of "random" asci will yield a high proportion of single-spore clones. The preparations can usually be enriched in asci by density gradient centrifugation.

In mass matings, hybrids between auxotrophic strains can be readily isolated by plating on selective (minimal) media. Zygotes obtained by mating of prototrophic strains can be isolated by micromanipulation and grown into diploid cultures. Pure colonies of the diploid hybrid strains are obtained within 2 or 3 days, saving several days in obtaining new hybrids.

Palleroni (1962) devised an ingenious method for obtaining hybrids from prototrophic, sporulating, homothallic strains. He dissected asci from sporulated strains and placed the spores on acetate-containing medium to reduce the growth rate and self-diploidization. The clones obtained could be mixed and mated while still in the haploid state. Hybrid zygotes were isolated by micromanipulation. Another method is to isolate spores from the two parental strains and pair them by micromanipulation. The pairs should be examined at frequent intervals in the first few hours, to see if fusion and zygote formation have taken place.

1.3 Rare Matings

Rare mating can be used to obtain hybrids between prototrophic industrial yeast strains and laboratory auxotrophs, to introduce markers into the former. One of the strains must be an auxotrophic, haploid, or diploid mating strain. The prototrophic, industrial strain is converted to the petite form, and the cultures are mixed and allowed to mate. Mating occurs between the auxotrophic mating strain and cells of the prototrophic petite strain which have undergone a change from a/α to a/a or α/α. The hybrids are isolated on minimal medium, to select against the auxotrophic strain, and glycerol as sole carbon source, to select against the petite, prototrophic strain, plus 3% ethanol to suppress sporulation during isolation. The method originated with Gunge and Nakatomi (1971) for use with laboratory yeasts, and was adapted by Spencer and Spencer (1977) for construction of hybrids with industrial yeasts. Since the hybrids usually sporulated, genetic analysis could be performed on hybrids in which a significant part of the genome was derived from the industrial yeast.

The technique was used to construct a brewing yeast carrying the DEX1 (STA2) gene, from *S. diastaticus*. The original hybrid also carried the POF gene (phenolic off-flavor), but segregants which did not have this gene were easily obtained.

1.4 Use of the *kar1* Mutant in Constructing Hybrids of Industrial Yeasts: Single-Chromosome Transfer

The *kar1* mutation results in loss of the ability of the nuclei in a hybrid to fuse during mating (Conde and Fink 1976). However, in such hybrids, one chromosome (seldom more) may be transferred from a donor strain to the mutant. This phenomenon can be used to transfer desirable genes to industrial yeasts.

The *kar1* mutant has been used to transfer cytoplasmic elements to industrial yeasts (cytoduction), yielding heteroplasmons containing cytoplasm of both parental strains. Mitochondria and "killer" VLPs have been transferred in this way. The killer character confers immunity to invading "wild" strains of *Saccharomyces* species without altering the beer, and interchange of mitochondria may alter the character of both brewing and baking yeasts (Spencer et al. 1988). The effects of replacement of mitochondria from different yeasts on industrial processes using yeasts are not known.

Transfer of single chromosomes or parts of chromosomes by use of the *kar1* mutant is possible (Dutcher 1981). A commonly used strain carries the *ade2* and *his4* mutations, making it easy to identify. This strain can be mated or fused with a donor strain having the desired gene, and *kar1* strains carrying the gene can be isolated by colony color and the phenotype of the desired gene. These can then be mated or fused with the recipient, industrial strain. Clones carrying the gene which are white and do not throw off red sectors can then be isolated. The final strains do not appear to carry other genes from the *kar1* mutant. We have used the method to transfer the STA2 gene from a laboratory haploid to a brewing yeast, and the MEL1 gene from a strain of *Saccharomyces kluyveri* to a baker's yeast. Therefore the method can be used for interspecific as well as intraspecific transfer of genes to industrial yeasts. This technique may be useful for laboratories which are financially unable to construct new strains using plasmids or YACs.

The Danish investigators developed a strain of lager yeast which sporulated freely enough and yielded enough viable spores for genetic analysis. They used the *kar1* mutant to transfer chromosomes from the brewing strain 244 into a laboratory haploid, and found sequences in chromosome III which were not homologous with those in the laboratory strain, especially in the HIS4 region. The yeast carried two and three copies of chromosomes V and X, respectively.

1.5 Protoplast (Spheroplast) Fusion

Protoplast fusion was used first in making "hybrid" plant cells, and animal cells had frequently been fused using Sendai virus as a fusogenic agent. The procedure

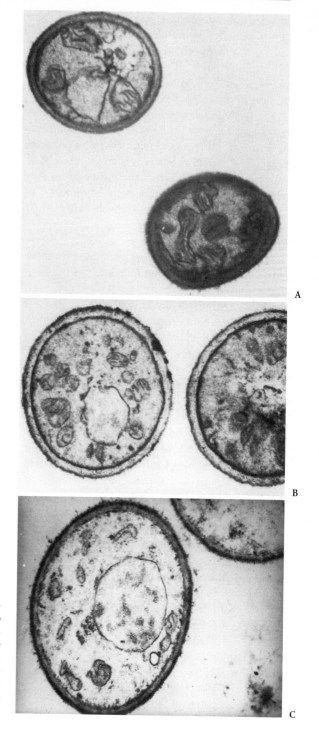

Fig. 13.1A–C. Products of protoplast fusion. Thin sections of two parental yeast strains and the hybrid obtained from them. Note the differences in the numbers of mitochondria and development of cristae within them. **A** *Hansenula capsulata*. **B** *Hybrid*. **C** *Saccharomyces diastaticus*

is applicable to yeast. Remove the cell walls enzymatically, mix the protoplasts obtained in stabiizing solutions of sorbitol or potassium chloride, add calcium ions and a fusogenic agent such as polyethylene glycol (PEG), incubate for 15–30 min, and plate out on osmotically stabilized selective media. The techcnique yields classes of hybrids, since there are no constraints on the recombination process in the units yielding colonies (Fig. 13.1).

Electrofusion is used more often to obtain hybrids of plant and animal cells, but is applicable to yeasts. The protoplasts are mixed in an osmotically stabilized medium of the proper conductivity, and subjected to an alternating electric field. This causes the protoplasts to line up in "pearl chains". They are then shocked with two square-wave pulses of direct current, which breaks down the protoplast membranes at the point of contact. The fusion products are transferred to regeneration medium and the hybrids are isolated after the cell walls have regenerated.

The hybrids obtained in intraspecific fusions possessed unwanted as well as desirable characters. Brewing yeast hybrids sometimes produced excess diacetyl. However, the ability to obtain hybrids between nonsporulating strains in brewing yeasts offsets this disdvantage. Hybrids obtained by fusion of protoplasts have desirable characteristics as well, and the technique has been used a good deal, especially in Japan. Protoplast fusion has been used to introduce the killer character into brewing and sake yeasts. Design of the section system is very important in isolation of desirable hybrids.

Hybrids obtained by interspecific fusions were frequently too unstable to be of practical use. Nevertheless, hybrids obtained between *S. cerevisiae* and *Zygosaccharomyces rouxii, Hansenula capsulata, Hansenula (Pichia) canadensis, Candida utilis, Pichia stiptis, Pachysolen tannophilus,* and *Kluyveromyces fragilis* mostly were stable under continued selection pressure.

We have found recently that auxotrophic, homologous mutations in a diploid strain of *S. cerevisiae* were complemented by fusion with a prototrophic strain of *Pichia canadensis*. The incoming DNA sequences were sufficiently homologous to the ones replaced to be able to function in the environment of the *S. cerevisiae* cell. To some extent, gene transfer by use of protoplast fusion can be directed.

The mechanism of formation of the genome of hybrids constructed by protoplast fusion was originally thought to be a simple fusion between the nuclei of the two parental strains. This is true for parental strains of *S. cerevisiae* only. When the genomes of hybrids between *S. cerevisiae* and *H. capsulata* and other interspecific hybrids were investigated by DNA reassociation, the genomes proved to consist mostly of the genome of the dominant species, plus some genes from the other (Spencer et al. 1985). This has some advantages, but the rules governing the introduction of foreign DNA into the genomes of industrial yeasts have not been determined. In fusions between *S. cerevisiae* and *Pa. tannophilus*, we observed two bands corresponding to chromosome-sized DNA, in pulsed-field gels of the genome of this hybrid. The chromosomes which were transferred have not been identified.

The dead-donor technique is a variant of the standard procedure for protoplast fusion. One of the parental strains is inactivated by UV irradiation. This reduces the problems in selecting the fusion products. The use of DMSO (dimethyl sulfoxide), 15% in the PEG solution for fusion, increases the number of fusants obtained. Treatment of the cells before or after protoplasting with hydroxyurea or 8-hydroxyquinoline causes G1 arrest of cell division. All three of these variants increase the fusant yield.

1.6 Nucleus-Protoplast Fusions

Viable nuclei are isolated from the donor strain and fused in the conventional manner with protoplasts from the recipient strain. So far, the method has been tested only on strains of *S. cerevisiae* and fusions between yeast protoplasts and nuclei from filamentous fungi (*Fuarium moniliforme* and *Trichoderma reesei*). It should have the advantage of avoiding mitochondrial-nuclear interactions, which occur in fusion of protoplasts of *Kluyveromyces lactis* and respiratory-competent strains of *S. cerevisiae*.

Interkingdom fusions have not been investigated to the same extent. Fusions of yeast and bacterial protoplasts have been obtained. Hen erythrocytes were successfully fused with yeast protoplasts, in 1975, by Cocking's group in Nottingham University. Yeast protoplasts have also been successfully fused with cells of *Xenopus* and other animal cells (Ward 1984). Small, slow-growing colonies were obtained. Protoplasts of petite mutants of *Saccharomyces cerevisiae* have been fused with human blood platelets, which contain mitochondria but not nuclei. Small, slow-growing, respiratory-competent colonies of yeast were obtained, the respiratory deficiency of the yeast cells apparently being complemented by the mitochondria from the platelets (H. Heluane, unpubl. data 1991). I.J. Bruce (unpubl. data), working in Wilkie's laboratory, fused rat liver nuclei with yeast protoplasts and detected DNA sequences from the rat liver in the regenerated yeast cells. The rat genes were not expressed.

Protoplast membranes of osmotolerant yeasts such as *Zygosaccharomyces rouxii*, grown on glucose, are rather fragile and the protoplasts burst easily. This was remedied by growing adapted yeasts in a medium containing galactose as sole carbon source. Adaptation can be done by two or three serial transfers on a galactose medium.

Pachysolen tannophilus forms protoplasts readily, but the viability of these is very low. Whittaker and Kavanagh (1990) attribute this to complete removal of the cell wall. If removal of the wall is halted at the spheroplast stage, the spheroplasts are viable and fusion can take place, yielding viable hybrids.

The walls of *Schizosaccharomyces pombe* are more difficult to remove than those of *S. cerevisiae*, but protoplasts can be obtained by standard procedures, and hybrids obtained by fusion.

It is more difficult to make protoplasts from basidiomycetous yeasts. Protoplasts have been obtained from *Rhodosporidium*, *Cryptococcus*, and *Phaffia* spp., but the incubation time required to digest the walls with Novozyme 234 (from *Trichoderma harzianum*) is much longer than for most ascomycetous yeasts, up to 4h. An enzyme from *Paecilomyces lilaceum* is more effective and will digest the walls of *Rhodotorula* spp. and *Phaffia rhodozyma*. Some strains of *Cryptococcus laurentii* can be protoplasted in an hour. Isolation of fragile mutants (Venkov et al. 1974; Mehta and Gregory 1981) or mutants having more easily digested cell walls may be another solution.

2 Genetic Engineering of Industrial Yeasts: Transformation, Plasmid Construction, Gene Expression, Secretion of Heterologous Proteins. (The Heavy Stuff)

Protein synthesis begins with the chromosomes, through transcription, synthesis of messenger RNA (mRNA) to the acquisition of first the 40S ribosomal unit, and then the 60S unit. From there, actual protein synthesis proceeds via amino acids transferred from charged tRNAs according to the message on the mRNA. Finally, the polypeptide chain separates from the ribosomal-mRNA complex. If the protein is marked with a signal sequence for excretion, it is passed from the polysome through the rough endoplasmic reticulum into the lumen of the ER. Here, the signal sequence is removed from the polypeptide chain by a peptidase. The polypeptide passes to the Golgi body, where it is sorted into the correct pathway and glycosylated. It passes from the Golgi body to the secretory vesicles or to the vacuole, and from the secretory vesicles, through the cytoplasmic membrane and the cell wall to the external environment.

Production of heterologous proteins is mainly focused on transcription – reading the message from the chromosomal genes and production of the optimum amount of mRNA for translation into the desired heterologous protein. The original polypeptide chains are modified to form the mature protein, after which, the protein is secreted into the culture medium. Here, it is separated from the other yeast proteins and recovered (King et al. 1989; Kozlov et al. 1995).

Two steps precede the actual synthesis of the polypeptide chain: (1) construction of the plasmid carrying the desired heterologous gene, plus any necessary promoter and terminator sequences, (2) introduction of the plasmid, YAC, or linear fragment of DNA into the yeast cell by transformation.

2.1 Transformation in Yeasts (Fig. 8.1)

The first transformations of yeasts were done with plasmid DNA and protoplasts of *S. cerevisiae* as recipients (Hinnen et al. 1978). Whole cells treated with alkali cations as lithium compounds can also be used. Li acetate gives higher yields of transformants. Transformation of whole cells avoids fusions between the protoplasts and unknown ploidy in the transformants, but transformation of protoplasts generally gives a better yield of transformants. Also lithium salts may cause deletions in the plasmid DNA.

Transformation is done by converting the yeast cells to protoplasts, by removing the cell walls enzymatically in an osmotically stabilized medium, and washing the protoplasts free of enzyme. The transforming DNA is mixed with the protoplasts with a fusogenic agent, usually PEG. Electroporation using pulses of an electric field also gives good results. The fusion mixture is then plated out on selective or nonselective, osmotically stabilized medium for regeneration of the transformed protoplasts and isolation of the transformed cells. Selectable markers are required for identificantion and isolation of transformed clones.

2.2 Construction of Plasmids

Plasmid construction is often complex and time-consuming. Plasmids for use in production of heterologous proteins are made by modification of an earlier plasmid.

Construction of plasmids used in modifying yeast strains for production of proteins such as insulin, interferon, and many others, and some of the plasmids used in production of proteins such as human factor VIII, for treatment of hemophiliacs, is done by assembling several DNA sequences and is difficult and time-consuming. However, the principles are simple.

Many plasmids used for transformations in yeast are based on plasmid pBR322, derived from *E. coli*. Sequences from pBR322 are found in many others used in research and in commercial applications. These are included to enable the plasmid to be amplified in *E. coli*, recovered, and used to transform the yeast. Sequences from yeast DNA, complementing an auxotrophic mutation in the recipient strain, are used as selectable markers to isolate yeast strains which have received the plasmid.

First, clone the desired gene into the plasmid. The original plasmid contains the sequences from pBR322, a selectable marker from yeast DNA such as LEU2, HIS3, HIS4, or URA3, two selectable markers for antibiotic resistance in *E. coli* (ampicillin and tetracyclin), a yeast centromere sequence or an ARS sequence, and sites for restriction enzymes. It is digested with a restriction enzyme (BamH1), which cuts in one of the genes for antibiotic resistance (tetracyclin). This digest is mixed with a digest of total yeast DNA, using the same restriction enzyme. The mixture is allowed to anneal, reforming the plasmids, some of which will contain sequences containing the desired yeast gene. The reaction mixture is treated with DNA ligase, usually from T4 bacteriophage, and used to transform a strain of *E. coli* sensitive to the antibiotics. The transformed cells are plated on selective media, and colonies resistant to ampicillin and sensitive to tetracyclin are isolated. The cells which are sensitive to tetracyclin are assumed to contain DNA sequences from the yeast DNA inserted at the BamH1 cloning site in the disrupted gene for tetracycline resistance.

There strains are pooled and the plasmid DNA amplified and recovered. It is used to transform the recipient (auxotrophic) yeast strain and select strains prototrophic for the yeast markers (frequently LEU2). The transformed yeast cells are then tested further for the ability to produce the protein and the yields.

Erratt and Nasim (1986) constructed a plasmid (based on plasmid YEp13) carrying the STA gene from *Saccharomyces diastaticus*. They used it to transform a strain of *Schizosaccharomyces pombe* to introduce the STA gene. They fractionated the enzymatic digest of the *S. diastaticus* DNA on a sucrose gradient, recovered the 5–15-kb fraction, and used it for construction of the gene library. They treated the plasmid digest with alkaline phosphatase to remove the sticky ends and prevent reformation of the original plasmid on annealing and ligation. They transformed *E. coli* with the ligated mixture, amplified the plasmid, reisolated the plasmid, and transformed *Scizosaccharomyces pombe* with it. They selected approximately 2000 transformants, 7 of which utilized starch. However, the glucoamylase activity in *S. pombe* was not more than 10% of that observed in transformants of *S. cerevisiae* with the same plasmid construct.

The plasmid may contain other sequences, for promoters, activators, initiators, terminators and other functions. This makes construction of the plasmid more complex. Typical plasmids are shown in (Fig. 13.2).

There are four general types of plasmid used in yeasts: (1) YRp (yeast replicating plasmids) (2) YEp (yeast episomal plasmids) (3) YIp (yeast integrating plasmids) and (4) YCp (yeast centromere plasmids). YEps have a high copy number, up to 100–150/cell, but YIps have a copy number of one, the significant DNA sequences being integrated into one of the yeast chromosomes. YCps are more stable than plasmids lacking a centromere sequence, but YEps, having a high copy number, generally lead to higher yields of the heterologous protein. Sequences from the 2-micron plasmid, a natural resident in many *S. cerevisiae* strains, are often included in the plasmid for increased stability. YEp plasmids are sometimes more stable in diploid than in haploid yeasts.

2.3 Expression of Heterologous Genes and Secretion of Heterologous Proteins (Piper and Kirk 1991)

2.3.1 Expression of Heterologous Genes

Expression vectors for heterologous proteins may be of several types: 2-μm-based vectors, ARS and ARS-CEN-based vectors, and integrating vectors. All have sequences allowing selection and replication in *S. cerevisiae*. These may complement auxotrophic requirements in the recipient strain, (LEU2, HIS3, URA3, TRP1) or are dominant selectable markers such as CUP1, DHFR (dihydrofolate reductase), resistance to antibiotic G418, resistance to hygromycin B, resistance to chloramphenicol, resistance to sulfometuron-methyl, and resistance to heavy metals, herbicides, and related compounds. Sequences allowing replication in *E. coli*, and resistance to antibiotics (tetracyclin and ampicillin) are required.

1. Two-micron based vectors have high copy number (20-200 copies per cell) and good stability. They usually contain the origin of replication and the STB locus of the 2-μm plasmid. These plasmids interact with the REP1 and REP2 proteins in the endogenous plasmids, which must be present. The pJDB219 and pJDB207 plasmids, and YEp24, do not contain the gene encoding FLP recombinase, required for amplifying the plasmid, so that the endogenous 2-μm plasmid must be present for these plasmids to function.

2. ARS and ARS-CEN-based vectors contain autonomous replicating sequences of yeast chromosomal DNA, which may be origins of replication. These vectors usually use the TRP1 sequence as a selectable marker (in plasmid YRp7) and can have a high copy number. They are unstable and are not used in the production of heterologous proteins. Use of the CEN (centromere) sequences usually increases the stability of the plasmid, but reduces the copy number of the plasmid to one. The selectable marker is usually TRP1 or URA3 (in plasmids pYE(CEN3)11 and YCp50 respectively. CEN plasmids are not used in expression of heterologous proteins on an industrial scale.

2.3.2 Integrating Vectors

These vectors induce insertion of single (sometimes multiple) copies of a gene into a specific site on the chromosome. One type of plasmid contains the expression unit and a selectable yeast gene, such as HIS3 (YIp1) or URA3 (YIp5). In the second type, the gene is flanked by sequences, homologous to sequences at the site where it is desired to insert the heterologous gene. Double-stranded breaks occur in the chromosomal DNA and the expression unit is guided into the site by the homologous sequences one-step gene replacement (Rothstein 1983).

2.3.3 Chromosomal Amplification of Genes

The gene may be incorporated into the chromosomal DNA by placing the foreign genes in a Ty element. For example, under the control of the GAL1 promoter, induction may lead to integration of at least ten copies of the hybrid Ty element into the genome. This system has not been tested for expression and production of heterologous proteins by yeast.

2.3.4 Intracellular Expression

Where secretion of the protein into the medium is not required, the method is: the desired gene is linked to a strong yeast promoter at one end and to a terminator of transcription at the other. Promoters from glycolytic cycle genes are often used, to give high-level production of the heterologous protein. The phosphoglycerate kinase (PGK) promoter gives very high levels of PGK itself, up to 80% of the total cell protein, when the PGK gene is introduced on a multicopy plasmid. Similar yields of heterologous proteins encoded by other genes are not always produced.

Yields of any heterologous protein are greatly influenced by the nature of the protein itself. Yields are also affected by plasmid copy number, plasmid stability, the culture medium, toxicity of the heterologous protein for the yeast cell, and the presence or absence of a secretion signal. Human serum albumin, α-interferon, α-1-antitrypsin, and hepatitis B surface antigen have been produced as intracellularly expressed proteins.

External factors also influence expression. Expression of a heterologous protein may be regulated by growing the cells to the required density, and then changing the composition of the medium (changing the carbon source to galactose, altering the phosphate level, altering the temperature, etc. The object is to activate the transcription of the gene, either directly or by inhibition of a repressor. The system should improve plasmid stability and/or avoid problems of toxicity of the protein.

Heterologous proteins expressed in this way include human leukocyte interferon D, which has been expressed at levels of up to 5% of the cell protein; hepatitis B core antigen (40%); and human superoxide dismutase (30–70%). Using a yeast strain carrying the pep4-3 mutation sometimes prevents degradation of the heterologous protein. Absence of proteinase A, encoded by PEP4, prevents activation of vacuolar proteases which may degrade heterologous proteins and reduce yields.

Several heterologous proteins (prochymosin, human gastric lipase, hepatitis B surface antigen) are expressed in an insoluble form. The latter is produced as insoluble, 22-nm, virus-like particles, which retain their original antigenic properties and can be used to produce a vaccine (Cregg et al. 1987). They can be coexpressed with other antigens as hybrid particles (Herpes simplex; HIV).

2.3.5 Posttranslational Modification

These processes include glycosylation, acetylation, phosphorylation, attachment of lipids, uptake of cofactors, and removal of precursors and leader and signal sequences. Glycosylation occurs after the protein enters the secretory pathway in the Golgi body. Superoxide dismutase produced in yeast is acetylated at the amino terminus, though natural yeast SD is not. The human c-myc protein is phosphorylated, and the p21 protein, encoded by the human Ha-ras gene, is palmitylated. Phosphorylation is a very common means of regulation of numerous processes in living organisms.

Rat liver cytochrome P-450, expressed in yeast, took up heme and was assembled as functional membrane-bound enzyme. Rat liver NADPH:cytochrome P-450 reductase was also expressed in yeast. With cytochrome P-450, it formed an electron transport chain with the same specificity for oxidation of drugs as the natural enzyme. Chimeric cytochrome P-450 enzymes from different cytochrome P-450 genes are expressed in yeast.

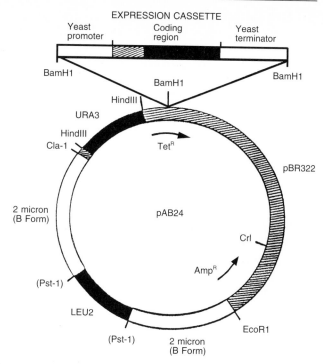

Fig. 13.2. Typical plasmid for introduction of genes encoding heterologous proteins (pAB24). The plasmid carries sequences for selection of vectors resistant to inhibitors (tet, Amp), for requirements for amino acids (URA3 and LEU2), and has restriction sites for various restriction enzymes. It also contains sequences from the yeast 2-micron circle and from the bacterial plasmid pBR322. The expression cassette contains a yeast promoter, a coding region for the heterologous protein, and a yeast terminator. For manufacture of heterologous proteins using *Pichia pastoris*, HIS4 is a useful selective marker

3 Secretion

The requirements and steps in secretion of a protein to the extracellular environment, after the protein has been translated from the mRNA chain by the ribosomes, (Fig. 13.3) are:

1. The protein precursor requires a leading "signal sequence", or secretion sequence, 20–30 amino acids long.
2. The secretion sequence binds the polysome to the outside of the rough endoplasmic reticulum and passes through it, bringing the polypeptide chain proper into the lumen of the ER.
3. The signal sequence is removed by a peptidase in the lumen of the ER.
4. The inner-core glycosylation is also done in the lumen of the ER.
5. The glycosylated polypeptides are transferred to the Golgi body, and outer core glycosylation occurs. The proteins are sorted into those which will be transferred to the vacuole and those which will be secreted.

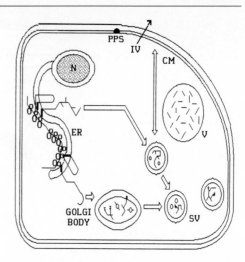

Fig. 13.3. Pathway of protein secretion in yeast. Polypeptide chains secreted on the ribosomes attached to the mRNA (polysome) pass through the endoplasmic reticulum to the Golgi body, where they are passed to the secretory vesicles. From there they are routed to the vacuole or to the periplasmic space, depending on their ultimate destination. *IV* Invaginations (in cytoplasmic membrane); *CN* cytoplasmic membrane; *ER* endoplasmic reticulum; *N* nucleus; *PPS* periplasmic space; *SV* secretory vesicle; *V* vacuole

6. The proteins to be secreted are packaged into secretory vesicles, which are transported to the inner surface of the cytoplasmic membrane.
7. The vesicles fuse with the membrane, and the proteins are released into the periplasmic space and may be diffused into the external medium.

S. cerevisiae does not normally secrete many proteins; invertase, acid phosphatase, killer toxin and mating-type pheromone (α-factor). Invertase and acid phosphatase are secreted only into the periplasmic space. Nevertheless, the signal sequences of all four of these proteins have been used to direct secretion of heterologous proteins from yeast.

Table 13.1. Polypeptides secreted from yeast using the a-factor pre-pro sequences as a promoter

Epidermal growth factor	β-Endorphin
Interferon	Interleukin-2
GM-CSF	Calcitonin
Insulin	Phospholipase A_2
Hirudin	Somatomedin C
Somatostatin	Prochymosin
Atrial natriuretic factor	α-Amylase
α-Amylase	β-Amylase
Invertase	

3.1 Pre-pro-α-Factor-Directed Secretion of Proteins

The MFα1 gene encodes the α-mating-type pheromone. It is first translated as a 165-amino acid precursor containing four copies of the α-factor itself, separated by spacer sequences, and preceded by a pre-pro signal sequence. The polypeptides are processed by three proteinases which release mature α-factor: yscF, encoded by the KEX2 gene, yscIV, a dipeptidyl aminopeptidase, encoded by the STE13 gene, and yscα, a carboxypeptidase encoded by the KEX1 gene. Processing may be rate-limiting to secretion during expression of heterologous proteins. Numerous peptides and proteins have been synthesized using the prepro-α-factor system (Table 13.1): the plasmid must be constructed correctly to ensure that the correct C- and N-termini are obtained. Calcitonin-glycine was secreted into the medium at levels reaching 97% of the total amount of the protein produced.

Some larger proteins using the α-factor leader sequence are not secreted correctly. An interferon was secreted but remained in the periplasmic space. Human gastric lipase was not efficiently secreted and the leader sequence was not properly removed. Nevertheless, a large Epstein-Barr virus envelope protein was reasonably well secreted, so large proteins can be expressed and secreted.

3.2 Use of Other Yeast Signal Sequences in Secretion

The signal sequences from the *Saccharomyces cerevisiae* killer toxin was used to direct synthesis of bacterial cellulase (endo-1,4-β-glucanase). The leader sequence of *Kluyveromyces lactis* killer toxin directed secretion of an interleukin-1β fragment, yeast acid phosphatase, α-interferon (but not tissue plasminogen activator), invertase, human interferon-α2 (which gave greater yields of the interferon than obtained using the natural sequence), calf prochymotrypsin, and α1-antitrypsin.

Secretion of heterologous proteins by their own leader sequences is possible, with varying results. Animal and plant signal sequences exist which function in yeast for secretion of their corresponding proteins. Wheat α-amylase has been expressed and secreted when its own signal peptide was used. The sweet plant protein thaumatin has been secreted when it was expressed as preprothaumatin. The fungal enzymes glucoamylase from *Aspergillus* and cellobiohydrolase from *Trichoderma reesei* were expressed and secreted under the control of their own signal sequences, and high levels of enzymes were recovered.

Carter's view (Piggott et al. 1990) is that the only function of the signal sequence is to insert the protein into the ER at the beginning of the secretory pathway. The actual secretion of the protein may be governed by the nature of the protein itself.

4 Yeast Artificial Chromosomes (YACs)

A new and valuable development. Plasmids, i.e., covalently closed circles of DNA, which may be chimeric creations able to replicate in both yeast and bacteria, are widely used to introduce heterologous genes into yeasts. DNA sequences which can be cloned into a plasmid are limited in size, and there are often problems of stability of the plasmid after being transformed into the yeast cell. Although yeast chromosomes are also unstable to a certain extent, they are far more stable than most plasmids. (Integrating plasmids become essentially part of the yeast genome.)

Problems of size and instability can be overcome to a great extent by using a construction having the same characteristics as normal yeast chromosomes; an artificially constructed yeast chromosome (Heslot and Gaillardin 1992). Such a structure must have the same structure as a naturally occurring chromosome: two telomeres, one attached to each end, a centromere, ARS sequences, and all other sequences necessary for the proper functioning of a normal yeast chromosome. If all of these structures are present, any structural genes on the chromosome should be expressed, and secreted if the elements required are present. The structure should replicate in the same way as the rest of the yeast chromosomes.

Such structures have now been synthesized. The initial structure is a circular, covalently closed plasmid, containing one centromere, ARS sequences, promoters, enhancers, and others, and the two telomeres, perhaps from *Tetrahymena* sp., with the outside ends "facing" each other and separated by a short sequence of DNA, having a restriction site (for BamH1, for instance) at each end of this throwaway sequence. The YAC vector contains a SUP4 gene having a Not1 cutting site at one end and an SfiI site at the other. The SUP4 gene contains an SmaI site which is used as the cloning site, and also a suppressor, so that cells expressing the suppressor are white, and unsuppressed cells, where foreign DNA is inserted at the smaI site and the SUP4 gene is disrupted, are red.

The vector obtained is digested with BamHI and SmaI, treated with alkaline phosphatase, so that three fragments are obtained, a right arm, a left arm, and a throwaway sequence. The two arms are ligated with the digested source DNA. The red colonies are tested for the presence of marker genes on the right (URA3) and left arms (TRP1). YACs containing 200–800 kb of human DNA have been constructed.

The recipient yeast, usually as a protoplast, is transformed with the linear DNA, which then functions as a normal chromosome. Within the yeast cell, it replicates at the same time and in the same way as the other chromosomes and any genes on it are expressed, if the proper control sequences are present.

The stability of the YACs increases as the length of the chromosome increases, short YACs being rather unstable. YACs can be used for storage and/or transfer of much larger DNA sequences than is possible in circular plasmids, and as model systems for study of chromosome segregation. So far, YACs have proven most useful for this function and for mapping of very large genes – as in the Human Genome Project. Cloning megabases of DNA in one vector greatly reduces the number of clones required to include the tremendous amount of genetic information in animal cell genomes.

4.1 Stabilization and Expression – Examples

The general method is illustrated by these examples. In the first, the gene for the killer toxin is included in the DNA sequences on the plasmid. This confers immunity to the toxin on cells carrying the plasmid. If the plasmid is lost, the cell becomes sensitive and is killed. In this way, the presence of the plasmid in 100% of the cells is ensured.

The cells can be stabilized by including a yeast centromere sequence in the plasmid, though the copy number is reduced to one.

The level of expression of the included gene can be increased by choice of the proper expression sequence, which gives efficient transcription and translation of the gene. Expression vectors from the yeast species, such as ADH1, CYC1, and PGK, encoding alcohol dehydrogenase, cytochrome c, and phosphoglycerokinase, usually give the best results. The nature of the foreign protein to be produced influences the yield. An episomal plasmid, copy number 50–100, carrying the PGK gene, may induce yields of phosphoglycerokinase of 80% of the cell protein. If the gene encoding interferon-α is included on the plasmid instead of the PGK gene, the percentage of heterologous protein may be less that 1%.

The gene encoding the desired protein may be switched on by including an upstream activation site in the plasmid. *Gal10* requires galactose to begin transcription from the linked promoter sequence. When the gene encoding human serum albumin was included on a plasmid transforming a brewing yeast, the yeast could be used to produce beer. After the fermentation, the yeast could be harvested, transferred to a galactose-containing medium, and the gene encoding HSA was expressed.

If the product was to be secreted, the DNA sequence encoding the amino acid signal sequence could be linked to the coding sequence of the desired gene, leading to secretion of the protein.

4.1.1 Yeast-Based Industries

These may be divided roughly into two groups: traditional, in which the product is the yeast cell or a metabolic product, and the new industries, which produce a "foreign", heterologous protein. This class includes the numerous hormonal pharmaceuticals such as insulin, interferon, somatostatin, tumor necrosis factor, and many others, plus numerous enzymes having industrial applications. The traditional industries, as well as the newer ones, can benefit from the application of the methods of molecular biology. Systems for control of such fermentations are important, as was discussed in the previous chapter (Perry et al. 1990).

Improvement of yeast strains for traditional industries begins with classical genetics, and merges into molecular biology. The techniques of crossing of known, haploid strains, carrying well-defined markers, sporulating the resulting hybrids, and analyzing the products of meiosis, becomes inadequate when applied to industrial yeast. These strains are often diploid or of higher ploidy; they may also be aneuploid, and seldom sporulate. If they produce spores, their viability is often low, only a few spores yielding new strains. Mating strains are difficult to obtain. If viable spores are obtained, they are usually homothallic and self-diploidize shortly after germination. They are often aberrant in other ways, sometimes having "defects" which lead to the formation of four spores of the same mating type, usually α in our own experience, or else being very slow to diploidize, so that some strains appear to be bisexual a few days after crossing with mating-type tester strains.

Baker's yeasts and distiller's yeasts have been more amenable to genetic modification than brewing strains. Nevertheless, investigators in the Carlsberg Laboratory in Copenhagen have made great progress in the genetic analysis of some of their strains normally used in production of high-grade lager beers.

Genetic manipulation of industrial yeast strains begins with the methods of "classical" genetics, and proceeds from there, through the use of rare-mating techniques, protoplast fusion, and its associated technique of fusion of yeast and fungal nuclei with yeast protoplasts, transformation of yeasts using chimeric plasmids, carrying genes for introduction into the recipient yeasts. Yeast artificial chromosomes (YACs), while of great importance in many aspects of fundamental research, have not as yet been used in modification of industrial yeasts. There are many specialized works on these subjects which discuss these methods.

A great many of the improvements in industrial strains of *S. cerevisiae* have been made by the methods of classical yeast genetics. Improvements in brewer's, distiller's, and baker's yeasts have concentrated on improved fermentation and/or growth rates, substrate utilization, and other aspects of the production process. Improvements in product quality in brewing yeasts, and in performance and flavor in the baking yeasts have been made using the methods of classical genetics.

A few examples of improvement of industrial yeasts, in the traditional yeast industries, beginning with the brewing industry are as follows:

4.1.2 Improvements of Brewing Yeasts (Hinchliffe and Vakeria 1989;
       Johnston 1990; Kielland-Brandt et al. 1983)

1. Improved substrate utilization. Genes encoding α-amylases, glucoamylases, and debranching enzyme have been cloned on to plasmids and brewing yeasts transformed with them. The transformed yeast strains could utilize maltotetraose, wort dextrins, and starch more efficiently. The amylases produced by *Schwanniomyces* spp. are useful for this.

2. Utilization of barley β-glucans. Microbial β-glucanases (endo-β-1,3-1,4-glucanase from *Bacillus subtilis* and the EG1 gene of *Trichoderma reesei*) have been cloned and transformed into production strains, giving good degradation of wort β-glucans and reduction of precipitates and hazes in the beer.

3. Construction of strains able to hydrolyze proteins. A gene from a wild yeast (*Yarrowia lipolytica*) encoding an acid protease was cloned and transformed into a brewing yeast. The enzyme degraded the beer proteins causing chill-haze. Unfortunately, the transformed yeast produced higher levels of diacetyl. *Saccharomycopsis fibuligera* and *Torulopsis magnoliae*, producing acid proteases, are being investigated as sources of genes encoding acid proteases.

4. Reduction of levels of vicinal diketones. These compounds, diacetyl and pentanedione, are produced via the isoleucine-valine pathway, and several strains forming lower levels of the compounds have been investigated. Mutants resistant to glucosamine produced lower levels of higher alcohols and esters. Mutants auxotrophic for isoleucine and valine, which lacked the enzymes threonine deaminase and α-acetolactate decarboxylase also produced reduced levels of propanol, butanol, diacetyl, and acetolactate. Strains carrying the ILV5 gene encoding the enzyme acetohydroxy acid reductoisomerase on a multicopy plasmid, produced a beer having a 60% reduction in levels of diacetyl and pentanedione.

Cloning the gene encoding α-acetolactate decarboxylase (ALDC) from *Enterobacter aerogenes* and *Aerobacter aerogenes* and introducing it into the yeast strain on a YEp vector may be more effective. The gene was also included in an expression cassette with PGK or ADC1 sequences in a YIp vector, and the gene was integrated into the yeast genome. Levels of vicinal diketones were reduced to negligible amounts, and the total production time was reduced from 5 to 2 weeks.

5. Other possibilities. These include increased ethanol tolerance, altered flocculation and adhesion, extended ranges of carbohydrate compounds utilized, and improved organoleptic properties. Improved ethanol tolerance is important in high-gravity fermentations, and is to a considerable extent dependent on an adequate nitrogen supply as free amino acids, and adequate supply of sterols and unsaturated lipids, and addition of the fermentable carbohydrate incrementally as the fermentation proceeds (Lonsane et al., this Vol.). Mutants having increased ethanol tolerance can be obtained in prolonged continuous culture. The mechanism of ethanol tolerance may involve genes encoding functions of the plasma membrane. Mitochondrial function is also related to both ethanol and temperature tolerance, both of which can be improved by transfer of mitochondria from a suitable donor.

Flocculation can be increased during brewing of lager beer using repitching with the residual yeast. This enriches the yeast in flocculent cells. FLO genes encoding this property have been mapped and cloned, but the study of the inheritance and control of flocculence is very difficult. The property is governed by polymeric genes and affected by the action of modifiers and suppressors (Stratford 1992, 1994). The control of adhesion to supports is also important, since the use of immobilized cells on solid supports has become common. The problems arising are similar to the factors influencing flocculation.

Genetic improvements of winery and distiller's yeasts are similar to those in modification of brewer's yeasts, and the methods are the same. Construction of strains suitable for use in manufacture of instant active dry yeast as starters in modern wineries is most important. These strains should give good growth and rapid fermentations. They should not introduce any undesirable flavors into the final product.

Requirements for a good baker's yeast are stringent (Evans 1990). This increases the difficulty of constructing improved yeast strains for the bakery. The yeast must grow rapidly and produce a good yield of biomass. It must also show good dough-raising properties in the bakery. It must have an active respiratory metabolism, i.e., a high content of cytochrome c and other cytochromes. It must also have a high fermentative activity when mixed into the dough. If it is produced for active or instant active dry yeast, it must remain active during drying and during mixing into the dough without previous growth in a "sponge". It should be resistant to freezing for mixing with doughs which are immediately frozen after mixing, and later thawed, allowed to rise, and baked. It should be osmotolerant and acid-tolerant for raising various types of specialist products. The task of the yeast geneticist constructing an improved strain of a baker's yeast is onerous and laborious.

There are few rapid test systems which can be used for screening strains of baker's yeasts. Laboratory determinations of growth rate and cell yield, acid- and osmotolerance give some indications of satisfactory strains. Good fermentative activity on maltose and low invertase activity, may indicate a potentially desirable strain. For the rest, good performance in doughs can be determined most reliably in a dough-raising test. Resistance to loss of viability during drying or freezing can be determined reliably by drying or freezing the yeast under conditions used in the bakery or kitchen. Statistical analysis of the results of a relatively large number of tests, by principal component analysis, may enable more accurate identification of desirable hybrids. Eventually, it is to be hoped that the actual characteristics which determine the nature of a good baker's yeast may be identified at the level of the genome and its regulation and expression by the methods of molecular biology, and that new strains may be constructed more easily and more rapidly by these methods.

The applications of molecular biology have brought great changes and improvements to the old and the new yeast-based industries. These achievements are the beginning, not the end, of the benefits of the discoveries in molecular biology. We do not stand on the last, great pinnacle of achievement in this field. We are probably on a ledge near the bottom of one of the nearer peaks of an almost unlimited range of future accomplishments.

We must not be misled into congratulating ourselves that the knowledge of the molecular biology of the yeasts is the be-all and end-all of our life with the yeasts. We will always need an understanding of the places and conditions in which yeasts live, their origin, evolution, and relationships, and their relationships with humans, if we are to profit from these relationships and the results of the metabolic activities of the simple unicellular fungi, the yeasts.

References

Conde J, Fink GR (1976) A mutation preventing nuclear fusion in yeast. Proc Natl Acad Sci USA 73:3651
Cregg JM, Tschopp JF, Stillman C et al. (1987) High level expression and efficient assembly of hepatitis B surface antigen in the methylotrophic yeast *Pichia pastoris*. Bio/Technology 5:479–484
Dutcher SK (1981) Internuclear transfer of genetic information in karl-1/KAR1 heterokaryons in *Saccharomyces cerevisiae*. Mol Cell Biol 1:245–253
Erratt JA, Nasim A (1986) Cloning and expression of a *Saccharomyces diastaticus* glucoamylase gene in *Saccharomyces cerevisiae* and *Schizosaccharomyces pombe*. J Bacteriol 166:484–490
Evans IH (1990) Yeast strains for baking: recent developments. In: Spencer JFT, Spencer DM (eds) Yeast technology. Springer, Berlin Heidelberg New York, pp 13–54

Gunge N, Nakatomi Y (1971) Genetic mechanisms of rare matings of the yeast *Saccharomyces cerevisiae* heterogeneous for mating type. Genetics 70:41–58

Heslot H, Gaillardin C (1992) Molecular biology and genetic engineering of yeasts. CRC Press, Boca Raton, pp 245–277

Hinchliffe E, Vakeria D (1989) Genetic manipulation of brewing yeasts. In: Walton EF, Yarranton GT (eds) Molecular and cell biology of yeasts. Blackie, London; Van Nostrand Reinhold, New York, pp 280–303

Hinnen A, Hicks JB, Fink GR (1978) Transformation of yeast. Proc Natl Acad Sci USA 75:1929–1933

Johnston JR (1990) Brewing and distilling yeasts. In: Spencer JFT, Spencer DM (eds) Yeast technology. Springer, Berlin Heidelberg New York, pp 55–104

Kavanagh K and Whittaker PA (1990) Formation of spheroplasts and protoplasts in the xylose-fermanting yeast *Pachysolen tannophilus* Biotechnol and Appl Biochem 12:57–62

Kavanagh K and Whittak PA (1996) Application of protoplast fusion to the nonconventional yeast-Enzyme Microb Technol 18:45–51

Kielland-Brandt MC, Nilsson-Tillgren T, Petersen JGL, Holmberg S, Gjermansen C (1983) Approaches to the genetic analysis and breeding of brewer's yeast. In: Spencer JFT, Spencer DM, Smith ARW (eds) Yeast genetics: fundamental and applied aspects. Springer, Berlin Heidelberg New York, pp 421–437

King DJ, Walton EF, Yarranton GT (1989) The production of proteins and peptides from *Saccharomyces cerevisiae*. In: Walton EF, Yarranton GT (eds) Molecular and cell biology of yeasts. Blackie, London; Van Nostrand Reinhold, New York, pp 107–133

Kozlov DG, Prahl N, Efremov BD, Peters L, Wambut R, Karpychev IV, Eldarov MA, Benevolensky SV (1995) Host cell properties and external pH affect proinsulin production by *Saccharomyces* yeast. Yeast 11(8):713–724

Mehta H, Gregory KF (1981) Mutants of *Saccharomyces cerevisiae* and *Candida utilis* with increased susceptibility to digestivve enzymes. Appl Environ Micdrobiol 41(4):992–999

Palleroni NJ (1962) Hybridization of homothallic yeasts by Chen's technique. Nature 195:1021

Perry BF, Miles RJ, Beezer AE (1990) Calorimetry for yeast fermentation and control. In: Spencer JFT, Spencer DM (eds) Yeast technology. Springer, Berlin Heidelberg New York, pp 276–347

Piggott JR, Doel SM, Goodey AR, Watson MEE, Carter BLA (1990) Heterologous protein producdtion from yeast. In: Spencer JFT, Spencer DM (eds) Yeast technology. Springer, Berlin Heidelberg New York, pp 366–386

Piper PW, Kirk N (1991) Inducing heterolgous gene expression in yeast as fermentations approach maximal biomass. In: Wiseman A (ed) Genetically engineered proteins and enzymes from yeast: production control. Ellis Horwood, Chichester, pp 147–184

Rothstein RJ (1983) One-step gene disruption in yeast. In: Wu R, Grossman L, Maldave K (eds) Recombinant DNA, Part C. Methods Enzymol 101:202

Spencer JFT, Spencer DM (1977) Hybridization of non-sporulating and weakly sporulating strains of industrial yeasts. J Inst Brew 83:287–289

Spencer JFT, Spencer DM (1988) Yeast genetics. In: Campbell I, Duffus JH (eds) Yeast, a practical approach, IRL Press, Oxford, pp 65–106

Spencer JFT, Bizeau C, Reynolds N, Spencer DM (1985) The use of mitochondrial mutants in hybridization of industrial yeast strains. VI. Characterization of the hybrid *Saccharomyces diastaticus* x *Saccharomyces rouxii* obtained by protoplast fusion and its behavior in simulated dough-raising tests. Curr Genet 9:649–652

Spencer JFT, Spencer DM, Reynolds N (1988) Effects of changes in the mitochondrial genome on the performance of baking yeasts. Antonie Leeuwenhoek J Microbiol Serol 55:83–93

Stratford M (1992) Yeast flocculation: a new perspective. Adv Microb Physiol 33:1–72

Stratford M (1994) Another brick in the wall? Recent developments concerning the yeast cell envelope. Yeast 10(13):1741–1752

Venkov P, Hadjiolov AA, Battaner E, Schlessinger D (1974) *Saccharomyces cerevisiae* sorbitol fragile mutants. Biochem Biophys Res Commun 56:599–604

Von Borstel RC (1990) Mutagenesis principles and strategies applied to yeast. In: Spencer JFT, Spencer DM (eds) Yeast Technology, Springer, Berlin Heidelberg New York, pp 355–365

Ward M (1984) Fusion of plant protoplasts with animal cells. In: Beers RF, Bassett EG (eds) Cell fusion: gene transfer and transformation, Raven Press, New York, pp 189–207

CHAPTER 14

Genetics and Molecular Biology of Methylotrophic Yeasts

E. Berardi

1 Introduction

At least 31 yeasts are able to utilize methanol as a sole carbon source (J.A. Barnett, pers. comm.). There is now much interest in these organisms for basic biological and ecological studies and practical applications. The development of genetic tools is advanced enough that three species are amenable to genetic analysis and recombinant DNA systems for them are known. Details of the methylotrophic yeasts are available (Table 14.1), including reviews on their genetics and molecular biology (Cregg 1986; Sibirny et al. 1988; Sudbery and Gleeson 1989). We discuss here current knowledge of fundamental and biotechnological problems.

2 General Features

Methylotrophic yeasts (MYs) are often isolated from soils and plants. Their numbers are higher in the presence of methoxy groups (e.g., lignin) and pectin (often assimilable by MYs), which liberate methanol on hydrolysis of methyl esters (Komagata 1991). Methanol is metabolized through a complex pathway, beginning in microbodies (Fig. 14.1).

MY fermentation is typically weak and sometimes occurs only under oxygen limitation (Van Dijken et al. 1986; J.A. Barnett et al. 1990; W.A. Scheffers, pers. comm.). These species are petite-negative organisms and glucose does not repress the respiratory enzymes, as in *Saccharomyces cerevisiae*.

Most MYs are ascomycetous and fall into the genera *Pichia* and *Candida* (Barnett et al. 1990). The methanol-utilizing species of *Hansenula* were recently reclassified to the genus *Pichia* by Kurtzman (1984), Considering DNA base composition immunological and biochemical reactions, Komagata (1991) divided the methanol-utilizing yeasts into four major groups.

The *Candida* species lack a sexual phase and are less suitable for genetic analysis and manipulation. They have been used for biochemical and physiological studies, only occasionally involving mutants and parasexual analysis (Sects. 4 and 5). *Pichia* species have a sexual phase and form asci and ascospores. So far, three species are being studied genetically; *P. pinus*[1], *P. pastoris* and *P. angusta* (formerly

---

[1] The form *pini*, used originally for the basionym *Zygosaccharomyces pini*, is also used (J.A. Barnett, pers. comm.).

J.F.T. Spencer/D.M. Spencer (eds)
Yeasts in Natural and Artificial Habitats
© Springer-Verlag Berlin Heidelberg 1997

Table 14.1. Reviews of methylotrophic yeasts

A General
Anthony C (1982) The biochemistry of methylotrophs. Metabolism in the methylotrophic yeasts. Academic Press, London, pp 268–295
Jensen TE, Corpe WA (1991) Ultrastructure of methylotrophic organisms. In: Goldberg I, Stefan Rokem J (eds) Biology of methylotrophs. Biotechnology series, vol 18. Butterworth-Heinemann, Boston, pp 39–75
Gleeson M, Sudbery PE (1988) The methylotrophic yeasts. Yeast 4:1–15
Komagata K (1991) Systematics of methylotrophic yeasts. In: Goldberg I, Stefan Rokem J (eds) Biology of methylotrophs. Biotechnology series, vol 18. Butterworth-Heinemann, Boston, pp 25–37
Large PJ (1983) Methylotrophy and methanogenesis. Physiology and biochemistry of methylotrophic yeasts. American Society for Microbiology, Washington, pp 57–64
Wegner H, Harder W (1986) Methylotrophic yeasts. In: Van Verseveld HW, Duine JA (eds) Microbial growth on C1 compounds. Proc 5th Int Symp Martinus Nijhoff, Dordrecht, pp 131–138

B Microbodies and related topics
Anthony C (1991) Assimilation of carbon by methylotrophs. In: Goldberg I, Stefan Rokem J (eds) Biology of methylotrophs. Biotechnology series, vol 18. Butterworth-Heinemann, Boston, pp 79–109
Borst P (1989) Peroxisome biogenesis revisited. Biochim Biophys Acta 1008:1–13
Fukui S, Tanaka A (1979) Yeast peroxisomes. Trends Biochem Sci: 246–249
Harder W (1990) Structure/function relationships in methylotrophic yeasts. FEMS Microbiol Rev 87:191–200
Lazarow PB, Fujiki Y (1985) Biogenesis of peroxisomes. Annu Rev Cell Biol 1:489–530
Lazarow PB, Moser HW (1989) Disorders of peroxisome biogenesis. In: Scriver CR, Beaudet AL, Sly WS, Valle D (eds) The metabolic basis of inherited disease, 6th edn. McGraw-Hill, New York, pp 1479–1509

Hansenula polymorpha). Chromosomal maps of *P. pinus* have been developed via classical genetic techniques. *P. pastoris* has been widely used for heterologous expressions and is proposed as a model organism for studies on peroxisomes. *P. angusta* is used as a model to investigate peroxisome function, biogenesis, and related topics; it is the yeast with the highest growth temperature (48–50 °C) and therefore could be a model for study of high temperature adaptation and related subjects. It is also used to study heavy metal/cell interactions (Berardi et al. 1990; Zoroddu et al. 1991) and for expression of heterologous genes.

The familiar name *Hansenula polymorpha* will be used in this chapter instead of the present official name of *Pichia angusta*.

3 Life Cycle

The methylotrophic *Pichia* species are homothallic haplodiplonts, usually found in nature as haploids. There are three cell types for each species: two haploid forms of opposite mating type, which may function as gametes, and one diploid form which can undergo meiosis and similar phenomena.

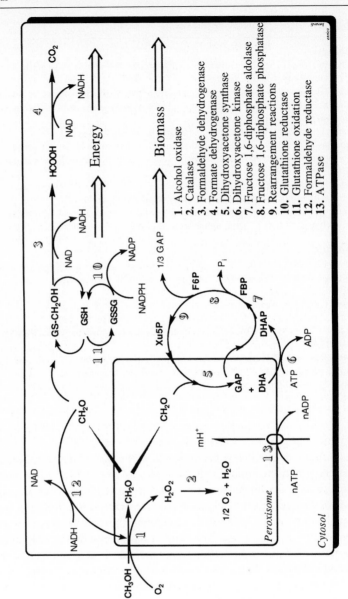

Fig. 14.1. Outline of the methanol metabolism in *H. polymorpha*. (After Van der Klei et al. 1991a,b)

Phase transitions are triggered by nutritional factors and can be controlled by the composition of the culture medium (Fig. 14.1); a haploid population containing both mating types remains in that state indefinitely in rich medium. Mating is induced by using a depleted medium. The resulting diploid phase can also be perpetuated in a rich medium. The haploid phase can again be obtained by using a sporulation medium. Clones derived from single spores can diploidize, so mating-type switching also takes place (Sudbery and Gleeson 1989).

4 Genetic Analysis

Nutritional control of mating and sporulation as above made possible the development of classical genetic techniques for analysis of MYs. A stable haploid phase enables a variety of mutants to be isolated and pairing of mutants cells to obtain diploids. Similarly, the ability to control the phases allows genetic markers to be followed through meiosis, which permits genetic analysis and linkage studies.

Out of 243 auxotrophic and temperature-sensitive (ts) mutants of *H. polymorpha*, isolated after random spore analysis, tetrad dissection, and complementation tests, 218 were assigned to 57 complementation groups (Gleeson et al. 1984; Gleeson and Sudbery 1988a,b). The remaining 25, which could not utilize methanol, were assigned to 5 complementation groups (De Koning et al. 1990b,c).

The first linkage map of this species obtained using these data included the genes *MUT2*, *MET6*, *ARG1*, and *ADE4*. More recently, genetic analysis of mutants unable to grow on methanol was also used to define various complementation groups (Fig. 14.1; Veenhuis et al. 1992).

Similarly, 84 auxotrophic mutants of *P. pinus* were assigned to 35 loci, by analyzing diploids which had undergone mitotic chromosome loss, coloss of a given pair of markers indicating linkage (Tolstorukov et al. 1977). Diploids of this species are fairly unstable, and lose chromosomes until stable haploids are formed. On rich medium, about 1% of the colonies have arisen from cells which had lost one or more chromosomes, and are easily recognizable by being smaller and slower-growing than normal. Gamma-irradiation of the cells increases the frequency of such colonies. In diploids heterozygotic for different marker genes, losses of chromosomes carrying dominant alleles are revealed by the expression of the recessive counterparts. Seventeen markers were assigned to four linkage groups and two fragments. Tetrad analyses of 164 crosses enabled 25 loci and 4 centromeres to be mapped (Tolstorukov et al. 1983; Tolstorukov and Efremov 1984). Later, 130 ethanol-negative mutants were assigned to 7 complementation groups (Tolstorukov et al. 1989).

Mutant isolation in *P. pastoris* has not been so extensive, but two collections of auxotrophic and methanol-negative phenotypes have been described (Gould et al. 1992; Liu et al. 1992). So far, a linkage map has not been constructed for this species (J.M. Cregg, pers. comm.).

Lahtchev et al. (1992) performed a parasexual analysis of *Candida boidinii*, based on spontaneous or induced mitotic segregation of markers in diploids obtained by protoplast fusion.

5 Transformation

If studies of gene isolation and characterization, chromosome engineering, and the expression of heterologous genes are desired, "foreign" DNA must enter the cell. High transformation frequencies allow gene isolation to be achieved via complementation assays.

The first report of transformation of an MY (*P. pastoris*) came from J.M. Cregg et al. (1985). Plasmid vectors containing the gene *HIS4* as a selectable marker were mixed with protoplasts of *P. pastoris*, auxotrophic for histidine, in the presence of polyethylene glycol (PEG) plus $CaCl_2$, and transformants were selected for histidine prototrophy. The efficiency was high, some plasmids yielding about 10^5 transformants per μg of plasmid DNA (Cregg et al. 1985).

Soon after, three groups succeeded in transforming *H. polymorpha*, also using complementing marker systems. They obtained transformation either by protoplasting the recipient strains, with approximate yields of 10^2–10^3/μg DNA, using PEG, or with LiCl, obtaining less than 10^2 transformants per μg DNA (Gleeson et al. 1986; Roggenkamp et al. 1986; Tikhomirova et al. 1986). The latter, similar to the one used for *S. cerevisiae*, was modified to yield 10^4–10^5 transformants per μg of DNA (Berardi and Thomas 1990). Electroporation of the cells is highly effective, yielding as many as 10^6/μg DNA (K.N. Faber, pers. comm.).

P. pinus cells have been transformed after treatment with LiCl, using a complementing marker system. Yields were 10^4 transformants per μg DNA (Tarutina and Tolstorukov 1992). *Candida boidinii* has been transformed with circular or linearized plasmids. The transformation efficiency depended on the strain, and was higher when linear DNA was used (Sakai et al. 1991).

Integrative and replicative transformation have both been demonstrated in methylotrophic yeasts. In the first, the transforming DNA becomes integrated in the host genome and is usually stable. In the second, the foreign DNA exists as an independent, mitotically unstable, replicating plasmid, in several copies.

Many aspects of plasmid behavior in methylotrophic yeasts resemble those already known for *S. cerevisiae*:

1. Plasmids which transform the host cells at low frequency usually integrate into the host genome.
2. Plasmids with sequences homologous to regions of the host genome also recombine homologously. In circular plasmids, there is a direct duplication of the sequence in question.
3. Plasmids transforming the host cells at high frequency usually contain an autonomous replicating sequence (ARS).
4. ARS-based plasmids are mitotically unstable and must be maintained under selective pressure.

However, replicating plasmids in methylotrophic yeasts may have characteristics which are different from those in *S. cerevisiae*.

1. In *P. pastoris*, plasmids harboring sequences homologous to regions of the genome may show high frequencies of integration, even though an ARS is present; homologous integration prevails over autonomous replication (Cregg et al. 1985). Possibly, this may be the result of a higher frequency of homologous recombination in this yeast than in *S. cerevisiae*. ARSs capable of maintaining plasmids autonomously must be relatively strong, considering the high growth rate of transformants under selection pressure and the low rate of plasmid loss in nonselective conditions. The time a replicating plasmid remains autonomous can be increased by using plasmids with low homology to the *P. pastoris* genome (by using heterologous selectable markers).

2. In *H. polymorpha*, plasmid multimerization may occur: During growth of transformants under selective conditions, replicating plasmids may exist in a head-to-tail configuration of monomers, leading to enhanced plasmid stability (Tikhomirova et al. 1986; Faber et al. 1992). This plasmid arrangement, and the consequent presence of more than one ARS per replicating molecule, may compensate for weak replicating sequences, improving growth under selective conditions. Or, multimers may compensate for low expression of heterologous marker genes. Multimeric integration of vector sequences may explain multiple integration events. Here, transformants are grown with no selection for several generations, and then stable prototrophic transformants are isolated. Although the vectors used have been designed for autonomous replication and bore ARSs, spontaneous integration occurs extremely frequently (Roggenkamp et al. 1986; Janowicz et al. 1991). In *H. polymorpha*, tandem, homologous integration of plasmids occurred more frequently than nonhomologous, single-copy integrations (Sierkstra et al. 1991).

It is not known whether multimer formation in *H. polymorpha* is restricted to particular strain/plasmid combinations. The factors governing them are not understood. Further investigation may develop a controllable system for generating useful multiple-copy transformants.

5.1 Homologous Integration

Integrating, single-copy vectors can be used to study stable, single-copy gene expressions (homologous and heterologous) and do chromosome engineering. The three aforementioned species of *Pichia* apparently undergo homologous recombination when foreign DNA with homology to regions of the host genome is introduced into the cells. Therefore, virtually any gene can be modified in vitro and replaced in the original location in the genome to investigate the effect on the phenotype, in vivo. This characteristic is shared by a few other organisms, primarily *S. cerevisiae*. It enables sophisticated manipulations of the chromosomes to be attempted.

These possibilities were first demonstrated in the methylotrophic yeast, *P. pastoris* (Cregg et al. 1989). The two alcohol oxidase coding sequences were disrupted in vitro and replaced at their respective loci, by cutting them within the flanking regions before transformation. As in *S. cerevisiae*, free dsDNA ends are highly recombinogenic and promote gene transplacements. Homologous recombination frequencies have reached 80%. A catalase-negative strain of *H. polymorpha* was constructed by gene transplacement (Didion and Roggenkamp 1992a,b).

Homologous integration also occurs in *H. polymorpha* and *Candida boidinii*. Beburov et al. (1990) targeted linear DNA fragments into the *H. polymorpha* genome, and obtained strains with disruptions in the AO and *HIS3* genes. Sakai and Tani (1992) described a model system for one-step gene disruption in *C. boidinii*. Targeted integration using replicating plasmids containing sequences sharing homology with regions in the genome only occasionally showed homologous integration of circular plasmids. Linearizing the plasmid by cutting within the homologous region resulted in an increase of homologous integration by 1–22%. Similar integration occurred in *S. cerevisiae* transformed with linearized replicating plasmids (Orr-Weaver et al. 1983). Lengthening of the homologous region in the plasmid increased the frequency of homologous integration. When there was no homologous integration, circular autonomous plasmids were found. This probably involved recombination-independent double-strand-breakage (DSB) repair, since DSBs in the nonhomologous regions of the plasmids were repaired as efficiently as those in the homologous regions. Homology-independent DSB repair also occurs in *H. polymorpha* (Gleeson et al. 1986). Although homology-independent DSB repair in *H. polymorpha* resembles that in *S. cerevisiae*, linearization of the plasmid always stimulated the transforming activity of replicating plasmids in *H. polymorpha*. In *S. cerevisiae*, linearization of this type of plasmid normally has a strong inhibitory effect on the frequency of transformation (Faber et al. 1992).

Although transforming with ARS-less fragments targeted into the chromosome should yield single-copy transformants, multicopy transformants are often formed in *P. pastoris*, indicating that events other than double crossovers occur (Clare et al. 1991; Fig. 14.2). Southern analyses and related determinations suggested that multicopy strains are not normally formed by multimerization followed by integration. Instead, in vivo intramolecular ligation of the fragment to be integrated apparently occurs, yielding circular forms which can then be repeatedly inserted into the chromosome by single crossover. Some ordinary, double-crossover transplacements were also observed, together with nondisruptive insertions at the homologous location on the chromosome. Insertions into different sites, homologous with other parts of the vector, were found.

Multiple integration also occurs in *H. polymorpha* transformed with replicating plasmids. Multicopy integrants could be obtained by growing transformants nonselectively for several generations, and then isolating any rare prototrophs found (Janowicz et al. 1991). The procedure eliminates cells carrying unstable

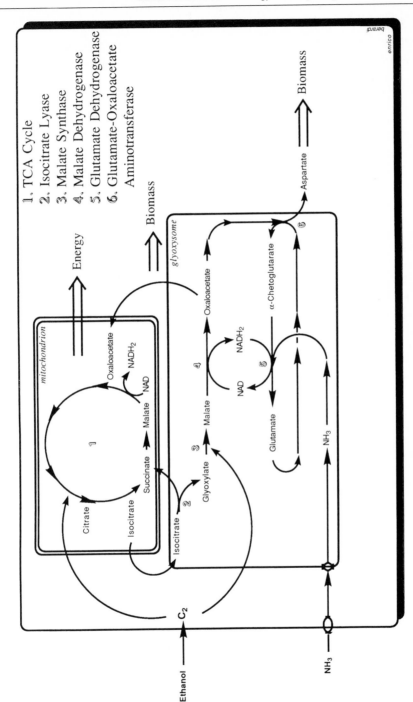

1. TCA Cycle
2. Isocitrate Lyase
3. Malate Synthase
4. Malate Dehydrogenase
5. Glutamate Dehydrogenase
6. Glutamate-Oxaloacetate
   Aminotransferase

Fig. 14.2. Outline of the ethanol metabolism in *H. polymorpha*. (After Sulter et al. 1991)

ARS-plasmids, so the surviving prototrophs were integrants. Plasmid copy numbers ranged from one to many. The mechanisms underlying these findings have not been elucidated. Stable integrants have been isolated from transformants harboring unstable plasmids, one of which carried three integrations (Fellinger et al. 1991). Two were tandemly inserted at the MOX locus (present in the plasmid) and the third somewhere else.

5.2 ARSs and Replicating Vectors

Replicating, multiple-copy vectors can be used to study homologous and heterologous, multicopy gene expression, and construct DNA libraries for isolation of genes by complementing host mutations.

Naturally occurring resident plasmids have not been found in methylotrophic yeasts. There were no DNA plasmids in 50 strains of *H. polymorpha* screened (E. Berardi, unpubl.). Development of episomal vectors such as those used in *S. cerevisiae* and *Kluyveromyces lactis* is not possible until such plasmids are found.

Sequences which confer autonomous replication ability and high transformation efficiency (*ARSs*) are usually required for establishing transformation systems. Normally, in methylotrophic yeasts the ability of an *ARS* to support autonomous replication and promote high transformation frequency in one host species is not strongly correlated with its *ARS* activity in another species. *ARSs* from *S. cerevisiae* do not function in *P. pastoris* and *H. polymorpha*, except that *ARS2* functions in the former species, and *ARS1* functions in the latter (Cregg et al. 1985; Roggenkamp et al. 1986).

Sequences with *ARS* activity in methylotrophic yeasts have been either discovered fortuitously, or have been isolated by their capacity of enhancing transformation frequencies. In the first case, the *S. cerevisiae LEU2* gene has *ARS* activity in *H. polymorpha*, *P. pastoris*, and *P. pinus*, and that the *cerevisiae HIS4* gene functions as an *ARS* in *P. pastoris* (Cregg et al. 1985; Berardi and Thomas 1990; Tarutina and Tolstorukov 1992). Two *ARSs* from *P. pastoris*, out of several isolated (*PARS1* and *PARS2*), have been characterized (Cregg et al. 1985). Their function was independent of other sequences and each shared some features with *ARSs* from *S. cerevisiae*, such as a high A + T content and homology with the *S. cerevisiae* consensus. They showed no *ARS* activity in *S. cerevisiae*. Two other *ARSs* have been isolated from *H. polymorpha*, *HARS1*, and *HARS2*, one of which has been sequenced (Roggenkamp et al. 1986). These, too, had high A + T contents and showed homology to the *S. cerevisiae ARS* consensus. Another 30 other *ARSs*, differing in stability, were isolated from *H. polymorpha*. Some of them supported the autonomous replication of plasmids in *S. cerevisiae* as well, and enabled shuttle vectors between *H. polymorpha* and *S. cerevisiae* to be constructed (E. Berardi, unpubl. data).

6 Gene Isolation

Genetic studies allow examination of the functional instructions of an organism, by isolating genes, characterizing their coding and controlling regions, and modifying the instructions they specify.

Genes of methylotrophic yeasts may be isolated by using several strategies:

1. Strategies based on gene products. Expression libraries can be screened using immunological methods if suitable antibodies are available.

2. Functional complementation. DNA from the source organism is cloned into a replicating vector to construct a gene library, and introduced into a mutant host with a known mutant phenotype. Transformants are tested for complementing DNA that restores the wild-type phenotype. Plasmids isolated from transformants and amplified in *E. coli* often carry the desired gene. This method results in both gene isolation and identification of gene function. It is widely utilized in investigations using *S. cerevisiae* and now for studies on methylotrophic yeasts. Efficient methods of transformation and replicating vectors are now available. The host strain and the gene under investigation are from the same species of organism. Heterologous complementation can be used if the gene functions in the recipient organism. The isolation of genes from methylotrophic yeasts by functional complementation of mutants of *S. cerevisiae* (Cregg et al. 1985) is extremely useful, since a great many mutants are available.

3. Strategies based on DNA homology. Gene products can be sequenced and the information obtained used to design suitable probes for screening of libraries. Heterologous probes can also be used at different stringencies, if the desired gene has some homology with them. The malate synthase and catalase genes from *H. polymorpha* have been isolated by these methods (Bruinenberg et al. 1990; Didion and Roggenkamp 1992a,b).

4. Strategies based on mRNA. Differential and subtractive screenings can be used to isolate differentially expressed genes. The dihydroxyacetone synthase gene of *H. polymorpha* and the two alcohol oxidase genes of *P. pastoris* have been isolated by this method (Cregg et al. 1985; Janowicz et al. 1985).

7 Microbodies

Microbodies are multifunctional organelles, bounded by a single membrane, found in all eukaryotes. Their importance in cellular metabolism was unrecognized until recently, when they were found to play a significant role in plants, fungi and mammals (Table 14.1b). The discovery of at least 11 human disorders related to aberrations in peroxisome function has stimulated interest in the biogenesis of microbodies and control of their proliferation. (Table 14.1b; Lazarow and Moser 1989; Moser et al. 1991).

Methylotrophic yeasts are excellent models for the study of microbodies. Their proliferation is easily controlled by means of the growth substrates, especially methanol. A great deal of physiological, biochemical, and cytological information is now available on structure-function relationships in microbodies and manipulation of growth conditions to control the degree of proliferation of the microbodies, and to obtain different enzymatic compositions. (Veenhuis and Harder 1991). The structure and function of microbodies has been studied in *H. polymorpha*. This species is an ideal eukaryote for investigation of microbody proliferation, biogenesis, assembly, maintenance, and regulation.

Microbodies invariably contain catalase. They may be specialized and contain other enzymes conferring functional identity. When they contain oxidases producing H_2O_2, they are referred to as peroxisomes, while if they participate in the glyoxalate cycle they are called glyoxysomes. If both sets of enzymes are present in the same organelle, the microbody is a glyoxyperoxisome.

7.1 Peroxisomes

Oxidases, producing H_2O_2, are either involved in carbon or nitrogen metabolism, and catalyze (usually) the initial reaction in the pathway.

Alcohol oxidase (AO). When MYs are grown on methanol, the cells contain high concentrations of a peroxisomal FAD-containing alcohol oxidase which uses O_2 as final oxygen acceptor, and catalyzes the oxidation of methanol to formaldehyde and hydrogen peroxide (Van der Klei et al. 1991a). Investigation of methanol-limited cultures of *H. polymorpha* showed that this matrix enzyme is an octameric molecule having identical subunits arranged in two alternating layers, which form a highly regular structure, crystalloid alcohol oxidase. The architecture of this crystalloid is cubical, with the unit cell containing six octamers in three mutually orthogonal orientations. Two holes, which can contain other perosixomal proteins, are connected by channels where the diffusion of relatively large molecules can take place (Vonck and Van Bruggen 1992). In the early stages of adaptation to methanol, protein monomers accumulate temporarily; in later stages, the pool of monomers is much smaller (Distel et al. 1988). In the presence of excess methanol, AO shows the so-called modification inactivation, which is associated neither with enzyme degradation nor with changes in the crystalline architecture of AO (Veenhuis and Harder 1991).

The amount of enzyme, and the features of the peroxisome (volume, numbers, shape) vary considerably according to growth conditions and the stage of development of the organelle. Maximum AO levels and peroxisome proliferation are observed in chemostat cultures at low dilution rates, when cells may contain more than 20 cuboid organelles, tightly packed and occupying a large proportion of the cell volume (up to 80%). These are crystalline peroxisomes.

The *H. polymorpha* AO gene (*MOX*) has been isolated and sequenced (Ledeboer et al. 1985). It has an intron-less coding region corresponding to a protein, 664 amino acids in length, having a putative FAD-binding motif and 7 potential protein-kinase-C phosphorylation sites. The gene is nuclear and present in one copy per haploid genome.

Pichia pastoris has two AOs, specified by two intron-less genes, *AOX1* and *AOX2*. Their coding regions are 1992 bp long and are 92% homologous (Ellis et al. 1985; Koutz et al. 1989). The predicted differences in the gene products involve amino acids having similar properties, resulting in the synthesis of very similar proteins having calculated molecular weights of 74 000. The contribution of each isoenzyme in supporting growth on methanol was determined (Cregg et al. 1989). First, mutants bearing disrupted *AOX* genes were grown using methanol as sole carbon source. Cells in which the *AOX1* gene was disrupted showed a two-thirds reduction of activity of AO, which increased the generation time considerably. If the *AOX2* gene was disrupted, the cells showed wild-type enzymatic activity and generation times. Mutants in which both genes were disrupted (double disruptants) had no AO activity and were unable to grow on methanol. Second, to determine whether the differences between the two genes were in the coding or the controlling regions, two strains having disrupted *AOX2* genes, and having at the *AOX1* locus either the *AOX1* coding region driven by the *AOX2* controlling regions or the *AOX2* promoter fused to the *AOX1* coding region. The promoter regions were responsible for the differences in behavior of the two strains. This was confirmed by Northern analysis of the *AOX* mRNAs, which showed much higher steady-state levels of the mRNA for *AOX1* as compared to *AOX2*. There are some selective advantages to the evolutionary conservation of two functional AOs in *P. pastoris*, but these have not been identified (Cregg et al. 1989).

The effect of gene dosage on peroxisomal import capacity was attempted after the *AO* genes were isolated by overexpression of the genes in both *H. polymorpha* and *Pichia pastoris*.

Overexpression of the cloned *MOX* gene in *H. polymorpha* was not achieved, and the contribution of the plasmid-borne gene, compared with the chromosomal one, could not be demonstrated, since the strain used had the wild-type *MOX* gene. (Roggenkamp et al. 1989). Out of 20 independently isolated strains deficient in alcohol oxidase, transformed with a high copy number, and replicating plasmid containing the *MOX* gene, only two showed *AO* overproduction (tenfold). Concomitantly, large, irregular peroxisomes were formed (Roggenkamp et al. 1989). These were very fragile when subjected to cell fractionation, and contained AO crystals which accounted for about two-thirds of the total cell protein. EM studies showed that increased growth of the crystals does not induce proliferation of the peroxisomes and may lead to their disruption. Whether these large peroxisomes retained their biological function was not determined, since the strains were catalase-negative (a defect used in isolation of the mutants, which impeded growth on methanol) and poor maters, so that meiotic segregation of catalase-positive haploids from +/– diploids was not possible. It is not known why there was overexpression of AO in only 2 out of 20 AO-defective mutants. Since all mutants arose spontaneously and a high percentage of strains (10%) was capable of overexpression, it is unlikely that additional mutants were present. The presence or absence of additional mutations could not be demonstrated genetically. The absence of mating prevented the use of genetic analysis to exclude segregation of the over-expression trait from other possible *MOX* lesions.

A strain of *P. pastoris* engineered to have one or two additional copies of the AO gene integrated in the genome was used to show an increase in AO activity (μmol H_2O_2/mg/min) from 0.85 to 1.47 (one additional copy) and to 1.89 (two additional copies) during growth in methanol (De Hoop et al. 1991b). Engineered cells showed the presence of immunoreactive material outside the peroxisome, often arranged as membrane-less, irregular, electron-dense structures. The organelle import capacity may not equal the AO oversynthesis.

Engineered forms of AO from *H. polymorpha* were used in *P. pastoris* to investigate the role of FAD binding on enzyme activity, octamerization, and intracellular location (De Hoop et al. 1991a). *MOX* was mutated in vitro at sites likely to be involved in FAD binding, and then inserted in the yeast genome to determine the relevant location, together with an evaluation of octamerization of the monomers in vitro. Four out of five amino acids tested appeared to be crucial for enzymatic activity. Some of the mutations were thought to prevent formation of an a-b-a-fold (Rosmann fold), impeding FAD binding via its ADP moiety. Others probably affected the actual binding of FAD by preventing formation of crucial hydrogen bonds. The mutations in the fifth amino acid (E42) allowed the production of functional enzyme, but appeared to destabilize the enzyme in vitro.

Acyl coenzyme a oxidase. This marker enzyme of peroxisomal fatty acid β-oxidation pathways was detected in *P. pastoris* growing on oleate (Liu et al. 1992). This pathway should be compared with similar systems in mammalian cells and *S. cerevisiae*. Utilization of more than one compound through the peroxisomal pathways may elucidate functional changes in the peroxisomes.

Enhanced levels of this and other enzymes of the β-oxidative systems were induced in *H. polymorpha* by oleate. This yeast is unable to use this compound. It responds by a rapid proliferation of membranes, resulting in the formation of multimembrane compartments, containing one or a few peroxisomes. Besides catalase, these organelles contained crystalline AO and thiolase only, of the β-oxidation pathway (Veenhuis et al. 1990).

Amine oxidase (AMO). Growth of *H. polymorpha* on primary alkylated amines (methylamine) as a nitrogen source induces a peroxisomal, copper-containing monoamine oxidase in high concentrations (Zwart et al. 1980; Veenhuis et al. 1983; Veenhuis and Harder 1991). This enzyme catalyzes the oxidation of methylamine to formaldehyde, ammonia, and hydrogen peroxide. Or it catalyzes the formation of acetaldehyde, ammonia, and hydrogen peroxide from ethylamine. Monoamine oxidase participates in the metabolism of poly-alkylated amines, probably via de-alkylations mediated by NADH-dependent mono-oxygenases, eventually yielding primary amines. These amines, in turn, are oxidized by AMO. Amine utilization is always accompanied by proliferation of peroxisomes. Their extent and characteristics depend on the growth conditions and the composition of the substrate. AMO and AO may coexist in the peroxisome, for instance when methanol + monomamine are present in the substrate, with the former utilized as a carbon source and the latter as a nitrogen source. AO occurs in the presence of methylated amines, although it has no apparent physiological role. Unless AO is present, AMO-containing peroxisomes are not crystalline.

The AMO gene from *H. polymorpha* includes an intron-less coding region corresponding to a protein of 77435 calculated molecular weight. The gene is nuclear and present in one copy per haploid genome (Bruinenberg et al. 1989).

Other oxidases. In methylotrophic yeasts, other nitrogen compounds induce proliferation of peroxisomes and enhanced levels of specific hydrogen peroxide-yielding oxidases, such as D-amino acid oxidase and urate oxidase (Veenhuis et al. 1983; Veenhuis and Harder 1991). The pathways have not been investigated by genetic methods.

Catalase. The hydrogen peroxide formed in the peroxisomes by oxidases is de-composed by a peroxisomal catalase the levels of enzyme increase with the levels of the enzymes producing H_2O_2. Catalase (CAT) probably moves freely through the spaces within the crystalloid structures (Veenhuis et al. 1983; Veenhuis and Harder 1991). The *Hansenula polymorpha* catalase is encoded by a 1521 bp ORF, corre-sponding to 507 amino acids. It has a high homology to other catalases (Didion and Roggenkamp 1992a,b).

Mutants lacking this enzyme cannot grow on methanol, and can only utilize this alcohol in the presence of another carbon source (Eggeling and Sahm 1980; Giuseppin et al. 1988a,b; Verduyn et al. 1988; Didion and Roggenkamp 1992a,b). This deficiency can be complemented by the peroxisomal catalase from *S. cerevisiae* (which is correctly processed and accumulated into peroxisomes), but not by the cytosolic catalase T, which is not imported into the microbodies (Hansen and Roggenkamp 1989). Microbody localization of this enzyme is essen-tial for growth on methanol. However, as discussed elsewhere, (cf. Sect. 7.5), CAT mislocation in methylotrophic yeasts does not abolish utilization of alkylated amines as nitrogen sources and similar functions (Sulter et al. 1990b). Other reducing systems in the cell (mitochondrial cytochrome-c peroxidase, CCP) may decompose the small amount of hydrogen peroxide formed.

A catalase-negative strain of *H. polymorpha* degraded hydrogen peroxide added to glucose-limited continuous cultures of this mutant (The enzyme was probably CCP, which was present at high activities.) Hydrogen peroxide may be an electron acceptor in the absence of O_2, in ethanol-grown cells (bypassing site III of the respiratory chain). This affects the metabolism of the cell adversely, so this finding is not very useful for estimating in vivo the relative contribution of this site to the overall energy metabolism of the yeast (Verduyn et al. 1991).

Transketolase. In MYs, formaldehyde formed in the peroxisomes by oxidation of methanol is assimilated via xylulose monophosphate cycles. Dihydroxyacetone synthase (DAS), the first enzyme in the pathway, is peroxisomal (Veenhuis et al. 1983; Veenhuis and Harder 1991). DAS is a special transketolase which transfers glycolaldehyde from xylulose-5-phosphate to formaldehyde. In *H. polymorpha* it is encoded by an intron-less gene corresponding to a protein of 702 amino acids, calculated MW 77 000 (Janowicz et al. 1985). The abundance of DAS in methanol-grown cells is probably the result of a high transcription rate, yielding up to 7% of the total poly-A mRNA (Roggenkamp et al. 1984; Janowicz et al. 1985). A DAS-negative mutant unable to grow on methanol as sole carbon source assimilates methanol in the presence of xylose under carbon-limiting conditions. The missing enzyme is probably replaced by the normal cytoplasmic transketolase (De Koning et al. 1990a).

In *P. pastoris*, clone P76, isolated by Ellis et al. (1985), contained the gene encoding DAS (Tschopp et al. 1987a). The mRNA which hybridized to it, translated in vitro, yielded a protein of MW 76 000 daltons.

Other enzymes. Enzymes such as α-hydroxyl-acid oxidase and other matrix enzymes may exhibit a low level of activity. Their metabolic role is obscure (Veenhuis and Harder 1991).

7.2 Glyoxysomes and Glyoxyperoxisomes

MYs have glyoxysomes (Zwart et al. 1983). Cells of *H. polymorpha* grown on methanol have key enzymes of the glyoxylate cycle, [isocitrate lyase (ICL) and malate synthase (MS)], in noncrystalline microbodies, along with catalase. The two other enzymes of the glyoxylate cycle, citrate synthase and aconitase, are in the mitochondrion, not the glyoxysome. Two other enzymes, malate dehydrogenase (MDH) and glutamate-oxaloacetate aminotransferase (GOT), are always present. Glyoxysomes are probably important in aspartate synthesis. MDH, unknown in other glyoxysomes, has a mitochondrial counterpart and catalyzes the formation of oxalacetate, used in gluconeogenesis as well as aspartate synthesis. *H. polymorpha* does not have a form of citrate synthase (CS) in the glyoxysomes. Electron microscopy shows that these organelles are closely associated with the mitochondria, so that there is probably a functional relationship. Unidirectional exchanges of isocitrate, succinate, and oxaloacetate interlock the metabolisms of the two organelles. If cells are grown on ethanol as a C source and methylamine as an N source, this induces both peroxisomal and glyoxysomal enzymes, which coexist in the same organelle (glyoxyperoxisome). Microbodies therefore respond functionally to nutritional changes (Zwart et al. 1983; Veenhuis and Harder 1991).

There are seven complementation groups of ethanol-negative mutants of *P. pinus*. The activities of the enzymes of the glyoxylate cycle, MS, ICL, and MDH were impaired. Mutants lacking acetyl-CoA synthase showed reduced activities of MS and ICL. Acetyl-CoA may be a positive effector, regulating enzymes of the glyoxylate cycle (Tolstorukov et al. 1989). The intracellular location of these enzymes has not been determined.

MAS from *H. polymorpha* is so far the only gene encoding an enzyme of the glyoxylate cycle which has been isolated. It contains an ORF encoding a protein of 555 amino acids, having a calculated MW of 63 254. The gene is homologous with all known MS genes (Bruinenberg et al. 1990).

Microbodies are unnecessary for C_2 metabolism. A peroxisome-deficient mutant which grows on ethanol but not on methanol has the enzymes usually found in the glyoxysomes in the cytosol (Sulter et al. 1991). The mutant shows reduced overall activity, particularly when ethanol is the limiting factor and the organism is grown, like wild-type cells, at high dilution rates in the chemostat. The mutant then accumulates glycogen, is unable to utilize the substrate completely, and shows decreased activities and amounts of alcohol dehydrogenase (ADH) and MS. Activity of MDH is increased, and of ICL decreased. The metabolic efficiency during growth on C_2 compounds in the absence of microbodies is reduced. Oxalacetate may be diverted to gluceonogenetic pathways rather than being converted to aspartate by GOT. This may account for the increased glycogen levels, as well as the activities of GOT and MDH when the mutant is grown at high cell density.

7.3 Membrane Proteins

The membrane proteins are involved in translocation and organelle assembly and other important functions. In *H. polymorpha* and *Candida boidinii*, most of these proteins are constitutive, the levels depending on growth conditions (Goodman et al. 1986; Sulter et al. 1990a).

Four genes encoding integral membrane proteins have been described; PMP70 from rat liver, PMP47 from *C. boidinii*, PAF-1 of CHO cells, and PAS3 from *S. cerevisiae* (Hohfeld et al. 1991). Their functions are still unknown.

7.4 Biogenesis

The biogenesis of microbodies is through direct transfer of proteins from the cytoplasm to the organelle, where they are integrated as membrane proteins or become components of the matrix. At least two groups of mechanisms are involved:

1. *Recognition mechanisms.* These operate through the interaction of topogenic signals (in the protein to be targeted) with receptor molecules on the cytoplasmic surface of the organelle.

Receptor molecules (or corresponding genes) have not yet been isolated in MYs, but a possible microbody receptor gene has been found in *S. cerevisiae* (Hohfeld et al. 1991).

There may be a cleavable leader sequence present, as in rat liver 3-ketoacyl-CoA thiolase (Hijikata et al. 1987), but most microbody proteins (MPs) have their topogenic information in the mature polypeptide sequence (Borst 1989). A group of microbody proteins (firefly luciferase and rat liver acyl-CoA) have a topogenic carboxy terminal signal apparently recognized by all eukaryotes and consisting of the tripeptide Ser-Lys-Leu-COOH (SKL) or of its conserved variants. If a cDNA luciferase clone, controlled by a yeast promoter, is expressed in *S. cerevisiae* and *H.*

polymorpha, peroxisomal translation is correct in both yeasts. A C-terminal segment of the peroxisomal protein PMP20 from *C. boidinii* functioned as a peroxisomal targeting signal in mammalian cells (Gould et al. 1990). Sometimes SKL is present, but not in regions involved in the targeting function. There are also some microbody proteins which do not contain the SKL motif or variants of it (Borst 1989). Many yeast MPs lack the SKL motif at the C-terminus, so different signals may be present. In *H. polymorpha*, the MAS topogenic region is unknown, and the SKL C-terminal motif is absent (Bruinenberg et al. 1990). The same applies to AMO; its unknown topogenic specification fails to target the enzyme to peroxisomes of *S. cerevisiae*, which shows that it is "nonuniversal". Instead, AMO occurs cytoplasmically as an active enzyme, at levels comparable to those found in *H. polymorpha*. Only C-terminal fusions of the SKL tripeptide force part of the enzyme into the heterologous microbodies. Peroxisomal AO and DAS from *H. polymorpha*, (also lacking the tripeptide motif) are routed into microbodies of *S. cerevisiae*, whether or not the growth conditions are appropriate for peroxisome proliferation. Therefore, the signals are recognized in the heterologous state (Distel et al. 1987). However, the AO found in *S. cerevisiae* microbodies does not assemble as active octamers, and DAS is present at low levels of activity.

2. *Import mechanisms.* Protein translocation across the microbody membrane may involve chaperones and transporters. None of these components has been identified in MYs, but biochemical evidence suggests that translocation into microbodies may require energy, as in other organelles. Energy can be provided either as high-energy bonds (ATP, GTP) or as an elecrochemical potential across the target membrane (Chap. 7; Membranes).

Hydrolysis of ATP is essential for import into microbodies (Borst 1989; Osumi and Fujiki 1990). ATP probably keeps the proteins translocation-competent, with floding after import. ATP-binding proteins concerned with microbody function (PAS1 from *S. cerevisiae*; Erdmann et al. 1991) and may be involved in the ATP-driven translocation of proteins and small molecules.

P^{32}-NMR studies have detected an in vivo pH gradient of approximately 1.2 across the microbody membrane (Nicolay et al. 1987), yielding an acidic peroxisomal environment (pH 5.8–6.0). This gradient is probably generated by a proton-translocating ATPase in the microbody membrane, since biochemical and immunochemical evidence indicates that *H. polymorpha* has a microbody membrane protein closely related to mitochondrial ATPase (Douma et al. 1988).

7.5 Microbody-Deficient Mutants

Microbody deficiency is being investigated in MY mutants (Cregg et al. 1990; Didion and Roggenkamp 1992a; Liu et al. 1992), and others, lacking microbodies (Zoeller and Raetz 1986; Brul et al. 1988; Erdmann et al. 1989; Zoeller et al. 1989; Tsukamoto et al. 1990; Shimozawa et al. 1992). Microbody deficiency is related to such serious human disorders as Zellweger's syndrome, a lethal autosomal recessive disease with complex clinical manifestations, and with fundamental biological

questions – organelle evolution, dispensability and relationships between structure and function. There is little information on yeast genes concerning these
mutants. Three sequences isolated from microbody-deficient mutants (MDMs) of
S. cerevisiae (PAS1, PAS2, and PAS3) encode a putative ATPase, a ubiquitin-
conjugating enzyme and a putative membrane receptor (Erdmann et al. 1991;
Wiebel and Kunau 1992; Hohfeld et al. 1991). Some essential component may be
altered or missing, in the reception, transport and unfolding/folding mechanisms
(Sect. 7.4). EM screening of methanol-negative MD mutants of H. polymorpha
yielded three classes of mutants (Cregg et al. 1990; Didion and Roggenkamp
1992a,b). Per⁻ had no visible microbodies, Pim⁻ had small organelles lacking most
of the matrix protein content, and Pss⁻ abnormal crystalline substrate (Veenhuis et
al. 1992). These mutants were not isolated as conditional lethals, so microbodies
are not vitally necessary for cell viability under all growth conditions. Unlinked
noncomplementathion was found between different per alleles, suggesting functional relationships between their gene products. These mutants abolished only
the methanol metabolism, while other microbody-associated pathways, such as
the glyoxysomal ones (Sect. 7.2) and those related to nitrogen metabolism, remain
functional, though the efficiency was sometimes reduced. Two H. polymorpha MD
mutants grow at rates comparable with wild-type if nitrogen is supplied as methylamine, ethylamine, or D-alanine (Sulter et al. 1990b). Both amine oxidase and D-
amino acid oxidase are correctly assembled and localized in the cytosol, where
they can form large protein aggregates close to the cell membrane.

Some MDMs, like normal wild-types, assembled active, octameric AO (Van der
Klei et al. 1991a), organized in one large crystalloid and having AO, CAT, and DAS
activities, located in the cytoplasm. Assembly of active crystalloids is independent
of protein transport across the microbody membrane. The other enzymic activities
of methanol metabolism were wild-type, and microbody membrane proteins were
present. Methanol was not metabolized, either alone or with glucose. Addition of
small amounts produced higher cell yields. Possession of intact microbodies is
crucial for C_1 metabolism. Possible explanations are impairment of formaldehyde
distribution between dissimilatory and assimilatory pathways, resulting in formation of S-hydroxymethylglutathione, the actual FDH (but not DAS) substrate,
leading to dissimilation, and alteration of H_2O_2 metabolism not mediated by catalase. Reduction of H_2O_2 concentration makes it unable to compete with cytochrome c peroxidase (CCP) or GSH-oxidizing systems which have a higher affinity
for this substrate (Van der Klei et al. 1991b). These systems consume reduction
equivalents (energy), which may explain the negative energy balance observed
with high concentrations of H_2O_2 (high concentrations of methanol in the medium). If rapid oxidation of GSH takes place, diverting free GSH from the dissimilation pathway may further reduce the efficiency of C_1 metabolism.

While these mutants resemble the MDMs of S. cerevisiae, (Erdmann et al. 1989),
others, from H. polymorpha and P. pastoris, are more like mutants of Chinese
hamster ovary cells and Zellweger fibroblasts. They do not show full activity of all
peroxisomal proteins in the cytoplasm. Such mutants were obtained from H.
polymorpha and P. pastoris.

Screening for mutants of *P. pastoris* was based on the ability to grow on methanol and oleic acid through pathways requiring the presence of the microbodies (Liu et al. 1992). Of ten methanol-negative and oleate-negative mutants, eight were microbody-deficient, as shown by electron microscopy and subcellular fractionation. Four other microbody-deficient mutants were obtained by similar procedures (Liu et al. 1992). These microbody-deficient mutants are currently in use for complementation experiments on gene isolation (M. Veenhuis, pers. comm.).

7.6 Regulation of Proliferation

Regulation of proliferation of microbodies in methylotrophic yeasts is complex, through growth and nutritional conditions. With catabolite repression, the mechanisms regulate the extent, volume fraction and organelle structure during proliferation. All are highly specific for different growth conditions (Veenhuis et al. 1983; Veenhuis and Harder 1991). Some components of the organelles are always present. Their amount varies only with the extent of proliferation. Others depend on the specific conditions and increase with proliferation. Certain constituents, such as catalase, are apparently regulated by more complex mechanisms. The content and function of the different enzymes in the microbodies appear to be independent.

Microbodies may arise from preexisting organelles by increase in volume and subsequent division. Methylotrophic yeast cells growing exponentially in glucose contain a single small microbody situated next to the cell wall. Compounds assimilated via oxidation in the peroxisomes, such as methanol and methylamine, induce proliferation. Alternative substrates (glucose and ammonia), often repress it, even in the presence of the inducer. Proliferation of glyoxysomes is induced by C_2 substrates, which are assimilated via glyoxalate and repressed by glucose. C_2 substrates repress proliferation of peroxisomes, which is induced by C_1 substrates. The order of preference for substrates is: (1) glucose, (2) ethanol, (3) methanol. Glyoxyperoxisomes proliferate when C_2 compounds are assimilated together with organic compounds which must be assimilated through the peroxisome system. Gratuitous induction of "C_1" peroxisome proliferation may occur during growth of *H. polymorpha* at low concentrations of glucose (but not of ethanol), or in the presence of nonrepressing substances (glycerol), even at high concentrations. Gratuitous induction may be strain dependent or species-specific. When gratuitous induction occurs, repression/derepression mechanisms rather than induction govern proliferation. Sensitivity to repression and induction are species- and even strain-dependent. Under simultaneous limitation of nitrogen and carbon, proliferation is also repressing. In conditions such as excess of methanol, AO shows the so-called modification inactivation. This process is not associated with either degradation of an enzyme, or changes in the crystalline architecture of AO.

Several classes of regulatory mutants of peroxisome proliferation have been isolated in MYs. Five complementation groups of mutants impaired in ethanol repression (ECR1), with defective glucose repression or showing partial or total constitutive synthesis of peroxisomal enzymes (CGR1, CGR2, CGR3, CGR4), have been isolated in *Pichia pinus* (Sibirny et al. 1987; Titorenko et al. 1990). Some of the CGR2 and CGR3 mutants showed impairments in catabolite repression triggered by dihydroxyacetone and malate. Whether the phenotypes are due to the presence of additional mutations is not known. The mutation *ecr1* impedes repression by ethanol of peroxisomal enzymes, and causes methanol-triggered repression of the key glyoxysomal enzymes. The allelic state of this gene may drive microbody biogenesis towards a peroxisomal or a glyoxysomal state (Sibirny et al. 1991). The monogenic recessive *adh1* lowers ADH activity and alters ethanol catabolite repression of C_1 peroxisomes (Sibirny et al. 1991).

In *P. pinus*, C_1-induced peroxisome proliferation is repressed by genetically distinct mechanisms. Four classes of carbon substrates repress the induction of AOX by methanol, probably through independent mechanisms (Sibirny et al. 1988).

In *H. polymorpha*, glucose and ethanol repress C_1 peroxisomes through distinct mechanisms (at least in part). Regulatory mutants exist, which retain their capacity for repression by ethanol, but not by glucose, of MOX transcription (C. Stanway and E. Berardi, unpubl. data). There are pleiotrophic mutants which are probably regulatory mutants with defective derepression, since they are unable to grow in methanol or on other carbon sources (ethanol, glycerol, maltose, and dihydroxyacetone; Sudbery and Gleeson 1989). They comprise three complementation groups and serve to isolate revertants having derepressed phenotypes in the presence of glucose analogs, independently of the substrate utilized. Therefore, the mechanisms governing glucose repression on different substrates have common steps.

Microbody proliferation is probably controlled by transcriptional regulation (Roa and Blobel 1983). The amount of mRNA from batch-grown cells parallels that of the protein under induction and repression conditions. (Roggenkamp et al. 1984). The AO gene is transcriptionally controlled and its regulation involves both repression-derepression and induction (C. Stanway and E. Berardi, unpubl.). In carbon-limited continuous cultures, using 4:1 mixtures of glucose and methanol, AO is formed at dilution rates (D) of <0.3/h and is then repressed by increasing levels of residual glucose. The amount of AO was relatively high and constant, from dilution rates of 0.05 to 0.14/h but then declined (Giuseppin et al. 1988a,b). Its activity in cell-free extracts peaked at rates of 0.1–0.15/h. These variations in AO activity had little effect on the rate of conversion of methanol or on the cell yield. Determination of the FAD/AO-octamer ratio, the levels of active AO, inactive AO, and AO messenger showed that the level of AO messenger increased with D up to approximately 0.15/h, and then remained high up to D = 0.3/h. Messenger levels correlated with AO activity only at low D values. Transcription may be important in controlling the formation of active AO at low values of D. At high D values other stages in synthesis, such as FAD incorporation and posttranscriptional events, may be crucial.

However, *AOX1*, *AOX2*, and DAS genes are regulated at the transcriptional level by induction and repression/derepression mechanisms (Tschopp et al. 1987a). All three genes are strongly repressed by glucose and induced by methanol. *AOX1* was activated on carbon starvation, but the DAS gene and *AOX2* are not.

The control of single peroxisomal proteins which are highly expressed does not account for all of microbody proliferation. If AO is expressed constitutively, glucose-grown cells assembled an active peroxisomal crystalline enzyme correctly. Proliferation did not occur, but the existing peroxisome enlarged, showing that the two events were governed separately (Distel et al. 1988). The AO import mechanism was controlled independently of proliferation. The components may be constitutively present in the cell, or they may be induced by the presence of AO. Investigation of the DAS gene heterologously expressed in *S. cerevisiae* gave similar results (Godecke et al. 1989).

Adaptation of MYs to conditions where microbody enzymes or whole microbodies are metabolically redundant involves elimination of the enzyme activity and of the microbodies themselves (Veenhuis et al. 1983; Veenhuis and Harder 1991). Most of the microbody enzymes studied in MYs are diluted out by new cell formation. However, AO-containing microbodies undergo selective inactivation triggered by glucose or ethanol. There is a rapid proteolytic degradation of individual organelles, which involves a vacuole-aided autophagic process, causing a complete elimination of the microbody enzymes. Neither formation of crystalloids nor their degradation occurs outside the microbody (Van der Klei et al. 1991b). Therefore, degradative inactivation is not directed against a single microbody protein but against the whole microbody. Possibly, degradation is mediated by microbody tagging, through phosphorylation of membrane proteins or ubiquitination. Mutations of a gene from *S. cerevisiae* impair microbody assembly. The gene is highly homologous to genes encoding enzymes which mediate ubiquitination (Wiebel and Kunau 1992).

8 Applications

MYs are of academic interest and increasingly important to biotechnology (Wegner 1990). They can grow in cheap substrates (methanol) and convert sugars with high efficiency. They are insensitive to glucose effects on the respiratory enzymes, so can be grown in media of high sugar concentrations and still produce high rates of conversion to biomass. Certain MY production processes are more efficient if continuous cultures are used. Mixed substrates can be used without interference from the diauxic effect. This increases the operative flexibility of these organisms. In particular, *H. polymorpha*, being thermotolerant, offers higher growth rates and reduced cultivation times, better cooling management, less viscous culture media, and fewer problems of contamination.

Potential and actual industrial applications of MYs include production of biomass, enzymes, metabolites, reducing power, and heterologous proteins.

8.1 Traditional Processes

Biomass (SCP) production from MYs (feed/food protein) is possible by high-cell-density processes (Phillips Petroleum Company), yielding up to 150 g dry cell weight/l, and utilizes a strain specifically selected for the process (Wegner 1983; Wegner and Harder 1986).

Mutagenesis has been widely used for strain improvement in processes using MYs. Mutants defective in glycerol catabolism could be used for development of glycerol-producing or dihydroxyacetone-producing fermentations (Kato et al. 1986; De Koning et al. 1987, 1990b,c). Metabolic blocks interrupt the flow of carbon through the cyclic assimilation pathway, driving DAS towards the production of DHA and GAP. Mutants blocked in glycerol kinase and dihydroxyacetone kinase accumulate trioses if xylose replaces xylulose-5-phosphate as a C_1-acceptor. Methionine production in *C. boidinii* has been improved using an ethionine-resistant mutant (Lim and Tani 1988), and branched-chain amino acid excretion has been achieved with retroinhibition-resistant mutants of *H. polymorpha* (Titorenko and Trotsenko 1983).

AO is an enzyme useful for bleaching, organic synthesis, and analytical probing. Catalase-negative mutants eliminate the need for removal of catalase, a major process contaminant (Giuseppin et al. 1988a,b). Glucose is used as a carbon source and formaldehyde (or formate) as an AO inducer, to avoid H_2O_2 formation. Mutants with increased AO activity have been desribed as well (see Sect. 7.1). These could be used in the production of formaldehyde from methanol. Catalase-enhanced mutants and strains insensitive to catabolite repression can both be used for this.

8.2 Foreign Gene Expression

An enzyme such as AO may constitute up to 40% of the cellular protein content (Giuseppin 1988). Heterologous protein production could become a commercial process, with highly transcribed regulatory sequences controlling the expression of a desired gene. Better understanding of the mechanisms of import into the microbodies may be required for control of production of heterologous proteins.

In any typical system for production of heterologous proteins, the required coding sequence is inserted into an expression cassettee, between the 5′ and 3′ controlling regions of a suitable gene. If secretion of the protein is required, a secretion sequence is added by in-frame fusion with the coding sequence. This construct is introduced on a vector into the host genome, usually by stable integration.

The first MY-based system for production of heterologous proteins was the result of joint projects between SIBIA and the Phillips Petroleum Corporation. The required genes were placed under the control of the highly expressed, tightly regulated *AOX1* promoter and on integrating vectors having high mitotic stability.

After the laboratory results (Tschopp et al. 1987a), high-level production of hepatitis B surface antigen (HBsAG) was attained (Cregg et al. 1987). Biomass was produced using glycerol as carbon source, and then heterologous expression was induced with methanol. High-level expression was obtained from strains containing a single expression cassette. Subsequently, unglycosylated HBsAG as properly assembled 22-nm particles was obtained, in yields of 90 g of particles from one run in a 240-l fermentor. The particles resembled those derived from human serum, which are immunogenic and used for actual vaccine production.

Later, efficient production of *S. cerevisiae* invertase in *P. pastoris* was reported – the first heterologous secretion in a MY (Tschopp et al. 1987b). The foreign gene, with its own leader sequence, was placed under the control of the *AOX1* promoter. Secretion of properly processed, glycosylated, active enzyme was obtained by induction of the yeast with methanol. Yields were up to 2.5 g/l, far higher than those reported for *S. cerevisiae*. These yields accounted for 80–90% of the total secreted protein, and only a small proportion of the total invertase synthesized remained in the periplasm, but the rate of secretion was low. This slow secretion is not a general characteristic of *P. pastoris* (Tschopp et al. 1987b).

The invertase produced by *P. pastoris* is less glycosylated than the natural *S. cerevisiae* enzyme, due to differences in the size of the N-asparagine-linked oligosaccharides (about eight mannose residues per chain for *P. pastoris*; longer for *S. cerevisiae*). The *sec18* phenotypes of *S. cerevisiae* and the large molecules of mannose-containing oligosaccharides of higher eukaryotes are similar.

The high yield of assembled HBsAG and the high yields of invertase were related to the use of an *aox1* strain which grew on methanol rather slowly, because of its less effective AO, encoded by *AOX2*. These *aox1* strains have advantages over *AOX1* strains. They are less exacting in their requirements, and scale-up from the laboratory to production is easier. However, *aox1* strains grow slowly on methanol. They require relatively long fermentation times and show low productivity (product/l/h). Therefore, *AOX1* strains (able to grow normally on methanol) were used for the first process for production of heterologous proteins in continuous culture (Digan et al. 1989). Biomass was produced using glycerol as carbon source, without methanol, slow induction was done by fed-batch additions of trace salts and increasing amounts of methanol, until full continuous production was begun, using a complete methanol medium. Growth was methanol-limited, and a mean cell residence time of 20 h was attained. The gene encoding bovine lysozyme (cDNA, with its own signal sequence, fused to the *AOX1* controlling sequences) was introduced into *P. pastoris* carrying an intact *AOX1* gene. Bovine lysozyme was produced in high volumetric yield throughout the fermentation, five times faster and in 30% less (v/v) fermentation volume than was required with *aox1* strains. Product concentration was >550 mg/l in the fed-batch phase and about 350 mg/l in the continuous phase. Volumetric productivity was approximately 13 mg/l/h. Thus, in 200 h of fermentation in a volume of 8 l, about 20 g of lysozyme was obtained, excluding that obtained during the phase of growth on glycerol.

Expression of tetanus toxin fragment-C, controlled by an *AOX1* cassette present at one copy per cell, was not affected by the chromosomal integration site or by the type of integrant, or by the ability of the strain to utilize methanol (Clare et al. 1991). Multicopy cassettes were required for high-level expression, which was shown with transplacement strains (Fig. 14.2). Expression levels of up to 12 g/l of fragment C were obtained in fed-batch fermentations, and accounted for as much as 27% of the total cell protein. The heterologous fragment C was as effective as the "natural" protein in immunizing test animals. The "*Pichia* yeast expression system*", by the Phillips Petroleum Company, is now widely used to produce many heterologous proteins of diverse origin.

Hansenula polymorpha has also been used successfully for expression of heterologous genes for production of "foreign" proteins, using two different methanol-regulated promoters, obtained from the MOX and FMD, formate dehydrogenase genes. The process resembles that used for *P. pastoris*; the production of biomass is followed by a strong methanol-triggered induction. Pre-S2-HBsAg was produced and assembled as 22-nm subviral particles and secreted across the plasma membrane into the periplasm (Shen et al. 1989). The subviral particles were released into the culture medium by permeabilizing the cell wall with β-1,3-glucanase. Pre-S2-HBsAG is transported and assembled as in mammals. The overall productivity of the synthesis and the ease of purifying the particles promise well for commercial application. The process was optimized to the fermentor scale. Biomass (35–40 g/l dry weight) was produced using glycerol as carbon source. This was then induced by changing the carbon source to methanol (de Roubin et al. 1991). The addition of yeast extract or other carbon source greatly increased the yield of antigen, which was only slightly less than using *S. cerevisiae*.

The coding region of the glucoamylase (and its leader sequence) from *Schwanniomyces occidentalis* was efficiently expressed and secreted under the control of the FMD promoter (Gellissen et al. 1991). Transformants carrying one to eight copies of integrated plasmids were isolated. The best yield of glucoamylase was obtained with a four-copy integrant, reaching 1.4 g/l of active enzyme, at cell densities of 100–130 g dry weight/l.

Balanced coexpression of S and L HBsAgs was achieved using strains carrying integrated copies of genes encoding both S and L antigens, controlled by the MOX promoter (Janowicz et al. 1991). Various transformants carrying different numbers of L and S cassettes yielded total amounts of L + S antigens ranging from 0.3 to 4.5 mg/100 mg toal protein. L/S ratios varied between 1:1 and 1:15. *H. polymorpha*, having a unique integration mechanism, yielded transformants with several L cassettes and up to 50 S cassettes, stably integrated in the genome.

The α-galactosidase from guar, a commercially valuable plant enzyme involved in seed germination, has been expressed in *H. polymorpha* (Fellinger et al. 1991). A construct containing an *S. cerevisiae* secretion sequence, fused to the MOX promoter, was stably integrated in the genome, and a three-copy integrant was used for determination of expression (see Sect. 5.1). Overall expression in shaken cultures reached 55% of the total protein. More than 85% of the enzyme was secreted after correct processing of the signal sequence. However, overglycosylation (9.5% as compared to <5% in the native enzyme) reduced the specific activity of the enzyme considerably. Enzymatic removal of the sugar groups restored the specific activity completely. Subsequently, a strain was developed which expressed the guar α-galactosidase gene more efficiently, giving yields of 22.4 mg/g dry cell weight, corresponding to more than 13% of soluble cell protein (Veale et al. 1992).

The many favorable properties of MYs and the many examples of high-level expressions described here and elsewhere make them highly attractive for production of foreign proteins in research as well as in industrial processes (Ratner 1989; Wegner 1990; Buckholtz and Gleeson 1991; Tschopp and Cregg 1991; Gellissen et al. 1992; Romanos et al. 1992).

9 General Perspectives

It has been shown here that MYs are interesting organisms for research and commercial application. The sophisticated genetic systems now available have made it possible to expand our knowledge considerably on topics as diverse as organelle biosynthesis, gene regulation, and metabolic pathways. Genetic approaches are increasingly valuable in the elucidation of the molecular mechanisms underlying the biogenesis, assembly, and maintenance of microbodies. Having MD mutants in microorganisms readily amenable to genetic procedures enables genes to be isolated by complementation experiments. These genes allow direct information to be obtained on the coding and controlling functions. Moreover, it is now possible to verify different hypotheses and models in vivo by chromosome engineering.

As was previously shown, some of the microbody-borne pathways are also functional in the cytoplasm, raising questions concerning microbody function and evolutionary origin. Microbody proliferation is the result of the interplay of numerous intricate phenomena. Construction of functioning models of biogenesis and assembly will require a vast amount of information on the regulatory mechanisms involved. In no case has the nature of the effectors been elucidated, nor have detailed transcription studies and analysis of controlling regions been reported.

H. polymorpha holds a leading role in studies of microbodies, but *P. pastoris* is the most highly developed methylotrophic yeast for heterologous expression. Both species are increasingly used for academic and technological purposes. This initial stage of biotechnological advancement, successfully demonstrating that industrial processes based on MYs work well, has been made possible by the dedicated work of several pioneers. It is hoped that more and more research will be carried out, to further the understanding and applications of these yeast.

Acknowledgments. The author wishes to thank C. Stanway for critical reading of the manuscript and J. Cregg, K.N. Faber, P. Haima, W. Scheffers, C. Stanway, P. Sudbery, V. Titorenko, I. Tolstorukov, and M. Veenhuis for sharing unpublished results. Work in the laboratory of the author is supported by C.N.R., the National Research Council of Italy.

References

Barnett JA, Paine RW, Yarrow D (1990) Yeasts. Characteristics and identification. Cambridge University Press, Cambridge

Beburov MY, Zlochewsky ML, Michailover VM, Semenova VD, Gracheva VD, Lahtchev K (1990) Transformation of methylotrophic yeast *Hansenula polymorpha* with linear DNA fragments. Yeast 6:S118

Berardi E, Thomas DY (1990) An effective transformation method for *Hansenula polymorpha*. Curr Genet 18:169-170

Berardi E, Meloni MG, Bonomo RP, Zoroddu MA (1990) Electron paramagnetic resonance study of retention of chromium (III), chromate or dichromate(VI) and copper(II) ions by thermotolerant *Hoansenula polymorpha*. J Chem Soc Faraday Trans 86:2579-2582

Borst P (1989) Peroxisome biogenesis revisited. Biochem Biophys Acta 1008:1-13

Bruinenberg PG, Evers M, Waterham HR, Kuipers J, Arnberg AC, Ab G (1989) Cloning and sequencing of the peroxisomal amine oxidase gene from *Hansenula polymorpha*. Biochim Biophys Acta 1008:157-167

Bruinenberg PG, Blaauw M, Kazemier B, Ab G (1990) Cloning and sequencing of the malate sythase gene from *Hansenula polymorpha*. Yeast 6:245-254

Brul S, Westerveld A, Strijland A, Wanders RJA, Schram AW, Heymans HSA, Schutgens RBH, Van den Bosch H, Tager JM (1988) Genetic heterogeneity in the cerebrohepatorenal (Zellweger) syndrome and other inherited disorders with a generalized impairment of peroxisomal functions. J Clin Invest 81:1710-1715

Buckholtz RG, Gleeson MAG (1991) Yeast systems for the commercial production of heterologous proteins. Bio/Technology 9:1067-1972

Clare JJ, Rayment FB, Ballantine SP, Sreekrishna K, Romanos MA (1991) High-level expression of tetanus toxin fragment C in *Pichia pastoris* strains containing multiple tandem integrations of the gene. Bio/Technology 9:455-460

Cregg JM (1986) Genetics of methylotrophic yeasts. In: Van Verseveld HW, Duine JA (eds) Microbial growth on C1 compounds. Proc 5th Int Symp. Martinus Nijhoff, Dordrecht, pp 158-167

Cregg JM, Barringer KJ, Hessler AY, Madden KR (1985) *Pichia pastoris* as a host system for transformations. Mol Cell Biol 5:3376-3385

Cregg JM, Tschopp JF, Stillman C, Siegel R, Akong M, Craig WS, Buckholz RG, Madden KR, Kellaris PA, Davis GR, Smiley BL, Cruze J, Torregrossa R, Velicelebi G, Thill GP (1987) High-level expression and efficient assembly of hepatitis B surface antigen in the methylotrophic yeast, *Pichia pastoris*. Bio/Technology 5:479-485

Cregg JM, Madden KR, Barringer KJ, Thill GP, Stillman CA (1989) Functional characterization of the two alcohol oxidase genes from the yeast *Pichia pastoris*. Mol Cell Biol 9:1316-1323

Cregg JM, van der Klei IJ, Sulter GJ, Veenhuis M, Harder W (1990) Peroxisome-deficient mutants of *Hansenula polymorpha*. Yeast 6:87-97

De Hoop MJ, Asgeirsdottir S, Blaauw M, Veenhuis M, Cregg J, Gleeson M, Ab G (1991a) Mutations in the FAD-binding fold of alcohol oxidase from *Hansenula polymorpha*. Protein Engin 4:821-829

De Hoop MJ, Cregg J, Keizer-Gunnink I, Sjollema K, Veenhuis M, Ab G (1991b) Overexpression of alcohol oxidase in *Pichia pastoris*. FEBS Lett 291:299-302

De Hoop MJ, Valkema R, Kienhuis CBM, Hoyer MA, Ab G (1992) The peroxisomal import signal of amine oxidase from the yeast *Hansenula polymorpha* is not universal. Yeast 8:243-252

De Koning W, Gleeson MAG, Harder W, Dijkhuizen L (1987) Regulation of methanol metabolism in the yeast *Hansenula polymorpha*. Isolation and characterization of mutants blocked in methanol assimilatory enzymes. Arch Microbiol 147:375-382

De Koning W, Bonting K, Harder W, Dijkhuizen L (1990a) Classical transketolase functions as the formaldehyde-assimilating enzyme during growth of a dihydroxyacetone synthase-negative mutant of the methylotrophic yeast *Hansenula polymorpha* on mixtures of xylose and methanol in continuous cultures. Yeast 6:117-125

De Koning W, Weusthuis RA, Harder W, Dijkhuizen L (1990b) Methanol-dependent production of dihydroxyacetone and glycerol by mutants of the methylotrophic yeast *Hansenula polymorpha* blocked in dihydroxyacetone kinase and glycerol kinase. Glycerol and dihydroxyacetone production. Appl Microbiol Biotechnol 32:693–698

De Koning W, Weusthuis RA, Harder W, Dijkhuizen L (1990c) Metabolic regulation in the yeast *Hansenula polymorpha*. Growth of dihydroxyacetone kinase/glycerol kinase-negative mutants on mixtures of methanol and xylose in continuous cultures. Yeast 6:107–115

De Roubin MR, Bastien L, Shen SH, Groleau D (1991) Fermentation study for the production of hepatitis B virus pre-S2 antigen by the methylotrophic yeast *Hansenula polymorpha*. J Ind Microbiol 8:147–156

Didion T, Roggenkamp R (1992a) Deficiency of peroxisome assembly in a mutant of the methylotrophic yeast *Hansenula polymorpha*. Curr Genet 17:113–117

Didion T, Roggenkamp R (1992b) Targeting signal of the peroxisomal catalase in the methylotrophic yeast *Hansenula polymorpha*. FEBS Lett 303:113–116

Digan ME, Lair SV, Brierly RA, Siegel RS, Williams ME, Ellis SB, Kellaris PA, Provow SA, Craig WS, Velicelebi G, Harpold MM, Thill GP (1989) Continuous production of a novel lysozyme via secretion from the yeast, *Pichia pastoris*. Bio/Technology 7:160–164

Distel B, Veenhuis M, Tabak HF (1987) Import of alcohol oxidase into peroxisomes of *Saccharomyces cerevisiae*. EMBO J 6:3111–3116

Distel B, Van der Ley I, Veenhuis M, Tabak HF (1988) Alcohol oxidase expressed under nonmethylotrophic condition is imported, assembled and enzymatically active in peroxisomes of *Hansenula polymorpha*. J Cell Biol 107:1669–1675

Douma AC, Veenhuis M, Waterham HR, Harder W (1988) Immunocytochemical demonstration of the peroxisomal ATPase of yeasts. Yeast 6:45–51

Eggeling L, Sahm (1980) Regulation of alcohol oxidase synthesis in *Hansenula polymorpha*: oversynthesis during growth on mixed substrates and induction by methanol. Arch Microbiol 127:119–124

Ellis SB, Brust PF, Koutz PJ, Waters AF, Harpold MM, Gingeras TR (1985) Isolation of alcohol oxidase and two other methanol-regulatable genes from the yeast *Pichia pastoris*. Mol Cell Biol 5:1111–1121

Erdmann R, Veenhuis M, Mertens D, Kunau WH (1989) Isolation of peroxisome-deficient mutants of *Saccharomyces cerevisiae*. Proc Natl Acad Sci USA 86:5419–5423

Erdmann R, Wiebel FF, Flessau A, Rytka J, Beyer A, Frohlich KU, Kunau WH (1991) PAS1, a yeast gene required for peroxisome biogenesis, encodes a member of a novel family of putative ATPases. Cell 64:499–510

Faber KN, Swaving GJ, Faber F, Ab G, Harder W, Veenhuis M, Haima P (1992) Chromosomal targeting of replicating plasmids in the yeast *Hansenula polymorpha*. J Gen Microbiol (in press)

Fellinger AJ, Verbakel JMA, Veale RA, Sudbery PE, Bom IJ, Overbeeke N, Verrips CY (1991) Expression of the α-galactosidase from *Cyamopsis tetragonoloba* (guar) by *Hansenula polymorpha*. Yeast 7:463–473

Gellissen G, Janowicz ZA, Merkelbach A, Piontek M, Keup P, Weydemann U, Hollenberg CP, Strasser AWM (1991) Heterologous gene expression in *Hansenula polymorpha*: efficient secretion of glucoamylase. Bio/Technology 9:291–295

Gellissen G, Melber K, Janowicz A, Dahlems UM, Weydemann U (1992) Heterologous protein production in yeast. Antonie Leeuwenhoek 62:79–93

Giuseppin MLF (1988) PhD Thesis, University of Delft, Delft

Giuseppin MLF, Van Eijk HMJ, Bes BCM (1988a) Molecular regulation of methanol oxidase activity in continuous cultures of *Hansenula polymorpha*. Biotechnol Bioeng 32:577–583

Giuseppin MLF, Van Eijk HMJ, Verduyn C, Bante I, Van Dijken JP (1988b) Production of catalase-free alcohol oxidase by *Hansenula polymorpha*. Appl Microbiol Biotechnol 28:14–19

Gleeson MA, Sudbery PE (1988a) The methylotrophic yeasts. Yeast 4:1–15

Gleeson MA, Sudbery PE (1988b) Genetic analysis in the methylotrophic yeast *Hansenula polymorpha*. Yeast 4:293–303

Gleeson MA, Waites MJ, Sudbery PE (1984) Development of techniques for genetic analysis in the methylotrophic yeast *Hansenula polymorpha*. In: Crawford RL, Hanson RS (eds) Microbial growth on C1 compounds. Proc 4th Int Symp. American Society for Microbiology. Washington DC, pp 228–235

Gleeson MA, Ortori GS, Sudbery PE (1986) Transformation of the methylotrophic yeast *Hansenula polymorpha*. J Gen Microbiol 132:3459–3465

Godecke A, Veenhuis M, Roggenkamp R, Janowicz ZA, Hollenberg CP (1989) Biosynthesis of the peroxisomal dihydroxyacetone synthase from *Hansenula polymorpha* in *Saccharomyces cerevisiae* induces growth but not proliferation of peroxisomes. Curr Genet 16:13–20

Goodman JM, Maher J, Silver PA, Pacifico A, Sanders D (1986) The membrane proteins of the methanol-induced peroxisomes of *Candida boidinii*. J Biol Chem 261:3464–3468

Gould SJ, Keller GA, Schneider M, Howell SH, Garrard LJ, Goodman JM, Distel B, Tabak H, Subramani S (1990) Peroxisomal protein import is conserved between yeast, plants, insects and mammals. EMBO J 9:85–90

Gould SJ, McCollum D, Spong AP, Heyman JA, Subramani S (1992) Development of the yeast *Pichia pastoris* as a model organism for a genetic and molecular analysis of peroxisome assembly. Yeast 8:613–628

Hansen H, Roggenkamp R (1989) Functional complementation of catalase-defective peroxisomes in a methylotrophic yeast by import of catalase A from *Saccharomyces cerevisiae*. Eur J Biochem 184:173–179

Hijikata M, Ishii N, Kagamiyama H, Osumi T, Hashimoto T (1987) Structural analysis of cDNA for rat peroxisomal 3-ketoacyl-CoA thiolase. J Biol Chem 262:8151–8158

Hohfeld J, Veenhuis M, Kunau WH (1991) PAS3, a *Saccharomyces cerevisiae* gene encoding a peroxisomal integral membrane protein essential for peroxisome biogenesis. J Cell Biol 114:1167–1178

Janowicz ZA, Eckart MR, Drewke C, Roggenkamp RO, Hollenberg CP (1985) Cloning and characterization of the DAS gene encoding the major methanol assimilatory enzyme from the methyltrophic yeast *Hansenula polymorpha*. Nucleic Acids Res 13:3043–3062

Janowicz ZA, Melber K, Merckelbach A, Jacobs E, Harford N, Comberbach M, Hollenberg CP (1991) Simultaneous expression of the S and L surface antigens of hepatitis B, and formation of mixed particles in the methylotrophic yeast *Hansenula polymorpha*. Yeast 7:431–443

Kato N, Kobayashi H, Shimao M, Sakasawa C (1986) Dihydroxyacetone production from methanol by a dihydroxyacetone kinase-deficient mutant of *Hansenula polymorpha*. Appl Microbiol Biotechnol 23:180–186

Komagata K (1991) Systematics of methylotrophic yeasts. In: Goldberg I, Stefan Rokem J (eds) Biology of methylotrophs. Butterworth-Heinemann, Boston, pp 25–37 (Biotechnology series, vol 18)

Koutz P, Davis GR, Stillman C, Barringer K, Cregg J, Thill G (1989) Structural comparison of the *Pichia pastoris* alcohol oxidase genes. Yeast 5:167–177

Kurtzman CP (1984) Synonymy of the yeast genera *Hansenula* and *Pichia* demonstrated through comparisons of deoxyribonucleic acid relatedness. Antonie Leeuwenhoek 50:209–217

Lahtchev K, Penkova R, Ivanova V, Tuneva D (1992) Genetic analysis of methylotrophic yeast *Candida boidinii* PLD1. Antonie Leeuwenhoek 61:185–194

Lazarow PB, Moser HW (1989) Disorders of peroxisome biogenesis. In: Scriver CR, Beaudet AI, Sly WS, Valle D (eds) The metabolic basis of inherited disease, 5th edn, McGraw-Hill, New York, pp 1479–1509

Ledeboer AM, Edens L, Maat J, Visser C, Bos JW, Verrips CT, Janowicz ZA, Eckart M, Roggenkamp R, Hollengerg CP (1985) Molecular cloning and characterization of a gene coding for methanol oxidase in *Hansenula polymorpha*. Nucleic Acids Res 13:3063–3082

Lim WJ, Tani Y (1988) Production of L-methionine-enriched cells of a mutant derived from a methylotrophic yeast, *Candida boidinii*. J Ferment Technol 66:643–647

Liu H, Tan X, Veenhuis M, McCollum D, Cregg JM (1992) An efficient screen for peroxisome–deficient mutants of *Pichia pastoris*. J Bacteriol 174:4943–4951

Moser HW, Bergin A, Cornblath D (1991) Peroxisomal disorders. Can J Biochem Cell Biol 69:463–474

Nicolay K, Veenhuis M, Douma AC, Harder W (1987) An ^{31}P NMR study of the internal pH of yeast peroxisomes. Arch Microbiol 147:37–41

Orr-Weaver TL, Szostak JW, Rothstein RJ (1983) Genetic applications of yeast transformation with linear and gapped plasmids. In: Wu R, Grossman L, Moldave K (eds) Methods in enzymology, vol 101. Academic Press, New York, pp 228–245

Osumi T, Fujiki Y (1990) Topogenesis of peroxisomal proteins. BioEssays 12:217–222

Ratner M (1989) Protein expression in yeast. Bio/Technology 7:1129–1133

Roa M, Blobel B (1983) Biosynthesis of peroxisomal enzymes in the methylotrophic yeast *Hansenula polymorpha*. Proc Natl Acad Sci USA 80:6872–6876

Roggenkamp R, Janowicz ZA, Stanikowski B, Hollenberg CP (1984) Biosynthesis and regulation of the peroxisomal methanol oxidase from the methylotrophic yeast *Hansenula polymorpha*. Mol Gen Genet 194:489–493

Roggenkamp R, Hansen H, Eckart M, Janowicz Z, Hollenberg CP (1986) Transformation of the methylotrophic yeast *Hansenula polymorpha* by autonomous replication and integration vectors. Mol Gen Genet 202:302–308

Roggenkamp R, Didion, T, Kowalik KV (1989) Formation of irregular giant peroxisomes by overproduction of the crystalloid core protein methanol oxidase in the methylotrophic yeast *Hansenula polymorpha*. Mol Cell Biol 9:988–994

Romanos MA, Scorer CA, Clare JJ (1992) Foreign gene expression in yeast: a review. Yeast 8:423–488

Sakai Y, Tani Y (1992) Directed mutagenesis in an ascosporogenous methyltrophic yeast: cloning, sequencing and one-step gene disruption of the 3-isopropylmalate dehydrogenase gene (*LEU2*) of *Candida boidinii* to derive double auxotrophic marker strains. J Bacteriol 174:5988–5993

Sakai Y, Kazarimoto T, Tani Y (1991) Transformation system for an asporogenous methylotrophic yeast, *Candida boidinii*: cloning the orotidine-5′-phosphate decarboxylase gene (*URA3*), isolation of uracil auxotrophic mutants, and use of the mutants for integrative transformation. J Bacteriol 173:7458–7463

Shen SH, Bastien L, Nguyen T, Fung M, Slilaty SN (1989) Synthesis and secretion of hepatitis B middle surface antigen by the methylotrophic yeast *Hansenula polymorpha*. Gene 84:303–309

Shimozawa N, Tsukamoto T, Suzuki Y, Orii T, Shirayoshi Y, Mori T, Fujiki Y (1992) A human gene responsible for Zellweger syndrome that affects peroxisome assembly. Science 255:1132–1134

Sibirny AA, Titorenko VI, Efremov BD, Tolstorukov II (1987) Multiplicity of mechanisms of carbon catabolite repression involved in the synthesis of alcohol oxidase in the methylotrophic yeast *Pichia pinus*. Yeast 3:233–241

Sibirny AA, Titorenko VI, Gonchar MV, Ubiyvovk VM, Ksheminskaya GP, Vitvitskaya OP (1988) Genetic control of methanol utilization in yeasts. J Basic Microbiol 28:293–319

Sibirny AA, Titorenko VI, Teslyar GE, Petrushko VI, Kucher MM (1991) Methanol and ethanol utilization in methylotrophic yeast *Pichia pinus* wild-type and mutant strains. Arch Microbiol 156:455–462

Sierkstra LN, Verbaki JMA, Verrips CT (1991) Optimization of a host/vector system for heterologous gene expression by *Hansenula polymorpha*. Curr Genet 19:81–87

Sudbery PE, Gleeson MAG (1989) Genetic manipulation of methylotrophic yeasts. In: Walton EF, Yarranton GT (eds) Molecular and cell biology of yeasts. Blackie, Glasgow, pp 304–329

Sulter GJ, Looyenga L, Veenhuis M, Harder M (1990a) Occurrence of peroxisomal membrane proteins in methylotrophic yeasts grown under different conditions. Yeast 6:35–43

Sulter GJ, Van der Klei IJ, Harder W, Veenhuis M (1990b) Expression and assembly of amine oxidase and D-amino acid oxidase in the cytoplasm of peroxisome-deficient mutants of the yeast *Hansenula polymorpha* during growth on primary amines or D-alanine as the sole nitrogen source. Yeast 6:501–509

Sulter GJ, Van der Klei IJ, Schanstra JP, Harder W, Veenhuis M (1991) Ethanol metabolism in a peroxisome-deficient mutant of the yeast *Hansenula polymorpha*. FEMS Microbiol Lett 82:297–302

Tarutina M, Tolstorukov II (1992) Mechanism of transformation of the methylotrophic yeast *Pichia methanolica* (*P. pinus*) MH4 with the LEU2 gene of the yeast *Saccharomyces cerevisiae*. Yeast 8:S622

Tikhomirova LP, Ikonomova RN, Kuznetsova EN (1986) Evidence for autonomous replication and stabilization of recombinant plasmids in transformants of yeast *Hansenula polymorpha*. Curr Genet 10:741–747

Titorenko VI, Trotsenko YA (1983) Selection of mutants sythesizing branched-chain amino acids in the methylotrophic yeast *Hansenula polymorpha*. Microbiology 52:774–780

Titorenko VI, Khodurskii AB, Teslyar GE, Sibirny AA (1990) Identification of new genes controlling catabolite repression of the synthesis of alcohol oxidase and catalase in the methylotrophic yeast, *Pichia pinus*. Genetika 27:625–635

Tolstorukov II, Efremov BD (1984) Genetic mapping of yeast *Pichia pinua*. II. Mapping by tetrad analysis. Genetika 20:1099–1107

Tolstorukov II, Dutova TA, Benevolenskii SV, Soom YO (1977) Hybridization and genetic analysis of the methanol yeast *Pichia pinus*. Genetika 13:322–329

Tolstorukov II, Efremov BD, Bliznik KM (1983) Construction of a genetic map of the yeast *Pichia pinus*. I. Determination of linage groups using induced mitotic haploidization. Genetika 19:897–902

Tolstorukov II, Efremov BD, Benevolensky SV, Titorenko VI, Sibirni AA (1989) Mutants of the methylotrophic yeast *Pichia pinus* defective in C₂ metabolism. Yeast 5:179–186

Tschopp JF, Cregg JM (1991) Heterologous gene expression in methylotrophic yeasts. In: Goldberg I, Rokem JS (eds) Biology of methylotrophs. Butterworth-Heinemann, Boston, pp 305–322 (Biotechnology series, vol 18)

Tschopp JF, Brust PF, Cregg JM, Stillman CA, Gingeras TR (1987a) Expression of the *lacZ* gene from two methanol-reguated promoters in *Pichia pastoris*. Nucleic Acids Res 15:3859–3876

Tschopp JF, Sverlov G, Kosson R, Craig W, Grinna L (1987b) High-level secretion of glycosylated invertase in the methylotrophic yeast, *Pichia pastoris*. Bio/Tchnology 5:1305–1308

Tsukamoto T, Yokota S, Fujiki Y (1990) Isolation and characterization of Chinese hamster ovary cell mutants defective in assembly of peroxisomes. J Cell Biol 110:651–660

Van der Klei IJ, Harder W, Veenhuis M (1991a) Methanol metabolism in a peroxisome-deficient mutant of *Hansenula polymorpha*: a physiological study. Arch Microbiol 156:15–23

Van der Klei IJ, Harder W, Veenhuis M (1991b) Selective inactivation of alcohol oxidase in two peroxisome-deficient mutants of the yeast *Hansenula polymorpha*. Yeast 7:813–821

Van Dijken JP, Van der Bosch E, Hermanns JJ, Rodriguez de Miranda L, Scheffers WA (1986) Alcoholic fermentation by "non-fermentative" yeasts. Yeast 2:123–127

Veale RA, Guiseppin MLF, Van Eijk HM, Sudbery PE, Verrips CT (1992) Development of a strain of *Hansenula polymorpha* for the efficient expression of guar α-galactosidase. Yeast 8:361–372

Veenhuis M, Harder W (1991) Microbodies. In: Rose AH, Harrison JS (eds) The yeasts, vol 4, 2nd edn, Academic Press, London, pp 601–653

Veenhuis M, Van Dijken JP, Harder W (1983) The significance of peroxisomes in the metabolism of one-carbon compunds in yeasts. In: Rose AH, Gareth Morris J, Tempest DW (eds) Advances in microbial physiology. Academic Press, London, pp 1–82

Veenhuis M, Kram AM, Kunau WH, Harder W (1990) Excessive membrane development following exposure of the methylotrophic yeast *Hansenula polymorpha* to oleic acid-containing media. Yeast 6:511–519

Veenhuis M, Van der Klei IJ, Titorenko VI, Harder W (1992) *Hansenula polymorpha*: an attractive model organism for molecular studies of peroxisome biogenesis and function. FEMS Microbiol Lett (in press)

Verduyn C, Giuseppin MLF, Scheffers WA, Van Dijken JP (1988) Hydrogen peroxide metabolism in yeasts. Appl Environ Microbiol 54:2086–2090

Verduyn C, Van Wijngaarden CJ, Scheffers WA, Van Dijken JP (1991) Hydrogen peroxide as an electron acceptor for mitochondrial respiration in the yeast *Hansenula polymorpha*. Yeast 7:137–146

Vonck J, Van Bruggen EFG (1992) Architecture of peroxisomal alcohol oxidase crystals from the methylotrophic yeast *Hansenula polymorpha* as deduced by electron microscopy. J Bacteriol 174:5391–5399

Wegner EH (1983) US Patent 4,414,329. Nov 8, 1983

Wegner GH (1990) Emerging applications of methylotrophic yeasts. FEMS Microbiol Rev 87:279–284

Wegner EH, Harder W (1986) Methylotrophic yeasts. In: Van Versefeld HW, Duine JA (eds) Microbial Growth on C1 Compounds: Proceedings of the 5th International Symposium. Martinus Nijhoff, Dordrecht, pp 131–138

Wiebel FF, Kunau WH (1992) The Pas2 protein essential for peroxisome biogenesis is related to ubiquitin-conjugating enzymes. Nature 359:73–76

Zoeller RA, Raetz CRH (1986) Isolation of animal cell mutants deficient in plasmalogen biosynthesis and peroxisomal assembly. Proc Natl Acad Sci USA 83:5170–5174

Zoeller RA, Allen LAH, Santos MJ, Lazarow PB, Hashimoto T, Tartakoff AM, Raetz CRH (1989) Chinese hamster ovary cell mutants defective in peroxisome biogenesis. J Biol Chem 264:21872–21878

Zoroddu MA, Bonomo RP, Di Bilio AJ, Berardi E, Meloni MGL (1991) EPR study on vanadyl and vanadate ion retention by a thermotolerant yeast. J Inorg Biochem 43:731–738

Zwart K, Veenhuis M, Van Dijken JP, Harder W (1980) Development of amine oxidase-containing peroxisomes in yeasts during growth on glucose in the presence of methyamine as the sole source of nitrogen. Arch Microbiol 126:117–126

Zwart K, Veenhuis M, Plat G, Harder W (1983) Characterization of glyoxysomes in yeasts and their transformation into peroxisomes in response to changes in environmental conditions. Arch Microbiol 136:28–38

CHAPTER 15

Yeasts in Food Fermentations and Therapeutics

G. Vijayalakshmi, B.K. Lonsane, and M.M. Krishnaiah

1 Introduction

With the rapid increase in world population, the need to find ways of satisfying hunger becomes more urgent, almost day by day. Therefore, food technology becomes a more and more important discipline as the urgency increases.

There are two aspects to the problem. First, the need to provide the necessary quantities of foodstuffs. Second, the preservation of existing foods and maintenance or improvement of their quality. That is, preservation of foods so that the flavor and nutritive characteristics are retained.

Fermented foods are widely used in Japan, China, Indonesia, and neighbouring regions, Africa and India (Wang and Hesseltine 1982; Wood and Hodge 1985) Fermented foods in India, unlike those of Japan, usually do not have a high salt content. They are foods fermented by mixed cultures of fermentative yeasts and lactic acid bacteria. Sourdough bread, popular the world over, even in North America and Europe, is a representative of this group. This bread is raised by a mixed culture of lactic acid bacteria and the yeast, *Candida milleri*, and has a pleasant tangy flavor, much liked by gourmets and all others who eat it.

Over the last 10 000 to 15 000 years, man, or probably woman, has been developing practical methods for food preservation. One of these is the use of fermentation by naturally occurring yeasts and bacteria, which impart a moderately acid flavor to the foods, and form small quantities of ethanol as well. Unfortunately, the process has often been corrupted to the point where much of the food value has been lost and converted to alcohol, to the detriment of the users.

2 Yeasts in Food Fermentations

2.1 Origin of Fermented Foods

Fermentation of foods probably originated in China, for preservation of cereals and legumes. In Indonesia and India, vegetable products having a meat-like texture, and methods of leavening batters of cereal-legume mixtures were developed. Acid fermented vegetables (pickles) and bread leavened by yeast were developed in Korea and Egypt, respectively. The practice has been extended to many other

J.F.T. Spencer/D.M. Spencer (eds)
Yeasts in Natural and Artificial Habitats
© Springer-Verlag Berlin Heidelberg 1997

foods. Sauerkraut, produced by a lactic acid fermentation of shredded cabbage, has been used for centuries in most of Europe. Fermented foods have a tremendous potential in the world today (Reddy et al. 1982; Rose 1982; Chan 1983; Reed 1983; Steinkraus 1983; Wood 1985).

2.2 Fermented Foods

Fermented foods are prepared by microbial action on one or more components, under relatively controlled conditions, causing changes in their physical, biochemical, and nutritional qualitites. These foods include cereals, legumes, roots, vegetables, fruits, edible parts of plants, fish, milk, and meats. Normally, the microorganisms are present naturally on the foods, though sometimes inocula containing specific bacteria, yeasts, molds, and/or actinomycetes are used. Substances containing microbial enzymes may be added. Contamination with undesirable microorganisms results in a spoiled or toxic product (Ko 1982).

Fermented foods are usually classified according to the nature of the original material and of the fermentation (alcoholic, or lactic, or both). They may also be classified by the region of origin (Japanese sake, Indian idli). Fermented foods are regional in nature, and their use is influenced by local preferences. The demand for these products is increasing and spreading to other countries.

2.3 Major Changes in Foods Imparted by Microorganisms

During fermentation of foods, complex materials (carbohydrates, proteins, fats) are metabolized to yield others such as ethanol, organic acids, and pectic hydrolysates. Polyunsaturated fatty acids and phospholipids may be oxidized or otherwise metabolized, organic acids decarboxylated, fatty acids degraded, alcohols oxidized to organic acids, amino acids deaminated, desaturated, or decarboxylated, or decomposed anaerobically. Compounds contributing to the flavor of the foods may be formed through condensation of fatty acids and alcohols into esters, and oxidation of fatty acids to carbonyl compounds (Rose 1982; Reed 1983; Wood and Hodge 1985). Melanoidins may be formed through the Maillard reaction. Vitamins of the B-complex may be formed.

Pigments such as mono-ascorubin and mono-ascoflavin are formed by *Monascus purpureus*. Anthocyanins are also formed. *Neurospora* sp. produce pigmented conidiospores which color the foods, and the mucin-like polysaccharides formed by some microorganisms give the characteristic texture to others (*natto*). *Rhizopus oligosporus* produces antimicrobial and antioxidant compounds during manufacture of *tempe*. Phenols, acids, and alcohols may also prevent the microbial degradation of some food products.

Undesirable compounds such as trypsin inhibitor, hemagglutinins, phytates, and flatulence factors are degraded during fermentation. The overall effect of fermentation is to reduce the pH of the product, solubilize some of the complex polysaccharides and other polymers, modify the texture, and contribute a desirable flavor to the food.

2.4 Advantages of Food Fermentation

The shelf life of fermented foods is frequently better than that of a comparable frozen food. Food fermentation is an efficient and simple method of food preservation, using low-cost equipment and little energy. The cost of the raw materials may be lower; broken rice used in manufacture of idli is cheaper than whole rice. The cost of preservation of fermented foods is much less than preservation by canning or freezing. It is therefore economically more feasible in the developing countries.

Fermented foods have a higher nutritive value and improved digestive properties. In idli, the combination of cereals and legumes gives a better balanced food. The fermented foods have better organoleptic properties; taste, color, texture, mouth-feel, and crispness (Lonsane 1989). Their dietary fiber content is relatively high. They are very desirable foods for infants, expectant and nursing mothers, and invalids. Fermented foods are less likely to contain disgestive inhibitors, toxic compounds, and pathogenic microorganisms, and are less liable to spoilage. Their increased content of vitamins, proteins, and energy makes them valuable in preventing malnutrition.

2.5 Importance of Yeasts

The importance of yeasts in the food and beverage industries was only realized about 1860, when their role in food manufacture became evident. Yeasts grow on a wide range of substrates, can tolerate extreme physicochemical conditions, are easily handled in the plant and in the laboratory, and can be manipulated genetically.

They produce ethanol, CO_2, proteins, vitamins, pigments, and flavoring compounds, and can contribute useful mechanical qualities to foods. Yeasts contribute significantly to recommended daily allowances for calories (energy), proteins, calcium, phosphorus, and iron, and the vitamins of the C and B groups and niacin in the human diet.

2.6 Interactions of Yeasts and Lactic Acid Bacteria

Yeasts and lactic acid bacteria participate in the fermentation of kefir, koumiss, kvass, and the raising of some specialized breads (Wood 1985). The fermentations are all anaerobic and produce alcohol and CO_2. This inhibits the growth of spoilage organisms, including filamentous fungi and toxin-producing bacteria. These fermentations are carried out to obtain a product of the desired flavor and texture, and the ethanol content of the beverages, rather than for preservation (Wood 1985).

The yeasts and lactic acid bacteria are firmly embedded in the kefir grains, which are bound together by microbial polysaccharides. The production of "bees' wine" and in the fermentation using the "ginger beer plant" are similar (Wood 1985).

The relationship between the yeasts and the bacteria is quasisymbiotic. In sourdough bread, the bacteria hydrolyze maltose to glucose, which the yeast utilizes, and produces essential nutrients required by the bacteria. The yeasts are tolerant of acetic acid and antibiotic agents produced by the bacteria. The bacteria do not utilize glucose, so the two species do not compete (Wood 1985).

In the soy sauce and miso fermentations, the role of the yeasts and bacteria is less understood. The lactic acid fermentation apparently occurs before the yeast fermentation, and adds to the flavor of the product. The interaction between the yeasts and bacteria is complex (Wood 1985).

2.7 Yeasts in Indian Fermented Foods

Most of the wide variety of Indian fermented foods are produced in the home, not commercially. There are some prepared dry mixes and ready-to-eat preparations which have appeared recently (Lonsane 1989). Some of the foods (idli and dosa) are served very hot, and many of those which are not (for e.g.: dhokla) must be consumed within a few hours, to retain their organoleptic properties. They are difficult to produce on a commercial scale. Foods and drinks produced commercially include papad, warries, kurdai, sondige, uppumensinakai, pickles, dahi, ghee, wine, feni (an alcoholic drink made from cashew apples), and liquors from *Mahua* flowers (Lonsane 1989).

The microorganisms associated with most of these foods and beverages have not been investigated.

2.7.1 Rice-Based Foods and Drinks

Indian fermented foods can be classified according to the materials used in their preparation. Rice is a very commonly used cereal. Drinks prepared from rice include *apong* (*laopani* or *mod*), nogin apong, ruhi, madhu, puchwai, kanji, jand, fermented rice extract, sonti annam, and torani. Other rice-based preparations are fermented doughs which are deep fried or cooked as pancakes, and include kurdai, sondige, seke papad, and anarse.

Nogin apong, one of the fermented drinks from rice, is prepared by cooking and grinding the rice, and inoculating it with a starter of rice batter or flour containing yeasts and lactic acid bacteria. This is fermented for 5 days and diluted with water, and the liquid phase is filtered off. Po-ro apong is prepared by adding rice straw ashes to the cooked rice, and fermenting the inoculated mixture for 10–12 days before diluting and filtering off the liquid.

The other drinks (sonti annam) are prepared similarly. The starters often contain ginger and other spices, which may inhibit undesirable microorganisms. Microorganisms in starters for preparation of puchwai may contain filamentous fungi (*Rhizopus arrhizus, Rhizopus cambodja, Mucor fragilis, Mucor rouxianus, Mucor javanicus, Mucor prainii, Amylomyces rouxii,* and *Amylomyces oryzae*), and the yeasts *Hansenula anomala, Endomycopsis (Pichia) burtonii,* and *Endomycopsis (Saccharomycopsis) fibuligera,* as well as the lactic acid bacteria. The yeasts occurring in the kanji fermentation were *Candida tropicalis, Candida guilliermondii, Hansenula anomala* var. *ciferri (H. anomala),* and *Geotrichum candidum*. The same yeasts were found in the torani fermentation. Sonti annam is fermented by unidentified yeasts and *Rhizopus sonti*.

Rice-based solid foods include kurdai, sondige, seke papad, and anarse. The procedure is like that for manufacture of fermented rice drinks, except the solids content is higher and the liquid is not filtered off. Fermentation occurs before the product is finally cooked for serving.

2.7.2 Foods and Drinks Based on Sorghum Grain

Bibdi, the only fermented food in this category, is made by soaking and grinding sorghum grain (*Sorghum bicolor*) (Lonsane 1989). Fermentation, by yeasts and lactic acid bacteria, takes place during the 3-day soaking under microaerophilic conditions. The grain is dried for 1 day, milled, and the bran removed, the meal boiled in water containing several spices, and rolled into balls and roasted. Microorganisms associated with the process have not been identified.

2.7.3 Foods Based on Barley

Chhang is a drink made from huskless barley (*Hordeum nudum*), though wheat, rice, or ragi millet are sometimes used. The barley is cooked, a mixture of a starter containing the yeasts *Candida krusei* and *Saccharomyces uvarum* (and probably other organisms) and wheat bran is added, and the moist grain is fermented for 8 days. The liquid is extracted with water. For a stronger liquor, the product can be distilled to yield arrack, containing 40% ethanol. Other similar drinks can be manufactured from Chhang. The solids fraction (lugri or busa) is eaten.

Thumba is a mild alcoholic drink made by fermenting millet (*Elusine coracana*) with *Endomycopsis (Saccharomycopsis) fibuligera* (Beuchat 1978a,b).

2.7.4 Wheat-Based Foods and Drinks

Jilebi is a fermented food of Arabic or Persian origin, which reached India about 1450 A.D. A batter of wheat flour and water is inoculated with baker's yeast and a yoghurt-like fermented milk preparation (dahi), containing a mixed culture of lactic acid bacteria, and fermented overnight. The fermented material is then squeezed through a cloth to obtain strands of dough which are then deep-fried in ghee, a melted butter preparation (Batra 1981) and immediately soaked in sugar syrup.

Microorganisms isolated from jilebi fermentations include *Saccharomyces bayanus*, *Saccharomyces cerevisiae*, *Hansenula anomala*, *Lactobacillus bulgaricus*, *Streptococcus thermophilus*, and *Streptococcus faecalis* (Batra 1981).

Nan bread is a flat, leavened bread, widely used as a dietary staple in northern India and in many other countries. A dough of wheat flour, sugar, salt, and water is mixed with a starter of batter from the previous day, or inoculated with dahi (fermented milk) and baker's yeast (Batra 1981). The dough is fermented for 12–24 h, worked by hand, and baked. Caraway or other spices may be added a few hours before baking.

Bathura and kulcha are products made from the same dough as nan bread. The doughs are deep-fried (bathura) or cooked on a griddle (kulcha) (Batra 1981). *Candida* spp., *Saccharomyces* spp., *Hansenula* spp., *Kluyveromyces* spp., *Pichia* spp., *Debaryomyces* spp., *Torulopsis* spp., *Trichosporon* spp., and *Rhodotorula* spp. were found in the fermented doughs, as well as bacteria (*Streptococcus*, *Leuconostoc*, *Lactobacillus*, and *Pediococcus* spp.). The content of amylases, proteinases, reducing sugars, soluble proteins, total proteins, and vitamins B_1 and B_2 increased during the fermentation.

Mangalore bonda is made by fermenting a wheat flour batter, using a small amount of dahi as starter. Cumin seeds, green chillies, grated coconut, and other condiments are also added at the beginning of the fermentation. The pH of the fermentation is lowered by the action of the lactic acid bacteria present in the dahi.

Phool warries are made by washing, soaking, and crushing wheat kernels, straining out the insoluble solids, adding salt and spices, fermenting for 4–8 h, thickening the batter with wheat flour or mung bean paste, and forming into strands or small balls and sun-drying. The preparation is deep-fried before eating. *Candida krusei*, occasionally *Rhodotorula glutinis*, and lactic acid bacteria are present.

Gulgule is a sweet food prepared from wheat flour, formed into balls and fermented probably by the action of yeasts and lactic acid bacteria. The dough is made of a mixture of wheat flour, mild, powdered jaggery, crushed ripe banana and cardamon powder, and is fermented for about 8 h before frying in ghee.

2.7.5 Legume-Based Fermented Foods

Punjabi warries are made from dehulled black gram, soaked and ground, mixed with spices, molded into balls, fermented, and dried (Soni and Sandhu 1990). Microbial species involved include *Saccharomyces cerevisiae*, *Zygosaccharomyces rouxii*, *Pichia membranaefaciens*, *Candida vartiovaarai*, *Candida krusei*, *Candida famata, Candida parapsilosis, Candida aquatica,* other *Candida* spp., *Kluyveromyces marxianus, Trichosporon beigelii, Hansenula anomala, Rhodotorula lactosa, Wingea robertsii, Debaryomyces hansenii, Cryptococcus humicola,* and *Geotrichum candidum, Leuconostoc mesenteroides, Lactobacillus fermentum, Streptococcus faecalis, Bacillus subtilis, Enterobacter* sp., and *Flavobacter* spp. (Batra et al. 1974; Batra 1981; Reddy et al. 1982). During the fermentation, total acids, soluble solids, nonprotein nitrogen, soluble nitrogen, free amino acids, proteinase activity, and water-soluble B vitamins increased, while the pH, reducing sugars, and soluble protein decreased. Amylase activity increased and then decreased.

Chana Ki warries are made from seeds of Bengal gram (Batra et al. 1974) and Maharastrian warries, from mung dal and moth dal, or from mung dal only, otherwise the processes are the same as for Punjabi warries.

Vadai is made from fermented black gram (*Phaseolus mungo*), without spices. The seeds are soaked for up to 12h, ground, prepared immediately or fermented for 18–24h at room temperature. The fermented batter is formed into balls, deep-fried, and eaten with coconut or dal chutney. The microflora include *Leuconostoc mesenteroides, Lactobacillus fermentum, Leuconostoc delbrueckii, Streptococcus faecalis, Bacillus subtilis, Achromobacter* sp., *Debaryomyces hansenii, Trichosporon beigelii, Hansenula anomala, Saccharomyces cerevisiae, Kluyveromyces marxianus, Pichia memranaefaciens, Candida curvata,* and *Candida krusei.* The microbial species are much the same as those found in Punjabi warries, as are the biochemical changes in the product.

Bhallae is made from Bengal gram, in the same way as Vadai, though the fermentation time is shorter, 6–24h. A similar group of microorganisms takes part in the fermentation, including *Candida famata, Candida membranaefaciens, Candida curvata, Candida parapsilosis, Trichosporon pullulans, Trichosporon beigelii, Wingea robertsii, Hansenula polymorpha, Hansenula anomala, Rhodotorula polymarina, Rhodotorula glutinis, Kluyveromyces marxianus, Debaryomyces hansenii, Pichia membranaefaciens, Saccharomyces cerevisiae, Lactobacillus fermentum, Leuconostoc mesenteroides, Streptococcus faecalis, Bacillus subtilis, Achromobacter* spp., and *Flavobacter* spp. The physicochemical and nutritional changes in the material are similar to the foods previously described (Soni and Sandhu 1990).

Papad is a thin tortilla-like wafer made from fermented black gram. Other leguminous grains may be added. The black gram flour is mixed with salt, spices, and water to make a thick dough, and fermented for 4–6h. It is then formed into cakes and sun-dried for several days. *Saccharomyces cerevisiae, Hansenula anomala, Trichosporon beigelii, Candida krusei, Pichia membranaefaciens, Debaryomyces hansenii, Leuconostoc mesenteroides, Lactobacillus fermentum,* and *Streptococcus faecalis* were found (Batra and Millner 1974; Reddy et al. 1982).

Khaman is also prepared from black gram, the batter being fermented for 12–15h at ambient temperature. The microorganisms involved are mainly *Saccharomyces cerevisiae* and *Candida* spp.

2.7.6 Foods Manufactured from Mixed Grains and Other Substrates

Sandoi is a type of papad, prepared in the same way as bibdi, from a mixture of sago and fermented sorghum (Lonsane 1989).

Ambus ghari is a fried food, made from a batter of Bengal gram and wheat flours, and fermented overnight. The microorganisms involved in the fermentation are those found in dahi (fermented milk).

Ambali is fermented by yeasts and lactic acid bacteria, and is used in liquid diets. Ragi millet (*Eleusine coracana*) or sorghum flour is fermented as a batter for 14–16 h and mixed with partly cooked rice and the cooking continued until complete. It is mixed with sour milk and eaten immediately, or buttermilk may be added to the ground millet before the fermentation step.

2.7.7 Foods Manufactured from Mixtures of Rice and Legumes

Idli is made from a fermented batter of rice and black gram (Fig. 15.1). It is often inoculated with a starter of fermented batter from a previous batch, or with sour buttermilk or baker's yeast. It is a mixed fermentation of yeasts and lactic acid bacteria, like many of the other fermented foods. The fermentation and the final product have been investigated extensively (Reddy et al. 1982). The yeasts

Table 15.1. Microflora involved in idli fermentation

Bacterla	Yeasts
Leuconostoc mesenteroides	*Candida parapsilosis*
Lactobacillus fermentum	*C. pelliculosa*
L. delbruckii	*C. cacaoi*
L. coryneformis	*C. fragicola*
L. lactis	*C. kefyr*
L. fermentii	*C. glabrata*
Pediococcus cerevisiae	*C. tropicalis*
Streptococcus lactis	*C. sake*
Str. faecalis	*C. krusei*
Bacillus spp.	*Torulopsis candida*
B. amyloliquefaciens	*T. pullulans*
Aerobacter cloacae	*T. holmii*
	Saccharomyces cerevisiae
	Debaryomyces hansenii
	D. tamarii
	Hansenula anomala
	Trichosporon beigelii
	T. pullulans
	Rhodotorula graminis
	Wingea robertsii
	Issatchenkia terricola
	Geotrichum candidum

Fig. 15.1. Flow diagram for preparation of idli

Trichosporon pullulans and *Torulopsis candida* (*Candida famata*) contribute acidity and flavor to the preparation (Mukherjee et al. 1965). Lactic acid bacteria lower the pH further. The microorganisms involved in the fermentation of idli batter are listed in Table 15.1.

Other types of idli are prepared using mixtures of soybeans and rice, northern bean and rice, and common bean and rice.

Rajmah idli is made from bhagar millet (*Panicum milliaceum*) and rajmah (*Phaseolus vulgaris*) instead of rice and black gram.

Bhagar idli is made with bhagar millet instead of rice, and wheat idli, with wheat flour. Several methods of preparation of idli use variations in the method of fermentation and cooking. Vaggaranai idli is made from batter which is left over from the morning and thus fermented longer and higher in acidity. Kolakattai are made by spreading fermented batter between two chapatti cakes and cooking by steaming.

Dhokla resembles idli, but is made from bengal gram rather than black gram, and rice. The batter made with the ground grain is fermented overnight in the presence of chopped leaves of fenugreek, green chillies and curry leaves, and steamed. The microorganisms in dhokla are those present in dahi. The bacterial species present included *Leuconostoc mesenteroides*, *Lactobacillus fermentii*, *Streptococcus faecalis*, and *Bacillus* spp. Yeasts are probably also present. Changes in the properties of the batter were the same as in the previous foods.

Dosa is made from fermented batter of black gram and rice, inoculated with a starter of batter from a previous batch or autofermented (Desikachar 1984). Sometimes 1–2% sucrose is added to the batter. It is cooked as a form of pancake. The microorganisms involved in the fermentation are the same as in warries, vadai, and bhallae. Yeast populations ranged from zero to ten cells per gram (Desikachar 1984), and bacterial populations from ten cells per gram (Beuchat 1978b) (Evans and McAthey 1991). The yeasts contribute, among other things, flavor compounds.

Besides plain dosa, there are numerous specialized types of dosa: masala dosa, with and without potatoes, paper dosa, set dosa, onion dosa, menthe dosa, containing fenugreek; Uttapam, bele dosa or parapu dosa, avalakki dosa, soybean dosa, mung bean dosa, koyee dosa or appam (a sweet dosa), and adai.

2.7.8 Fermented Foods Based on Milk

Dahi is made from cow's milk, buffalo milk, or sometimes sheep or goat's milk, and is a thick, sour product resembling yoghurt (Batra and Millner 1974). It forms a major part of the diet everywhere in India and neighbouring countries. About 70% of the milk produced in India is converted into dahi. The milk is sterilized by boiling, cooled, and inoculated with a starter of dahi from a previous batch.

The microflora of dahi includes *Streptococcus thermophilus*, *Streptococcus faecalis*, *Streptococcus lactis*, *Streptococcus dextranicum*, *Lactobacillus bulgaricus*, *Lactobacillus casei*, *Lactobacillus brevis*, *Lactobacillus helveticus*, *Lactobacillus plantarum*, *Candida krusei*, *Trichosporon* sp., and *Torulopsis* sp. The yeasts produce the flavoring compounds. The properties of the dahi depends upon the milk used, its solids content and fat distribution. The acidity of the product varies with time and temperature of preperation. Dahi has antibacterial activity against *E. coli* in the human intestine.

Sweet dahi is sweetened with cane sugar and flavored with nutmeg.

Buttermilk is diluted dahi with water and salt added. It is drunk as a beverage in summer. Spices such as ginger and coriander leaves may be added.

Lassi is dahi mixed with iced water, sweetened with sugar, and churned with spices.

Dahi Vadai and Dahi Bhalle are Vadai or bhalle, soaked in dahi, with salt, red chilli powder, garam masala, and other spices and sometimes cumin added.

Srikhand is a sweetened fermented milk paste. The starting material is sour dahi, produced by fermenting milk for 12–18 h. The whey is drained off the dahi and the solids retained are termed chakka. This is sweetened and flavored with cardamom, saffron, and dried fruits. The final product contains 40–43% moisture, 45% sugar, 5–6% protein, and 51–66% fat. It carries viable microorganisms which can colonize the human intestine.

Ghee is an edible fat used in sweets and for cooking of vegetable, rice, and legume curries. It is produced by fermenting milk fats (cream) which rise to the top of boiled milk. The cream is inoculated with sour dahi and fermented for 2–3 days at room temperature. The cream is then churned, the butter is separated and melted. The solid residues are filtered off. The microorganisms (yeasts and lactic acid bacteria) are the same as in dahi.

Kadhi is prepared from buttermilk mixed with bengal gram flour and salt. Ghee is heated and seasoned with mustard seed, cumin, turmeric powder, curry leaves, garlic and ginger. The mixture of bengal gram and salt is stirred into the ghee. The product is garnished with coriander leaves.

Pakoras or vegetables (bottle gourds, bitter gourd, and others) can be cooked in the kadhi. The microorganisms are those found in dahi.

2.7.9 Fermented Foods Having a Fruit or Vegetable Base

These include alcoholic beverages, chillies and pickles.

Cashew apple feni is an alcoholic beverage obtained by fermenting the juice of the cashew apple (the enlarged peduncle), which contains about 17.6–21.9% sugar, and is very perishable. The juice is fermented with *Saccharomyces cerevisiae* var. *ellipsoideus* in a normal fermentation. The product may be drunk as wine or distilled to yield *feni*. It may be produced in batch or continuous fermentation.

Mahua daru is an alcoholic drink distilled from fermentation of mahua (*Madhuca longifolia*) flowers, which contain 56–62% sugar. The fermentation is due to *Saccharomyces cerevisiae* and the alcohol recovery is by normal distillation. About 1.5 million tonnes of mahua flowers are available annually, equivalent to 375 million liters of alcohol.

Toddy (kallu) is an alcoholic beverage made by fermenting palm saps (sendi toddy, tari toddy, nareli toddy, from *Phoenix sylvestris*, *Borassus flabellifer*, *Coco nucifera*, respectively). The sap as collected contains numerous bacteria and yeasts, and is often used as a starter for preparation of nan. The fermentation is a mixed lactic acid and alcoholic type. The yeast microflora include *Kloeckera japonica, Endomycopsis monospora, Pichia farinosa, Pichia membranaefaciens, Saccharomyces chevalieri, Saccharomycoides ludwigii, Schizosaccharomyces pombe* and *Candida* spp. *S. pombe* is also found in palm wine fermentations in parts of Africa. *S. pombe* is seldom, reported as a constituent of the microflora of other fermented foods and drinks in India (Batra and Millner 1974).

Carrot kanji is a mild alcoholic beverage made by extracting carrot juice (*Daucus carota* var. *sativa*) with water, adding spices, ginger, salt, and lemon juice, and fermenting it. Beet juice is sometimes used. The juice is fermented by yeast, usually *Hansenula anomala* var. *anomala*, and consumed immediately (Batra and Millner 1974). The fermentation is inoculated with *kanji* from the previous fermentation.

Kharkya are fermented ber fruits (*Zizipus mauritiana* Lam.). They are pricked with a fork, dipped in buttermilk, and sun-dried to about 8–10% moisture. The fruits are stored in closed containers for about a year and used as a condiment in vegetable curries and dal. They are often used as candies by children.

Uppumensinakai is a fermented green chilli prepared by using dahi as inoculum. The stalks are removed from the chillies, they are cut into large pieces and put into sour dahi diluted 1 : 1 with water, and held overnight. Next morning, the pieces of chilli are removed and sun-dried. This is repeated until all of the dahi has been absorbed by the chillies. The product is fried in oil and used as a condiment. The microorganisms are those found in dahi.

Masala mirchi is prepared from green chillies, which are fermented by mixed cultures of yeasts and lactic acid bacteria. Split green chillies are stuffed with a mixture of coriander powder, dry mustard, fenugreek seed powder, turmeric powder, salt, vegetable oil, and lemon juice. The stuffed chillies are held overnight at room temperature and sun-dried.

Tomato pickle is made by mixing pieces of tomato and mango ginger with salt and turmeric and holding overnight. Tamarind is soaked in the juice released from the tomato pieces. Mustard and fenugreek seeds are roasted and ground. The tomatoes and mango ginger are ground and mixed with the tamarind extract. These ingredients, plus red chilli powder and a trace of asafoetida, are cooked in a little vegetable oil until thick.

Lemon pickle is made by cutting up lemons and adding extra lemon juice, plus salt and turmeric. After 24 h, the pieces of lemon are removed and sun-dried. In the evening of the same day, the pieces of lemon are replaced in the juice and allowed to absorb more of it. This is repeated two to three times, and in the final stage, the pieces are placed in vegetable oil in which spices (mustard, fenugreek, asafoetida) have been cooked. Tomato and lemon pickles are fermented by yeasts and lactic acid bacteria.

Mango, green chillies, and mixed fruits and vegetables can be pickled in this way. Fruits and vegetables can also be stored in concentrated brine, for 6–9 months.

Table 15.2. Yeasts in some fermented foods from countries other than India

Species	Rice	Wheat	Corn	Sorghum	Rye	Other Cereal	Cassava	Vegetable	Fruit	PlantPr	Milk	Meat	Fish	Misc	Product	Region
Zygosaccharomyces_ rouxii	Y		Y					Y	Y						Sake, ogi, shoyu, cucumber, wine	Japan, Africa
Saccharomyces_sake	Y														Sake	Japan
Torulopsis sp.	Y									Y					Sake, lao-chao, kokja, cocoa,	Japan, China, Korea, cocoa,
Endomycopsis sp.	Y														Laco_chao	China
Hansenula sp.	Y														Chiu-miang, kokja, cocoa	China, Korea
Saccharomycopsis fibuligera	Y														Lao-chao	China
Saccharomyces cerevisiae	Y	Y	Y	Y	Y				Y	Y				Y	Puto, hopper, tapuy, tape, crackers. Bread, beer, ogi, bussa, chincha, Whisky, burukuttu, rye_bread, wine, Vinegar, basi, palm_wine, tibi, vodka,	Philippines, Sri-Lanka, Indonesia, USA, Kenya, South_America Nigeria, Europe, Mexico, France, Africa, Malaysia,
Saccharomyces_ uvarum	Y	Y													Cheese, rum Tapuy, sourdough_ bread	Philippines, USA
Saccharomyces_ sp.	Y			Y		Y		Y		Y			Y		Brem, tapai pulut, takju, kokja, Kisra, kawal, kecap, tapaiubi Takuan-zuki, cocoa, jeotkal	Indonesia, Malaysia, Korea, Sudan Japan Korea
Endomycopsis_ burtonii	Y														Tapuy, kapeketan	Philippines, Indonesia
Candida_lactosa	Y														Tapuy	Indonesia
Pichia sp.	Y					Y				Y					Kokja, miso, cocoa	Korea, Japan

Table 15.2 (*Contd.*)

Species	Rice	Wheat	Corn	Sorghum	Rye	Other Cereal	Cassava	Vegetable	Fruit	PlantPr	Milk	Meat	Fish	Misc	Product	Region
Candida sp.	Y		Y	Y			Y				Y				Kokja, pito, kisra, hulu-mur, gari, Fufu, Tapaiubi, cocoa, gariss	Korea, Nigeria, Sudan, West Africa
Saccharomyces_busae_asiaticae	Y									Y					Busa	Turkestan
Hansenula_anomala	Y							Y							Tape, takuan-zuke, cucumber brine,	Indonesia, Japan,
													Y		Fish_roe	Korea
Endomycopsis_fibuligera	Y														Tapuy, tape	Philippines, Indonesia
Torulopsis_holmii		Y			Y			Y			Y				Sourdough bread, olives, cucumber, Koumiss, sour rye bread	USA, Spain, Russia, Europe
Saccharomyces_exiguus		Y			Y				Y						Panettone, pandoro, rye_bread, wine	Italy, Germany, Finland
Saccharomyces_inusitatus		Y													Sourdough bread	USA
Candida_krusei		Y	Y		Y	Y		Y	Y						Sourdough bread, beer, bussa, rye Olives, wine	USA, Uganda, Kenya. Europe, Sudan, Spain,
Candida_tropicalis		Y													Sourdough_bread	USA, France
Saccharomyces_minor		Y													Sour bread	Russia
Torulopsis_candida		Y													Sangak bread	Iran
Torulopsis_colliculosa		Y								Y					Sangak break, basi	Iran, Philippines
Candida_mycoderma			Y												Ogi	Africa
Rhodotorula_sp.			Y					Y							Ogi, cucumber, cocoa	Africa,
Geotrichum_candidum			Y							Y	Y				Pito, poi(taro), cheese, vilia	Nigeria, Hawaii, France, Finland
Candida_quilliermondii			Y												Injera,	Ethiopia
Saccharomyces_chevalieri			Y												Burukutu	Nigeria

Species	(1)	(2)	(3)	Product	Origin
Pichia_saitoi	Y			Rye_bread	Europe
Torulopsis_halophilus		Y		Shoyu	Japan
Torulopsis_versatilis		Y		Shoyu, miso, brined cucumbers	Japan
Torulopsis_famata		Y		Shoyu	Japan
Torulopsis_etchellsii		Y	Y	Shoyu, miso, cucumbers	Japan
Torulopsis_nodaensis		Y		Shoyu	Japan
Torulopsis_mannitofaciens		Y		Shoyu	Japan
Torulopsis_halonitratophila		Y		Shoyu	Japan
Saccharomyces_acidifaciens		Y		Shoyu	Japan
S. a. var. halomembranis		Y		Shoyu	Japan
Zygosaccharomyces_rouxii		Y		Shoyu	Japan
Z._rouxii var._halomembranis		Y	Y	Shoyu, miso, cucumber brine, wine	Japan
Pichia_farinosa		Y		Shoyu	Japan
Candida_polymorpha		Y		Shoyu	Japan
Trichosporon_behrendii		Y		Shoyu	Japan
Zygosaccharomyces_nukamiso				Shoyu	Japan
Candida_tenuis			Y	Takuan-zuke	Japan
Candida_solani			Y	Olives	Spain
Candida_rugosa			Y	Olives	Spain
Candida_valida			Y	Olives	Spain
Torulopsis_sphaerica			Y	Olives	Spain
Pichia_membranaefaciens			Y	Olives, cheese	Spain, France
Hansenula_subpelliculosa			Y	Olives	Spain
Torulopsis_lactis			Y	Olives, brined cucumber	Spain
Saccharomyces_rosei			Y	Cucumber	

Table 15.2 (Contd.)

Species	Rice	Wheat	Corn	Sorghum	Rye	Other Cereal	Cassava	Vegetable	Fruit	PlantPr	Milk	Meat	Fish	Misc	Product	Region
Zygosaccharomyces_bailii								Y							Cucumber	
Saccharomyces_delbrueckii								Y	Y						Cucumber, wine	
Debaryomyces_membranaefaciens								Y	Y						Cucumber, wine	
Pichia_ohmeri								Y							Cucumber	
S. cer. var. ellipsoideus								Y							Cucumber	
Torulopsis_delbrueckii									Y						Wine	
Candida_stellata									Y						Wine	
Candida_vini									Y						Wine	
Metschnikowia_pulcherrima									Y						Wine	
Schizosaccharomyces_pombe									Y						Wine	
Schizosaccharomyces_sp.									Y	Y					Vinegar, palm wine	
Saccharomyces_bayanus										Y					Cocoa, coffee beans, palm wine	
Endomycopsis vini										Y					Basi	Philippines
Candida_brumptii										Y					Basi	Philippines
Candida curiosa										Y					Basi	Philippines
Candida_capsuligena										Y					Basi	Philippines
Torulopsis_globosa										Y					Basi	Philippines
Torulopsis_ernobii										Y					Basi	Philippines
Saccharomyces carlsbergensis										Y					Basi	Philippines
Saccharomyces_carbajali										Y					Beer	
Saccharomyces_marxianus										Y					Pulque	Mexico

Organism			Substrate	Origin
Schizosaccharomyces_sp.	Y		Coffee beans	
Kloeckera_sp.	Y		Coffee beans	
Hanseniaspora_sp.	Y		Cocoa beans	
Saccharomyces_pyriformis	Y		Cocoa beans	
Saccharomyces-kefir	Y		Ginger beer	
Kluyveromyces_marxianus	Y		Kefir, yoghurt	Russia
Kluyveromyces_fragilis	Y		Cheese	France
Kluyveromyces_bulgaricus	Y		Cheese, vinegar	France
Yarrowia (Candida) lipolytica	Y		Cheese	France
Candida_sake	–		Cheese	France
Candida_intermedia	Y		Cheese	France
Debaryomyces_hansenii	Y		Cheese	France
Candida pseudotropicalis	Y		Cheese	France
Saccharomyces majortaette	Y		Vinegar	France
Debaryomyces_kloeckera	Y		Taette	Europe
Debaryomyces_phaffii		Y	Dry sausage, cured_meat	
Saccharomyces_specialis		Y	Dry sausage, cured_meat	
Candida_clausenii		Y	Dry sausage, cured_meat	
		Y	Patis	Philippines

2.8 Yeasts in Fermented Foods from Other Countries

These include soy sauce, miso paste and sake (rice wine) from Japan, and similar fermented foods from China, Indonesia, and other countries of the region (Table 15.2). Many of these are high in salt content, and are fermented by osmotolerant yeasts and, often, by filamentous fungi. *Zygosaccharomyces rouxii*, *Pichia miso* (*Pichia farinosa*), *Aspergillus oryzae*, and *Rhizopus oligosporus* are often found in these products (Onishi 1990). Sauerkraut and whole pickled cabbage are used from the Ukraine to Germany, Holland, Belgium, and France, and are produced by lactic acid fermentation, by bacteria occurring naturally on the surface of the leaves. Yeasts are also present and may produce esters which contribute to the aroma of the product (Reddy et al. 1982; Rose 1982; Reed 1983; Steinkraus 1983; Wood 1985). Occasionally, yeasts (*Pichia membranaefaciens*) may dominate the microflora and become spoilage agents. Sauerkraut is widely used in North America and parts of Latin America, especially in areas where colonists from Northern Europe have settled. Sourdough bread has also become extremely popular in North America recently.

In Africa, kaffir beer is used as a beverage and a food (Odunfa 1985). It is the product of fermentation by a mixed culture of lactic acid bacteria and yeasts. Sorghum (*Sorghum caffosum* or *Sorghum vulgare*), 180–360 g/l of beer, is malted by soaking in water for 1 or 2 days, dried, and sprouted. The sprouted grain is dried, the malt is ground, made into a thin gruel, boiled, and cooled. Uncooked malt is added as inoculum and the mash is fermented overnight. It is boiled, cooled, and let stand for 2 days, when more uncooked malt is added. After fermentation for another day, the husks of grain are filtered off and the beer is consumed.

In large-scale production, the mash is soured before fermentation and inoculated with a brewing yeast (*Saccharomyces cerevisiae*). The husks are filtered off and the mash is fermented. The beer has a relatively high starch and B-vitamin content and is quite nutritious. The process is like that used in fermentation of other cereal-based drinks in India, and the mash is fermented by a similar microflora.

3 Yeasts in Therapeutics

Yeasts are a possible source of a wide range of therapeutic products, from vitamins to heterologous proteins. Yeast, dried or as extracts (Marmite), is an excellent source of the B-vitamin complex, including riboflavin and thiamin; carotenoids are produced by *Rhodotorula* species, saccharocin, an aminoglycoside antibiotic by *Saccharopolyspora* spp., Vanoxonin by *Saccharopolyspora hirsula*, ergosterol, and analog of vitamin D3, by *Saccharomyces cerevisiae* and the closely related species, *Saccharomyces carlsbergensis* (*S. pastorianus*), and other sterols or steroid or steroid precursors by *Hansenula*, *Kloeckera*, *Pichia*, *Rhodotorula*, and *Schizosaccharomyces* species. Glycan, a by-product of the manufacture of yeast extract, is a useful proteinaceous additive to foods. Glutathione and nucleic acids are also obtainable from yeasts. RNA, from yeast, has been reported to have a number of therapeutic uses. Skin respiratory factor, which is reported to have wound-healing properties, is a yeast-derived product. The coenzyme A protein synthesizing complex may be used in detection of cancer at an early stage.

Yeasts of several species are used as vehicles for the production of heterologous proteins, some of which, of great therapeutic value, are interferon, hepatitis B surface antigen (used in vaccine production), epidermal growth factor, used in hastening wound healing in corneal transplants and in similar situations, fibroblast growth factor, platelet-derived growth factor, human insulin, tissue plasminogen activator, and many others. Numerous enzymes may also be produced in yeasts as heterologous proteins, some of which have therapeutic value.

Yeasts may also be used in special diets, such as to supply chromium to diabetic patients who lack the glucose tolerance factor.

Yeast protein is of reasonably high quality, though it is deficient in sulfur-containing amino acids (methionine). Although yeast proteins are not equal in value to animal proteins, the deficiencies can be made up in properly mixed diets. *Candida utilis* (*Hansenula jadinii*) has been used as a food and feed supplement since World War II. The content of nucleic acids (RNA) must be reduced by heat treatment, or the amount of yeast in the diet limited, to avoid toxic effects due to the accumulation of uric acid in the blood plasma and in the joints (gout).

Yeasts of the genus *Kluyveromyces* (and their anamorphs in the genus *Candida*) form the enzyme β-glalactosidase, which can be used to remove lactose from milk concentrates of use to some individuals who show lactose intolerance.

Milk or ice cream from which the lactose had been removed does not form crystals when in frozen storage.

Thus, we have seen some examples of the utility of yeasts, in preservation of foods cheaply, increasing their nutritional value, and in production of valuable therapeutic agents, which are and will be of great value to humans everywhere.

References

Batra LR (1981) Fermented cereals and grain legumes of India and vicinity. In: Moo-Young M, Rovinson CW (eds) Advances in biotechnology, vol 3. Pergamon Press, Toronto, 547pp

Batra LR, Millner PD (1974) Some Asian fermented foods and beverages and associated fungi. Mycologia 66:942–948

Beuchat LR (1978a) Traditional fermenteds food products. In: Beuchat LR (ed) Food and beverage mycology. AVI, Westport, 224pp

Beuchat LR (1978b) Microbial alterations of grains, legumes and oilseeds. Food Technol 32(5):193–198

Beuchat LR (1983) Indigenous fermented foods. In: Reed G (ed) Biotechnology, vol 5. Food and feed production with microorganisms. Verlag Chemie, Weinheim, 477pp

Chan HT (1983) Handbook of tropical foods. Marcel Dekker, New York, 639pp

Desikachar HSR (1984) Fermented Cereal and Legume-Based Foods of India. In: Saono S, Winarno FG (eds) Proc Int Symp Microbial aspects of food storage, processing and fermentation in tropical Asia. Bogor Agricultural University, Bogor, Indonesia, 141pp

Evans IH, McAthey P (1991) Comparative genetics of important yeasts. In: Weiseman A (ed) Genetically engineered proteins and enzymes from yeasts: production control. Ellis Horwood, New York

Hesseltine CW (1984) Microorganisms involved in food fermentation in tropical Asia. In: Saono S, Winarno FG (eds) Proc Int Symp Microbial Aspects of Food Storage, Processing and Fermentation in Tropical Asia. FTDC, Bogor Agricultural University, Bogor, Indonesia

Ko SJ (1982) Indigenous fermented foods. In: Rose AH (ed) Economic microbiology, vol 7. Academic Press, London, 15p

Lonsane BK (1989) Potential of two sorghum-based fermented foods of Maharashtra for exploitation by the food industry. In: Souvenier, Silver Jubilee Celebrations of the International Food Technology Training Center, Central Food Technological Research Institute, Mysore, India, 45pp

Mukherjee SK, Albury MN, Pederson CS, van Veen AG, Steinkraus KH (1965) Role of *Leuconostoc mesenteroides* in leavening the batter of idli, a fermented food of India. Appl Microbiol 13:227–237

Odunfa SJ (1985) African fermented foods. In: Wood BJB (ed) Microbiology of fermented foods, vol 2. Elsevier, London 155pp

Onishi H (1990) Yeasts in fermented food. In: Spencer JFT, Spencer DM (eds) Yeast technology. Springer, Berlin Heidelberg New York, 167pp

Reddy NR, Pierson MD, Sathe SK, Salunkho DK (1982) Legume-based fermented food: their preparation and nutritional quality. CRC Crit Rev Food Sci Nutr 17:335–370

Reed G (ed) (1983) Biotechnology, vol 5. Food and feed production with microorganisms. Verlag Chemie, Weinheim, 631pp

Rose AH (ed) (1982) Economic microbiology, vol 7. Fermented foods. Academic Press, London, 337pp

Soni SK, Sandhu DK (1990) Indian fermented foods: microbiological and biochemical aspects. Indian J Microbiol 30:135–157

Steinkraus KH (1983) Handbook of indigenous fermented foods. Marcel Dekker, New York, 439pp

Wang HL, Hesseltine CW (1982) Oriental fermented foods. In: Reed G (ed) Prescott & Dunn's industrial microbiology. AVI, Westport, 492pp

Wood BJB (1985) Microbiology of fermented foods, vols 1, 2. Elsevier, London, 371pp, 292pp

Wood BJB, Hodge MM (1985) Yeast-lactic acid bacterial interactions and their contribution to fermented foods. In: Wood BJB (ed) Microbiology of fermented foods, vol 1. Elsevier, London, 263pp

CHAPTER 16

Yeasts and Energy. The Production of Fuel-Grade Ethanol

B.K. Lonsane, G. Vijayalakshmi, and M.M. Krishnaiah

1 Introduction

Life requires energy. With increasing population, the supplies of nonrenewable energy sources have become scarcer, and competition for them stronger. Concomitantly, the importance of renewable sources has increased. They are a partial solution to the energy crisis caused by the rising consumption of energy. Their efficient use is under urgent investigation. The use of yeasts for production of ethanol for energy is therefore under investigation.

"Renewable resources" are those materials produced by the activities of green plants in fixing carbon dioxide and simultaneously trapping solar energy, such as cellulose, starch, sugars, and other carbohydrates. Some of these are converted to fats and oils. The energy of these materials can be released by burning. Microorganisms, except for a few photosynthetic bacteria, do not utilize solar energy for growth and metabolism, but can convert the compounds synthesized by photosynthesis to forms utilizable by humans, particularly for transport. Yeasts and some anaerobic bacteria convert carbohydrates to ethanol, which is easily usable as a fuel. Other bacteria convert plant and animal material to methane and other gaseous hydrocarbons. Methanol is obtainable directly by distillation of wood. All of these materials are "renewable resources".

Wood now supplies much of the energy for developing countries. It can be used as such or converted to liquid fuels, which are more useful. Conversion of wood (cellulose) to liquid fuels may require the action of microorganisms, after hydrolysis to sugars, and fermentation by yeasts or bacteria.

The cost of recovery is the major economic factor influencing the production of ethanol as a fuel. The lower limit to the concentration of ethanol necessary in the mash is about 8%. This determines the dry matter content and the BOD of the stillage, which is normally in the range of 50000 to 75000 ppm. This must be reduced to approximately 30 ppm before disposal, which adds to the cost of production.

Other important factors in selecting a microbial culture for industrial use include (Maiorella et al. 1981):

1. Absence of metabolites other than ethanol. Additional metabolic products may increase recovery costs, lower product quality, or both.

J.F.T. Spencer/D.M. Spencer (eds)
Yeasts in Natural and Artificial Habitats
© Springer-Verlag Berlin Heidelberg 1997

Table 16.1. Yeasts capable of producing ethanol from different carbohydrates

Species of yeasts	Glucose + sucrose	Lactose	Starch	Inulin	Xylose	Cellobiose	Cellodextrins
Saccharomyces cerevisiae	+						
Sacch. ellipsoideus	+						
Sacch. fragilis	+						
Sacch. carlsbergensis	+						
Sacch. oviformis	+						
Schizosaccharomyces pombe	+						
Candida pseudotropicalis		+		+			
Torula cremoris		+					
Kluyveromyces fragilis		+		+			
Brettanomyces spp.		+					
Torulopsis spp.		+		+			
Trichosporon spp.		+					
Debaryomyces spp.		+					
Saccharomyces diastaticus			+				
Kluyveromyces marxianus			+	+	+		
Candida tropicalis			+		+		
Candida shehatae			+		+		
Schwanniomyces occidentalis			+	+			
Schwannimyces alluvius			+				
Schwanniomyces castelli			+	+			
Endomycopsis fibuligera			+				
Endomycopsis castellii			+				
Candida macedoniensis				+			
Candia kefyr				+			
Candida membranaefaciens				+			
Kluyveromycs thermotolerans				+			
Sacch. fermentati				+			
Sacch. cheresiensis				+			
Sacch. kluyveri				+			
Schizosaccharomyces malidevorans				+			
Zygosaccharomyces microellipsoides				+			
Debaryomyces castellii				+			

Table 16.1 (*Contd.*)

Species of yeasts	Glucose + sucrose	Lactose	Starch	Inulin	Xylose	Cellobiose	Cellodextrins
Torulaspora delbrueckii				+			
Torulaspora pretoriensis				+			
Torulaspora globosa				+			
Kluyveromyces cellobiovorus					+	+	
Clavispora spp.					+		
Pachysolen tannophilus					+		
Pichia stipitis					+	+	
Brettanomyces anomalus					+		
Brettanomyces claussenii					+		
Candida lusitaniae						+	
Candida molischiana						+	
Candida versatilis						+	
Candida wickerhamii						+	+

2. High osmotolerance and ethanol tolerance, which increases unit productivity and reduces recovery costs. It also allows use of crude feedstocks having a higher salt content and recycling of part of the stillage.
3. Rapid growth and fermentation rate of the culture.
4. Low pH optimum of the fermentation, which reduces the possibility of contamination.
5. High temperature optimum of the fermentation, which reduces cooling costs.
6. Resistance of the strain to physicochemical stresses, including high osmotic tensions.
7. Good flocculation characteristics, to aid in separation of the cells before distillation.
8. Ability to utilize a wide range of carbohydrates.

The feedstocks available may determine the yeast species used (Table 16.1). If the feedstock is glucose, sucrose, or hydrolyzed starch, *Saccharomyces cerevisiae* will probably be used. If lactose from whey is used, a strain of *Torula cremoris* (*Candida pseudotropicalis*; *Kluyveromyces marxianus* var. *lactis*) should be selected (Castillo 1990). *Pachysolen tannophilus* may be used for fermentation of xylose, or *Schwanniomyces castellii* (*occidentalis*) for unhydrolyzed starches (Slininger et al. 1982). Some species of bacteria and filamentous fungi also produce ethanol from carbohydrates (Table 16.2).

Table 16.2. Bacteria and fungi capable of producing ethanol from different carbohydrates

Microorganism	References
Bacterial cultures	
Zymomonas mobilis, Erwinia amylovora Sarcina sp., *Sar. ventriculi*	Jain and Nath (1987); Dawes (1963); Stephenson and Dawes (1971); Bauchop and Dawes (1959)
Clostridium thermosulfurogens, Cl. thermohydrosulfuricum, Cl. thermosaccharolyticum, Cl. thermocellum, C. thermocellulaseum, Cl. acetobutylicum, Cl. perfringens, Thermoanaerobium blockii, Thermoactinobacter ethanolicus	Enebo (1951); Wood (1961); Bryant (1976); Spivey (1978); Cooney et al. (1978); Zeikus (1979); Lamed and Zeikus (1980); Hyun (1986); Shirai and Yorifuji (1986); Kanno and Tariyama (1986)
Lactobacillus casei, Leuconostoc. sp. *Aeromonas* sp., *Aero hydrophila, Klebsiella* sp., *Escherichia coli, Aerobacter aerogenes, Aerob. indologenes,* various enteric bacteria	deVries et al. (1970); Rogers et al. (1979) Reynolds and Werkman (1937); Adams and Stainer (1945); Neish and Ledingham (1949); Altermatt et al. (1955); *Blackwood* et al. (1986); Stainer et al. (1976)
Bacillus macerans, B. polymyxa, Spirochaeta stenostrepta, Ruminococcus albus	Northrop et al. (1919); Adams and Stainer (1945); Bryant et al. (1958); Hespell and Canale-Parola (1970)
Fungal cultures	
Fusarium oxysporum, Fusa. lini Rhizopus sp., *Mucor* sp.	Gibbs et al. (1954); Gleason (1971); Batter and Wilke (1977); Margulies and Vishiac (1961)

2 Feedstocks and Their Processing

Raw materials – feedstocks – are 55–75% of the total cost of production of fuel ethanol (Maiorella et al. 1981). The choice of feedstocks is often dictated by the materials available. Sugar cane, molasses, cassava, babassu nuts, Jerusalem artichoke, and sweet potatoes are major carbohydrate sources in tropical countries, and corn, wheat, potatoes, and sugar beets in temperate regions. The cost of transport and processing of the feedstock is significant. Transportation costs for the ethanol produced are obviously much lower.

2.1 Feedstocks Containing Glucose and Sucrose

These feedstocks are fermented directly by yeast; the processing required is minimal. They include sugar cane, sugar beet, fodder beets, molasses, surplus fruits and fruit processing wastes, and sweet sorghum. Processes for their utilization are well known. Some of these materials are important foods and food products which must also be considered.

2.1.1 Sugar Cane

Sugar cane (*Saccharum* spp.) is an important crop in tropical countries. The ratio of energy output:input in its production may reach 22:1 (Joule:Joule). Global production was 56×10^6 tonnes per annum in 1979. Its yield is more than 200 tonnes/ha/year, and the ethanol yield can reach 70–90 l/t under efficient. The juice contains 12–16% sugar, so 1 t of cane yields about 125 kg sugar. The solid residue (bagasse) after sugar extraction can be used for paper manufacture or boiler feed. About 2 kg bagasse/kg sugar obtained, is produced, representing 19 700 kJ/kg of heating value.

The juice is extracted by cutting the cane into chips and passing through a hammer mill, and pressing the cane in a roller press. The solids are washed with hot water, and pressing and washing is repeated three to five times. Approximately 85–90% of the sugar is extracted when the total volume of wash water is about 25% of the weight of the cane. The juice is usually clarified with milk of lime and sulfuric acid, to precipitate inorganic fractions. The resulting liquid is fermented.

The Ex-Ferm process, developed by Rolz and coworkers, gives improved recovery of juice (Rolz et al. 1979). The sugar is extracted from the sugar cane segments and simultaneously fermented by yeast to ethanol. Dilute ethanolic broth is recycled to the fermentor containing fresh sugar cane segments.

A drum fermentor operated at a high solids content of sugar cane segments produced ethanol in high concentrations and minimum amounts of liquid (Er-El et al. 1981) Sugar cane pulp was inferior to sugar cane juice as a feedstock. Treating the pulp with crude cellulase, pectinase, α-amylase, and amyloglucosidase increased ethanol yields by 20%.

Using sugar cane for fuel ethanol production occurs in tropical and subtropical countries. In Brazil, about 43% of the fermentation plants use sugar cane directly. The juice and cane lack stability in storage, and sugar cane is not cheap, since it is also used for manufacture of food-grade sugar. Bagasse can be used as fuel and for paper production. Sugar cane wax and aconitic acid are by-products.

2.1.2 Sugar Beets and Fodder Beets

Sugar beets (*Beta vulgaris*) and fodder beets (a genetic cross between sugar beet and mangolds) grow in tropical and temperate countries. Yields are 40–50 t/ha and 50–105 t/ha, respectively. Sugar beets yield per tonne 140–190 kg of sugar and 70–100 kg of dry cellulosic pulp. The latter is used as feed for cattle, sheep, and other livestock. The sugar content of fodder beets is less, though this is offset by the higher total yields. The juice of both is rich in nutrients. It is unnecessary to add nutritional supplements (nitrogen) for efficient fermentation. Yields of ethanol per hectare may be two to five times as great as from corn. The same advantages and limitations to production of fuel alcohol from sugar cane apply to sugar beets and fodder beets. Crop rotations which include nonroot crops are necessary.

Beet roots are harvested and the tops and adhering soil removed. The roots are sliced into V-shaped wedges (cossettes) and the sugar extracted by continuous countercurrent diffusion in hot water. For fuel alcohol, the juice is fermented directly. Thin slices of beet may also be used. The beets may be ground coarsely (1.27–1.91 cm) and fermented directly, without prior sugar recovery, by solid state fermentation. The efficiency can be increased by addition of water, but dilution of the juice reduces the sugar concentration to less than 15% and reduces the final ethanol concentration, making vacuum distillation necessary and increasing the cost. The solid-state fermentation process is simplest, which may offset the lower yields (Lonsane et al. 1992).

2.1.3 Molasses

Molasses is a viscous, sugar-rich by-product obtained during the manufacture of sucrose from cane sugar or sugar beet (Paturau 1982). The sugar remains noncrystallizable due to the high viscosity of the juice after repeated evaporation during sugar recovery. It contains 50–55% fermentable sugars, inorganic salts, soluble nonsugar organic residues and nonfermentable reducing compounds. Blackstrap molasses from sugar cane contains sugar degradation products (1,3-fructopyranose, caramel residues, hydroxymethyl furfural, acetoin, formic acid, levulinic acid, and others), which are not fermented by yeasts. Molasses contains some of the compounds (sulfite, phosphoric acid, and sodium phosphate), added to clarify the juice during sugar manufacture. Some of these compounds and the high level of calcium salts in molasses are inhibitory to the yeasts. Only about 90% of the sugar in blackstrap molasses is fermentable.

Treatment of molasses is important. The molasses may be treated with $(NH_4)_2SO_4$ and calcium superphosphate, the pH adjusted to 4.5–5.2, and heated to 80 °C. The sludge of $CaSO_4$ and colloidal organic inhibitors is allowed to settle. Or the molasses is heated to 70–90 °C and H_2SO_4 added to reduce the pH to 3–4 and precipitate the sludge. Both methods cause some degradation of sugars and loss of yeast growth factors and nutrients, especially at high temperatures. Additional nutrients are required when clarified molasses is used. The molasses must be clarified, especially if the yeast or the stillage is recycled. Clarification reduces scale formation in the distillation column.

The composition of the molasses varies considerably. However, molasses is probably the cheapest feedstock available, and is easily shipped and stored at ambient temperatures.

2.1.4 Fruits and Fruit Wastes

Fruit juices (grape, peach, apricot, pear, pineapple, banana, etc.) contain 6–15% glucose or sucrose, are good substrates for ethanol production, and are used for production of beverage alcohol. They are used for human food and cost too much to be used for production of fuel alcohol. Dates, *Mahua* flowers, and rain tree pods from tropical and semiarid climates could be used in production of fuel alcohol, but are not used commercially.

The demand for fruit juices is increasing with increasing urban population. The wastes: apple and grape pomace, citrus crush, and peach peel and pits, are potential feedstocks for fuel ethanol production. Utilization of these wastes reduces the cost of treatment, though this may pose other problems. The water solutions resulting from extraction of the sugar from these wastes are too dilute for economic use in production of fuel ethanol. The waste may be fermented directly in the solid state, without extraction of the sugars. The fermentative production of fuel ethanol may be integrated into a waste disposal system (Maiorella et al. 1981). Seasonal fluctuation in production of fruit and fruit-processing wastes remains a problem in ethanol production from these feedstocks.

2.1.5 Sweet Sorghum

Sweet sorghum (*Sorghum bicolor*) contains sucrose and starch in the stem. The more complicated recovery process required makes it uneconomic to use it for ethanol production. The plant is easily cultivated on a wide range of soil and climatic conditions, so it is a potential feedstock.

2.1.6 Nipa Plant

The nipa plant (*Nipa fructicans*) grows well in estuarine swamps in southeast Asia and Pacific countries. The sap contains about 15% sugar and is a potential feedstock for production of fuel ethanol. The wild trees are difficult to reach, the sap is perishable, and the plant is not cultivated commercially.

2.2 Starchy Feedstocks

2.2.1 Potential Feedstocks

Cereals (corn, wheat, oats, barley, rice, grain, sorghum, and roots such as cassava, potato, and sweet potato) are some possible starch-containing feedstocks for production of fuel ethanol. Cereals are available throughout the year.

These feedstocks usually contain 50–55% and sometimes 90% starch. Corn, cassava, and sweet potato probably have the greatest potential for production of fuel ethanol, corn being the most suitable in the USA, cassava and sweet potato in tropical countries. All of these crops are used in great quantities for human food, but can be used for fuel ethanol production when there is a surplus.

2.2.2 Cull Potatoes

These may comprise 10–20% of the total crop. About 2 million gallons of fuel alcohol could be produced from these, plus more, from the fibrous waste and soluble wastes from potato processing. The high water content of potatoes (80–85%) makes it difficult to produce fuel ethanol from them economically. Small fermentation units may be economic where surplus potatoes are available.

2.2.3 Starchy Wastes

These and by-products from food, chemical, and related industries are also potential feedstocks. The fibrous residue from cassava contains 50–60% dry matter and is approximately 20% of the weight of tubers processed during manufacture of starch and sago. This can be utilized either as feedstock for ethanol production or in solid state fermentations (SSF) for enzyme production. Similar wastes (coffee husk/parchment, tamarind seeds, jackfruit seeds, sal seed residues from oil recovery), generated in food and related industries, are also potential feedstock.

2.2.4 Moldy Cereals

Corn and wheat, unfit for human consumption because of fungal contamination, are potential feedstocks for fuel ethanol production. Grain contaminated with such mycotoxins as aflatoxin, ochratoxin, zealraenone, patulin, citrinin, and Toxin T-2 must be detoxified, but could be used as feedstocks for fuel ethanol production first. The mass of residues is reduced during fermentation. Many of the mycotoxins are destroyed as well. Mycotoxins remaining in the spent grains can be destroyed by standard methods, at lower costs because of the reduced mass. The metabolism of the yeast is not affected by most of the mycotoxins, except for Toxin T-2. Ammonium hydroxide, which is added to the cereal hydrolysate as a nitrogen source, assists in detoxifying the mycotoxins during fermentation. Utilization of spoiled cereals allows production of a useful liquid fuel and reduces environmental pollution with various highly toxic agents.

3 The Starch Molecule

Starch, a major carbohydrate reserve in higher plants, is a heterogeneous polysaccharide which occurs as water-insoluble granules in the cells of grains or tubers. It consists of the glucose polymers amylose and amylopectin, arranged in a complex, folded structure. The ratio of amylose to amylopectin is normally 20–25:75–80. The amylose molecule contains about 800–1500 glucose residues and amylopectin, 5000–40000 residues.

The starch granules are practically insoluble in water and are not attacked by α-amylase or amyloglucosidase at ambient temperatures. During gelatinization and liquefaction, the granules swell when heated in water and then rupture gradually. The amylose and amylopectin unfold and the starch is dispersed in the solution. Gelatinization is normally initiated at 65–75 °C and is complete at 140–150 °C. Temperatures of 105–110 °C are sufficient for most starches. There is little gain in glucose formation by gelatinization at 140–150 °C. The gelatinized starch is highly viscous and can be liquefied with acid or bacterial α-amylases through conversion of starch to soluble dextrins. Amyloglucosidase converts liquefied starch into glucose.

3.1 Importance of Starch Saccharification

Most strains of *Saccharomyces cerevisiae* used for production of industrial ethanol by fermentation cannot utilize starch directly. Liquefaction and saccharification of starch to reducing sugars is essential for utilization of starch for ethanol production. Capital investment for starch liquefaction reaches 15–20% of the total plant cost (Maiorella et al. 1981). The equipment and technique must avoid formation of compounds which inhibit fermentation and efficiency of conversion of starch to reducing sugars. The use of dried chips instead of fresh tubers as starting materials alters the nature and economics of the process. Cereals contain high-value products which are lost if not recovered before fermentation. Isolated starch may be a better fermentation feedstock. Whole grain should be free of dirt and cracked hulls. The grains are ground to flour of 20–80-mesh particle size to facilitate wetting (Maiorella et al. 1981).

3.1.1 Starch Solubilization by Heating

Cereals have long been fermented for production of potable alcohol. Saccharification of the grains by batch cooking the slurry of mashed or pulverized feedstock at atmospheric pressure for gelatinizing and solubilization of the starch is done before enzymatic saccharification with malt. Cooking the slurry under pressure improves the process. The mash, containing approximately 0.27 kg dry solids/l, is heated to 130–150 °C at pH 5.5 by injecting steam at 690 kPa (100 psi) into a horizontal pressure vessel equipped with rake agitators for 10–30 min. The slurry is then cooled rapidly by pressure blowdown and vacuum evaporation to approximately 60 °C before addition of barley malt for diastatic saccharification of the starch. The slurry may be fed continuously through a jet heater into a cooking tube for a plug-flow residence time of 5 min at 180 °C, after which the slurry is flash-cooled to about 60 °C for saccharification with malt. The continuous process allows more uniform cooking, lower steam consumption, and reduced capital investment. Pressure cooking and continuous operation have replaced the traditional system of cooking at ambient pressure. The batch process is labor-intensive and requires high-pressure steam, while continuous pressure-cooking permits reduced cooking times. This prevents losses of sugar due to degradation, improves the yield by 3–5%, and sterilizes the mash (Maiorella et al. 1981).

The energy consumption (for both the batch and continuous processes) for cooking the starch at pH 5.5 is high. Cooking the starch at 140 °C requires about 700 kcal/kg of ethanol produced. The total enthalpy of combustion of the ethanol is 7000 kcal/kg (Maiorella et al. 1981). Solubilization of the starch by cooking at high pressure is appropriate for the production of beverage alcohol. Good taste and flavor bring better prices. The method is not feasible for production of fuel ethanol. The process has been improved in the past few years.

3.1.2 Acid Hydrolysis

In this method, the starch is hydrolyzed by acid treatment at high temperature. The maximum yield obtainable was about 75% of the theoretical. Other disadvantages were: (1) incomplete hydrolysis of acid-stable α-linkages; (2) acid-induced reversion of the glucose molecule to more heat-stable dimers/polymers; (3) formation of unfermentable degradation and condensation products which inhibit microbial growth; and (4) severe corrosion problems.

3.1.3 Acid-Enzyme Hydrolysis

Combined acid and enzymatic hydrolysis is used for liquefaction and saccharification of starch. Initial hydrolysis is at atmospheric or higher pressure. The acid concentration is less than in starch hydrolysis by acid treatment alone. Acid treatment produces liquefaction and partial saccharification of the starch. The pH of the acid-treated slurry is then adjusted to 4.5, the slurry is cooled to about 60 °C, and saccharified enzymatically at 60 °C for 24–48 h with amyloglucosidase. This step may be combined with the fermentation stage (Sect. 8.2.6). The hydrolysate is purer and contains fewer degradation products and produces a higher yield of reducing sugars.

3.1.4 Enzyme-Enzyme Process

This process represents a major advance in utilization of starch. The stages are: liquefaction and partial saccharification of the starch slurry at pH 6.5–7.0 by a thermostable bacterial α-amylase at 90–100 °C for 1–2 h, and saccharification of the liquefied starch slurry, pH 4.5, at 60 °C for 24–48 h by a fungal amyloglucosidase. High-purity glucose is obtained in high yield, and the amount of degradation products is negligible. The method has almost completely replaced the acid and acid-enzyme processes. It allows direct use of starch and ground grain, though with grain, the value-added products are lost. Other advantages of the enzyme-enzyme process are: lower capital and recurring expenses; significant reductions in energy requirements; no requirement for expensive corrosion-resistant equipment which is essential in acid-based methods; formation of fewer by-products; and better process control.

Batchwise enzyme-enzyme hydrolysis of starch uses slurries containing 15–20% starch. This produces sugar solutions and ethanolic fermentation liquors which do not affect the activity of the yeast. The steps are:

1. Coarse grinding of the grain.
2. Preparation of a slurry containing about 20% solids.
3. Steam heating in a closed, agitated cooker.
4. Addition of thermostable bacterial α-amylase before starch gelatinization.
5. Holding at 70–80 °C for 15–30 min for preliquefaction, to avoid excessive mash viscosity during cooking.
6. Heating the mash to 100–160 °C and holding for 15–30 min to complete gelatinization.
7. Cooling the mash by vacuum and/or heat exchanger, to 70–85 °C.
8. Addition of a postliquefaction dose of thermostable α-amylase.
9. Holding the mash for 15–30 min for complete liquefaction.
10. Adjusting the pH to 4.5.
11. Cooling the mash to about 60 °C for 24–48 h for complete saccharification.

The α-amylase is added in two steps. The process requires larger quantities of enzyme, since the residual enzyme from the first step is inactivated when the slurry is heated to 140–150 °C.

Saccharification of starch by amyloglucosidase and fermentation of glucose to ethanol by yeasts are done simultaneously. The mash is cooled to about 30 °C, the pH is adjusted to 4.5, and both amyloglucosidase and yeast inoculum are added along with the yeast nutrients. This reduces the time required significantly. The rate of saccharification decreases as the saccharification nears completion. The maximum yield of glucose is approximately 95%, but is increased to near 100% if glucose is continuously removed from the hydrolysate by fermentation to ethanol.

Alternatively, the mash is treated with α-amylase, cooled to 60 °C, the pH adjusted to 5.0, and the full dose of amyloglucosidase added. The mash is held at 60 °C or allowed to cool gradually for about 4 h to achieve rapid partial saccharification. It is then cooled to about 30 °C and incubated with yeast for simultaneous saccharification and fermentation.

3.1.5 Modified Enzyme-Enzyme Methods

Continuous culture requires an important modification of enzyme-enzyme methods. The slurry of ground grains or tubers, containing 20% solids, is cooked in a continuous process for starch saccharification. This begins with preliquefaction, high-temperature gelatinization and postliquefaction, using a jet cooker and tubular heat exchanger. α-Amylase is added in a single dose, in continuous processes, which requires less enzyme than when two-step addition is used. Other modifications include (1) cold-slurry preliquefaction, (2) hot-slurry preliquefaction and (3) the manioc process. In (1) the slurry of granular starch is mixed with α-amylase, amyloglucosidase, nutrients, and yeast inoculum for simultaneous saccharification and fermentation at 30 °C, without a liquefaction step. The fermentation rate in this process is about half the rate for the process which includes liquefaction. Increasing the starch content from 20 to 40% increases the rate. The rate-limiting step in this process is the breakdown of granular starch to dextrins by α-amylase.

Potatoes or grain may be suspended in water to a concentration of 20% solids, without previous milling of the starchy substrates. The suspension is heated in a Henze cooker (a specially designed autoclave) and the cooked mass is blown out through a strainer valve into a mash tub equipped with an agitator and cooling coil. A preliquefaction treatment with enzyme is not used.

Use of immobilized enzymes can reduce the quantity of enzyme required. Use of immobilized α-amylase and amyloglucosidase caused technical and operational problems in starch hydrolysis.

3.1.6 Improvements in Efficiency of the Enzyme-Enzyme Method

The efficiency of the enzyme-enzyme method can be improved by increasing the rate and extent of starch hydrolysis. Pullulanase, a debranching enzyme which splits α-1,6-glucosidic linkages in pullulan and amylopectin (Peppler and Reed 1987), can accelerate the rate of hydrolysis of α-1,6-linkages when used in combination with amyloglucosidase. Isoamylase, another debranching enzyme, is capable of hydrolyzing all the α-1,6-interchain linkages in amylopectin. It is very specific and cannot act on pullulans. It cannot completely debranch β-limit dextrins and a minimum of three α-D-glucose residues is required by this enzyme in the B or C chains of the substrate.

α-Amylase may be used with β-glucosidase for starch hydrolysis at lower temperatures for reduced gelation of the starch and beginning the saccharification much earlier than in the normal process. These enzymes may not yet be commercially viable for production of fuel alcohol.

3.1.7 Enzymatic Hydrolysis of Raw Starch Without Cooking

The enzymatic hydrolysis of native or raw starch at ambient temperatures without cooking has tremendous economic advantages over high-temperature methods. Uncooked glutinous rice starch can be completely hydrolyzed to glucose by a mold glucoamylase. A mixture of enzymes from hog pancreas and *Aspergillus oryzae* hydrolyzed uncooked starch completely, and an amyloglucosidase having stronger debranching activity hydrolyzed raw starch. All these enzymes can be adsorbed on the starch granules and act on raw, ungelatinized starch.

The rate of hydrolysis of raw starch by these enzymes is rather slow, but the hydrolysis rate can be improved significantly. The rate is increased threefold by the synergistic action of α-amylase.

Isoamylase, pullulanase, and β-amylase also have synergistic action. The hydrolysis of raw starch at room temperature has been tested on an industrial scale. The method may be useful for industrial production of fuel ethanol from starch.

3.1.8 Direct Fermentation of Starch

The cost of production of ethanol by fermentation of starch would be greatly reduced if prior saccharification with acid and/or enzyme were not required. The cost of raw materials and processing amounts to 50–70% of the total cost of fuel ethanol. The cost of thermostable bacterial α-amylase for starch hydrolysis is about 12% of the total raw material cost. The total batch time is rather high, because of the slow rate of enzymatic hydrolysis, which increases the cost of the process. This is because of the inability of most industrial strains of S. cerevisiae to hydrolyze starch.

Some fermentative yeasts produce amylolytic enzymes, and ferment the glucose formed to ethanol. The value of these yeasts in production of ethanol from starchy materials is known, and attempts were made to develop economical and practical technologies using these species are being developed.

The amylolytic fermentative yeasts (Table 16.1), unlike *Saccharomyces cerevisiae*, produce extracellular amylases (α-amylase, amyloglucosidase) and can utilize soluble or raw starches as sole source of carbon. Not all are suitable for commercial production of ethanol. *Saccharomyces cerevisiae* var. *diastaticus* does not produce α-amylase, and liquefied starch or dextrin must be used as substrate (Laluce and Mattoon 1984), or external α-amylase must be added to the medium. Its amyloglucosidase has little or no debranching activity (Russell et al. 1986), and *Saccharomycopsis fibuligera* produces little ethanol. Using mixed cultures for cofermentation with *Saccharomyces cerevisiae* or *Zymomonas mobilis* was unsatisfactory. However, Katsuragi et al. (1994) fused protoplasts of S. fibuligera and S. diastaticus, sorted the fusants with a flow cytometer, and obtained a true hybrid which consumed starch as fast as the parental S. fibuligera, grew as fast as S. diastaticus, and produced more ethanol.

Schwanniomyces occidentalis is one of the most promising of the amylolytic yeasts. Soluble starch, 2.5%, was converted to ethanol nearly quantitatively (Calleja et al. 1982). Higher concentrations were not converted so completely. It has also a lower ethanol tolerance. This species can be cocultured with *Saccharomyces cerevisiae*, with dextrins as substrate. Other limitations include (1) impaired activity of amylase at increased ethanol concentrations, (2) repression by ethanol even in nongrowing cells, (3) strong reduction in α-amylase production in anaerobic conditions, and (4) requirement for dissolved oxygen (10% of saturation) in the medium for active amylase biosynthesis.

Ethanol production by *Schw. castellii* has been investigated at ORSTOM, Montpellier, France, in a column fermentor packed with sugar cane pith bagasse in an SSF system. The medium contained soluble starch (10%). The first stage lasted 24 h, for production of biomass, amylases, and simultaneous partial saccharification of soluble starch under aerobic conditions. The second stage was an anaerobic fermentation for production of ethanol and continuous stripping of the alcohol produced. The maximum ethanol yields were 83.2–109.3 mg/g initial dry matter under the conditions used (Saucedo-Castenado et al. 1992).

The pH of the medium significantly affected ethanol yields. Yields obtained in a single-stage, single-vessel process for biomass buildup, starch hydrolysis, and ethanol production equalled those obtained in a similar, two-vessel, two-stage process. For process control, measurement of the CO_2 content of the exhaust air was an efficient criterion for obtaining equal biomass yields at different bed heights in a column fermentor under aerobic conditions. On-line monitoring and control systems for CO_2 and O_2 concentrations in aerobic and anaerobic solid state fermentations have been developed. A novel system for scaleup, based on maintenance of equal heat and water balances at all stages, was effective for production of ethanol by *Schw. castellii* from starchy substrates (Saucedo-Castenado et al. 1992).

3.1.9 Choice of Method

The physicochemical characteristics of the starch or starchy substrate may determine the method of hydrolysis. The enzyme-enzyme method was used at CFTRI for production of ethanol from cassava chips, and the acid-enzyme method for ground potato and cassava fibrous residue. CFTRI developed a novel type of cooker for the mild acid treatment of fibrous cassava mass and avoiding low sugar concentrations in the hydrolysate.

4 Cellulosic Feedstocks

Cellulose is a linear homopolymer of anyhydroglucose units through 1,4-β-glucosidic bonds. It is a major component of plant biomass. It is synthesized by green plants by capture of solar radiation, and is a renewable carbon source available in quantity everywhere. The total carbon fixed through photosynthesis is about 100 billion t/a, cellulose accounting for about half. Cellulose is a long-chained molecule consisting of hexose units and exists as fibrous crystals in plants. A matrix of lignins and hemicelluloses is associated with the cellulose microfibrils and gives rigidity and flexibility to the cell wall in plants.

Potential cellulosic feedstocks include plants and plant residues of all kinds (Maiorella et al. 1981). Ethanol was produced from wood in quantity during World Wars I and II. Yields were low, use of strong acids caused corrosion problems, and production costs from other processes were lower. There was renewed interest in these processes recently because of the world energy crisis.

The cellulose molecule is highly recalcitrant because of its high crytallinity, making it difficult to hydrolyze. Access of hydrolyzing agents to cellulose is reduced by the lignin matrix, so pretreatment of the feedstock is critical. For acid hydrolysis, chipping is sufficient. Enzymic hydrolysis requires mechanochemical pretreatment. Disadvantages of acid hydrolysis are degradation of the product, impurities in the hydrolysate, low yields, high capital costs, corrosion of the equipment, and inactivation of acid by noncellulosic materials in crude cellulosic feedstocks. Hydrolysis of cellulose by cellulolytic microorganisms or by cellulolytic enzymes (bacterial) lacks most of the limitations of the acid method and can be done at ambient temperature and pressure. This eliminates sugar decomposition, gives higher yields, reduces costs of downstream processing such as neutralization and purification of the hydrolysate, produces waste streams which are less expensive to manage, and has lower capital and operating costs (Maiorella et al. 1981). It has not been used industrially because of the high cost of enzymes and mechanochemical pretreatment.

4.1 Hydrolysis by Dilute Acid at High Temperature.
The Slow Acid Hydrolysis Process

Wood chips are hydrolyzed in a packed bed pressure vessel with 0.2–1.0% sulfuric acid or 40–45% hydrochloric acid. The acid percolates through the bed and is continuously removed as sugar solution. The reactor is heated by direct steam injection (350 kPa) initially, and gradually increased to 1135 kPa at the end of hydrolysis (2.5–3.5 h). The sugar solution from the reactor is continuously neutralized with lime, and flash-cooled to minimize sugar decomposition. The yield is about 500 kg of sugars from a ton of bark-free wood, about 70% being fermentable glucose. The solution contains about 5–8% fermentable sugar, which makes the ethanol content too low for economic recovery. Various decomposition products (furfural, methanol, levulininc acid, other sugar acids) reduce the sugar yield and inhibit the yeast fermentation.

The wood may be dried to 12% moisture, chipped, and hammer-milled to sawdust. Prehydrolysis is done in a pressurized vessel using hot water and steam at 150 °C with 0.5% H_2SO_4. After 30 min, a 2% acid solution is added to the mixture, the temperature raised to 121 °C with more steam, and held for 2 h. The process yields a slurry containing 5% sugar. A mechanochemical treatment reduces the hydrolysis time to 10 min. The rate of hydrolysis can be improved by short-time processing at 500 °C.

Limitations are: (1) lower yields, (2) slower rates of hydrolysis, (3) high energy consumption, (4) lower sugar concentration in the hydrolysate, and (5) presence of compounds inhibiting yeast fermentation. The process is no longer used commercially in North America and western Europe (Maiorella et al. 1981).

It is used industrially in Brazil and Russia, and, on a pilot scale, in New Zealand. Here, the cost of hydrolysis may be lower than in processes using concentrated acid (Maiorella et al. 1981).

4.2 Low-Temperature Hydrolysis by Concentrated Acid

Most of the limitations of dilute acid hydrolysis method are overcome to a large extent in this case. Crystalline cellulose is completely soluble in 72% H_2SO_4 or 42% HCl solutions at 10–42 °C, and is hydrolyzed to cellulotetraose. Diluting and heating the mixture to 100–200 °C completes saccharification. The final hydrolysate is neutralized to pH 4.0 with milk of lime.

4.3 Improved Acid Hydrolysis Techniques

In the high temperature-high pressure method, hydropulped newspaper or sawdust is cram-fed into a twin corotating screw extruder to cause mechanical shear. It is then compressed to a solid plug at 3450 kPa pressure with dewatering to about 50% cellulose. The material is heated to 240 °C by injection of high pressure steam and acidified by injection of 0.5% H_2SO_4. The process converts about 60% of the cellulose in 20 s before the plug is expelled from the screw through a high pressure valve. The blown slug, containing about 30% glucose, is flash-cooled before extracting sugars in a two-stage countercurrent washer. The process consumes about 2.1 kg steam and 0.75 kWh electricity per kg of sugar produced. Corrosion-resistant equipment is required.

Two-step hydrolysis may be used if the hemicellulose content of the feedstock is high, which gives faster hydrolysis at lower temperatures. Step 1 converts hemicelluloses and part of the cellulose to monosaccharides at lower temperatures. Step 2 gives hydrolysis of cellulose to hexoses, with higher temperatures or acid concentrations and longer fermentation times. A three-stage sequential method is also possible. In stage 1, hemicelluloses are hydrolyzed and the resulting sugars are separated. In stage 2, cellulose is hydrolyzed to oligosaccharides and some glucose, in more concentrated acid. In stage 3, the oligosaccharides are hydrolyzed. Or dried wood chips are hydrolyzed using gaseous hydrogen fluoride.

4.4 Hydrolysis by Enzymes

Cellulases are added to pretreated cellulosic feedstock, and the reaction is carried out at pH 4.5–4.8 and 45–50 °C, to complete saccharification. Three classes of enzymatic activities are required for complete hydrolysis of cellulose: endoglucanases cleaving internal β-(1,4)-glucosidic linkages in the cellulose polymer at random to produce shorter chains, exoglucanases which cleave cellobiose from the nonreducing ends of the chain, and β-glucosidases which split cellobiose and other short-chain oligosaccharides to fermentable glucose. The endo- and exoglucanases have a synergistic action in saccharification.

The enzyme is the most expensive part of the cellulose saccharification process. This prevents the use of the method on a commercial scale, though a few pilot plants have been established in some countries.

4.5 Direct Saccharification by Cellulolytic Microorganisms

Direct saccharification by cellulolytic organisms would eliminate the costs of production and recovery of cellullolytic enzymes. Microbial saccharification of cellulosic feedstock directly to convert the cellulose to glucose, and fermentation of the glucose to ethanol by yeast could perhaps be done in a single- or two-step process.

This might be done by: (1) coculture of mixtures of cellulase-producing and glucose-fermenting microorganisms, or (2) use of monocultures of species having both cellulolytic and fermentative properties. This would enhance the rate of hydrolysis, improve the ethanol yield, minimize β-glucosidase inhibition of saccharification, reduce the probability of contamination, and facilitate maximum hydrolysis of cellulose. The process has been tested in the pilot plant, but so far, the process is not economical.

4.6 Pretreatment of Cellulose

The cellulose molecule in wood has an outer sheath of lignin and is crystalline in nature. It is almost inaccessible to the action of enzymes, which are limited to the slow cleavage of bonds near the ends of the cellulose molecule and conversion reaches only about 25%. The cellulose feedstock requires mechanophysical and mechanochemical pretreatment to reduce the crystallinity of the cellulose and the degree of polymerization of both cellulose and lignins. The amorphous cellulose obtained is then rapidly hydrolyzed by the enzyme. Similar mechanophysical and mechanochemical pretreatments of cellulosic feedstocks are beneficial in on acid hydrolysis. Other pretreatment systems are listed (Table 16.3). The steam explosion method is efficient and economical for hardwoods and agricultural wastes but not for softwoods.

Table 16.3. Mechanophysical and mechanochemical pretreatment methods for improving cellulose hydrolysis by enzymes

Pretreatment method	Effect
Mechanophysical methods	
Ball or hammer milling	Reduction in crystallinity, depolymerization of cellulose and lignin, particle size reduction
Wet milling	Fibrillation and delamination of cellulose, no change in crystallinity and chain length
Steam explosion	Delignification, depolymerization of hemicellulose, increased surface area, breaking the substrate into wool-like amorphous fibers
Mechanochemical methods	
Dissolving in solvent and precipitation	Complete destruction of crystallinity
Extraction with solvent	Partial delignification
Swelling	Reduction in crystallinity and chain length, partial delignification
Pulping	Partial delignification, increased surface area

4.7 By-Product Utilization

This is important in the economics of ethanol production from celluloses. The lignin residue from the dilute acid method was blown down, filter-pressed to 50% moisture, and used for steam generation. The low-pressure steam from steam explosion pretreatment was recovered and used in the plant. Acids and solvents used in pretreatment and cellulose hydrolysis are recovered for recycling. The solid residue remaining after rapid hydrolysis of cellulose saccharification is burnt to produce process steam (Maiorella et al. 1981).

The pilot plant under construction at the Biochemical Engineering Research Center, Indian Institute of Technology, New Delhi, India, is designed for complete utilization of all by-products. It uses a novel process for conversion of rice straw to ethanol. The rice straw is broken down to lignin, hemicellulose, and α-cellulose through autohydrolysis and solvent extraction. The α-cellulose is converted to ethanol by simultaneous saccharification and fermentation. Pure lignin is a basic raw material for numerous aromatic compounds. An improved and easily digestible livestock fodder is also obtained.

5 Hemicellulosic Feedstocks

Hemicelluloses (polymers of pentoses and hexoses) constitute about 40% of the plant biomass. They are second only to celluloses as carbohydrate sources. They are short, branched-chain, easily hydrolyzable heteropolysaccharides of mixed pentoses and hexoses. Pentosans yield mainly D-xylose and L-arabinose and hexosans yield D-glucose, D-mannose, and D-galactose. Hemicelluloses are potential sources for production of fuel ethanol. Some hemicellulosic feedstocks include: stem, leaves, and fibers of monocotyledons; hard- and softwoods; news- and waste-papers; waste fibers; corn residues; wheat and flax straws; sweet clover hay; sugar cane bagasse; peanut hulls; and residues from sunflower and soybean plants. Hemicelluloses are hydrolyzed by acids or enzymes.

Xylose and glucose, the major sugars in the hydrolyzates of hemicelluloses, are fermented efficiently by many bacteria. *Saccharomyces cerevisiae* can utilize only glucose, and the residual pentoses make the process highly uneconomic. Fermentation of pentoses requires a different group of yeast species (Table 16.1), *Pachysolen tannophilus* being the most promising (Schneider et al. 1981). The fermentation is strongly inhibited by high concentrations of ethanol (Slininger et al. 1982). Pentoses may be converted into fermentable isomers; i.e., D-xylose to D-xylulose, in a two-step process for ethanol production.

Further investigation of D-xylose-fermenting yeasts to obtain strains for bioconversion of hemicelluloses and their component sugars to ethanol is required. Isolation of new microbial strains for degradation of pentoses and a knowledge of their regulatory metabolism is required if efficient utilization of hemicelluloses for fuel alcohol is to be achieved.

6 Lactose-Containing Feedstocks

Cheesemaking generates approximately 115 million tons of liquid whey, containing 6 million tons of lactose, during the manufacture of 12.7 million tons of cheese. Whey is an excellent substrate for production of fuel ethanol. It contains 4.5–5.0% lactose, about 70% of the whey solids. Lactose, a disaccharide of glucose and galactose, has a limited market. Whey is mostly discarded as waste except for a small fraction used for food and feed additives. Disposal of whey creates pollution problems, since the BOD may reach 35000–40000 ppm. Intensive waste treatment is necessary for its treatment. Utilization of the lactose of whey yields ethanol and reduces the pollution caused by untreated waste.

6.1 *Saccharomyces cerevisiae*

This yeast does not possess lactase (β-galactosidase) activity, and lacks a lactose transport system. Yeasts having lactase activity and a lactose transport mechanism can assimilate and ferment lactose (Table 16.1) (Castillo 1990). Ethanol is usually produced by fermenting whey directly with lactose-fermenting yeast. Sometimes, the whey is deproteinized or mixed with other sugars before fermentation. Or lactose in whey is first hydrolyzed with lactase (β-galactosidase) and fermented with nonlactose-utilizing yeasts (*Saccharomyces cerevisiae*). Processes using batch cultivation, continuous culture, cell recycle, and cell immobilization have been developed, and industrial plants are operating. Ethanol production occurs at about 90% conversion efficiency.

6.2 Whey

Whey represents a potential substrate for fuel ethanol production, but the capital investment required is high. Nevertheless, the discharge of whey as waste results in a loss of valuable substrate.

7 Feedstocks Containing Inulin

Many plants contain inulin (polyfructose), a polymer of fructose linked by β-1.2 bonds. It is the major polysaccharide in the roots and tubers of Jerusalem artichoke, chicory, and dahlia (Guiraud and Galzy 1990). Burdock, inula, dandelion, colonopsis, and great campanula roots contain about 40% inulin, so inulin is a possible feedstock for ethanol production. Inulin can be hydrolyzed to difructose anhydride, low-molecular-weight polyfructans, fructose, and glucose by microbial enzymes (inulinase, invertase, fructosidase). The composition of the hydrolysate is determined by the feedstock composition and the specificity of the enzyme (Guiraud and Galzy 1990). A number of microorganisms, including yeasts, can assimilate and ferment inulin (Table 16.1). However, *Saccharomyces cerevisiae* has invertase activity but cannot ferment inulin.

Jerusalem artichokes have been used for ethanol production on pilot and industrial scales. Methods of hydrolysis are like those for starch hydrolysis, except that (fungal) inulase is used instead of amylases. Direct fermentation was difficult, and acid hydrolysis was too expensive (Guiraud and Galzy 1990). Steam sterilization of artichoke extract yielded about 30% reducing sugars, which facilitated fermentation with *Saccharomyces cerevisiae*. Solid-phase fermentation of artichoke tuber mashes was evaluated in 400–500-lb batches using 125-gal stainless steel fermentors and a progressive cavity pump (Ziobro and Williams 1983).

Integrated processes for the production of fructose, ethanol, and biomass, to ensure efficient utilization of inulin-containing feedstocks, were investigated. Fermentation using yeasts having inulase activity, may be preferable.

8 Fermentation Aspects

Fuel ethanol must compete price-wise with other energy sources, so the cost of production must be kept at a competitive level at all stages of the production process. The factors affecting the fermentation of the feedstock into ethanol and strategies of fermentor operation must be thoroughly understood if the process is to be economic.

8.1 Factors Affecting Fermentation

Productivity and economics of microbial processes are determined by the growth, metabolism, and regulatory mechanisms of microbes, and by physical, chemical, nutritional, and cultural factors. The objective is maximum conversion of substrate into product at minimum cost. *Saccharomyces cerevisiae* is the species most commonly used in production of ethanol, and this and other yeast species used in industrial fermentations are discussed here and in Castillo (1990).

8.1.1 Ethanol Tolerance

Yeast metabolism produces ethanol and is inhibited by it. It affects the integrity of cell membranes and the activity of the glycolytic enzymes of the pathways leading to ethanolic fermentation. Cell growth and ethanol production are inhibited in yeast at ethanol concentrations of about 11%. Higher ethanol concentrations are reached in sake fermentation, and ethanol concentrations of 23% have been obtained in conventional yeast fermentations. Ethanol toxicity is also influenced by substrate concentration, fermentation temperature, and the presence of lipids in the medium. Tolerance of the yeast to ethanol determines the maximum substrate concentration to 15–20%, and the maximum ethanol concentration to 7–10%.

Higher concentrations of ethanol can be attained by control of nutritional factors, development of yeast strains tolerant of higher ethanol concentrations and elevated osmotic tensions, and by continuous removal of ethanol from the mash.

8.1.2 Substrate Concentration

The rate of ethanol production by yeasts is determined by the sugar concentration in the medium. Productivity is reduced at lower concentrations, and is maximum at about 15% sugar concentration. The glycolytic cycle enzymes are repressed by higher sugar concentrations (catabolite repression), resulting in lower conversion rates. Yeasts having higher ethanol tolerances can ferment media having higher sugar concentrations. Glucose inhibits transcription of the genes encoding enzymes of the oxidative pathways and the enzymes themselves (Chap. 7, this Vol.). The Crabtree effect makes it possible to carry out ethanolic fermentation at lower sugar concentrations.

8.1.3 Dissolved Oxygen

Aerobic metabolism in yeast uses sugars for cell growth, not for ethanol production, at low sugar concentrations. The presence of oxygen is unimportant at higher sugar concentrations, since the Crabtree effect ensures anaerobic metabolism and ethanol formation. Trace amounts of oxygen (0.5–0.7 mmHg oxygen tension) are required during anaerobic metabolism for biosynthesis of polyunsaturated fatty acids (oleic acid) and other lipids.

8.1.4 Fermentation Temperature

The optimum temperature for *Saccharomyces cerevisiae* is usually 30–36 °C. Lower temperatures reduce the fermentation rate. At higher temperatures, viability of the yeast cells, activity of a number of enzymes, and ethanol tolerance are reduced. Heat is produced during fermentation, about 11.7 kcal/kg of substrate consumed, and the temperature must be controlled within optimum limits. This is expensive in tropical countries. Yeast strains tolerating higher temperatures have been developed in India and Brazil.

8.1.5 Other Factors

Ethanolic fermentation is affected by many factors: (1) types and concentration of inorganic and organic compounds, (2) pH of the medium, (3) nature and concentration of cometabolites formed with ethanol, (4) impurities and yeast inhibitors in feedstocks, (5) inoculum ratio, and (6) stability of the culture during long-term operation. These factors affect the growth and metabolism of yeasts and may increase the cost of production.

9 Fermentor Operation Strategies

Many different fermentor designs and operational strategies have been developed for production of ethanol, to find the most economic system (Maiorella et al. 1981). The fermentor design and/or the operation strategy may determine the success of a plant producing fuel ethanol, and play a significant role in waste management, influence recovery and reuse of by-products, and minimize the volume and strength of effluent (Maiorella et al. 1981; Lonsane and Ahmed 1989).

The common features of these strategies are: inoculum development by conventional methods, involving serial culture in vessels of increasing size: growth of the pure culture from the stock culture on a fresh agar slant, subculture and growth in shaken flasks, successive growth in aerated, agitated fermentors of increasing capacity to obtain enough yeast cells to inoculate the ethanol production fermentor at the optimum rate. The inoculum may be grown by a semi-anaerobic method using a fed-batch system to obtain a three to four times greater cell density. Commercially available dehydrated starter cultures are obtainable.

Medium composition: optimized for all nutrients. Molasses not clarified or sterilized except in the most modern plants.

Cooling is by internal cooling coils, circulating the medium through external heat exchangers, jacketing the fermentation vessel, or spraying cooling water on the vessel.

Mechanical agitation is most effective, agitation by rising bubbles of CO_2 or sparging with air less so.

The pH of the medium falls from 4.5–5.5 to about 4.0 without pH control. The pH is often maintained artificially during the fermentation. The yeast cells are harvested by gravity or mechanical means.

The fermentors used in large plants are usually cylindroconical or bottom cone tanks. These facilitate better circulation, allow thorough drainage, and permit simpler cleaning operations. After harvesting the culture, the fermentors are cleaned with water, or sterilized by chemical means or steam.

Fermentor-operating strategies are shown in Table 16.4. The methodologies are as follows.

9.1 Batch Fermentation

The sequence is:

1. Preparation of the fermentation medium.
2. Charging the fermentor.
3. Addition of inoculum, 2–5%.

Table 16.4. Different fermentor operation strategies and their major advantages/limitations for production of ethanol

Strategy	Specificity	Advantages-	Limitations
1. Batch fermentation	Traditional technology, simplest and easiest	Process management by unskilled labor, easier management of feedstock, no requirement of strict sterilization	Poor productivity high, production cost, slower rate of reaction, long turnaround time, occasional contamination
2. Batch fermentation with cell recycle	Improved traditional technology with high cell density	Reduced fermentation time, no lag phase, facilitates use of up to 22% sugar	Productivity as like in batch process, additional expenditure on cell separation
3. Semicontinuous fermentation	Cyclic operation of fermentors to combine plus points of batch and continuous modes	Significant reduction in fermentation time, yeast reuse without resorting to mechanical separators, lower heat generation	Losses of substrate due to different residence times in fermentors
4. Conventional continuous fermentation	Simplest possible mode for continuous feeding of nutrients and hervesting of product	Higher reaction rate, lower medium volume and cell densities, High sugar in medium, Increased productivity	Frequent contamination, Inhibition of cell growth due to high sugar concentration and oxygen limitation, washout if sugars are low
5. Continuous fermentation with cell recycle	Continuous mode with yeast recycle for high cell density	Higher productivity, reduced residence time, higher rate of reaction	High capital cost for yeast separators, complex operation, Increased energy and labor requirement
6. Alcon's rapid continuous fermentation	High cell density for rapid continuous fermentation	Enhanced rate of reaction, cell density upto 45 g/l, high productivity	High capital investment and operating cost
7. Continuous tower fermentation	Retention of yeast in fermentor due to fermentor design, continuous mode of operation	High cell density, rapid fermentation rate, high productivity, Yeast recycle without using mechanical separators, Low by-product volume, no need for agitation	Oxygen limitation, long wating period to reach stable operation, reduced yeast settling, additional power for pumping
8. Ex-Ferm process	Simultaneous extraction of sugars in aqueous phase and ethanolic fermentation	High extraction efficiency, lower initial investment, greater conversion, operation in batch or continuous modes	Limitation of fermentation by slower rate of substrate diffusion, lower ethanol levels in batch fermentation, contamination problem

Table 16.4 (*Contd.*)

Strategy	Specificity	Advantages-	Limitations
9. Extractive fermentation	Integrated fermentation and product recovery using water immiscible solvent	High productivity, economy in distillation, overcoming of ethanol toxicity	Additional expenditure on solvent recovery and recycle, lack of ideal solvent of desired attributes
10. Membrane extractive fermentation	Use of diffusion membrane for selective separation of solvent	Complete sugar utilization due to nondiffusion through membrane, simplicity, energy reduction in distillation, marginally added equipments	Membrane fouling, high cost of membranes, frequent shutdowns
11. Selective product recovery membrane fermentation	Use of selective ultrafiltration membrane for diffusion of ethanol	No need for using solvent for ethanol extraction, one of the most promising strategy	Low flux rate, need for larger membrane surface, membrane fouling, costly membrane, shutdown during membrane replacement
12. Dialysis fermentor	Dialysis of ethanol and sugar through fixed membrane	High productivity at high rate, elimination of inhibitory effects of high sugar and ethanol concentrations	Slow diffusion of substrate through membrane
13. Pressure dialysis system	Use of high pressure feed pump for substrate feeding	Overcomes slow substrate diffusion problem	Membrane fouling
14. Rotor fermentation	Continuous pressure dialysis through rotating membrane	Continuous removal of ethanol without membrane fouling, retention of viable yeast cells	Highly complex nature of machinery and operation, leakage through seals of the membrane
15. Hollow fiber fermentation	Use of fine hollow fiber as membrane	Highly productive, larger membrane surface area, rapid diffusion, high reaction rate	Highly complex, cost intensive, membrane plugging, problem in venting of CO_2
16. Porous frit support plate system with a layer of Kieselguhr	Substitute for membrane	Retention of yeast in high density, rapid reaction rate, greater productivity, reduced fermenter volume	High capital investment, need for high pressure pumping system, gradients in O_2, decreased cell viability with time, frequent shutdowns

Table 16.4 (*Contd.*)

Strategy	Specificity	Advantages-	Limitations
17. Vacuum fermentation	Boiling off ethanol during fermentation	Overcoming ethanol toxicity, increased productivity, simplified distillation	Need for specially designed fermenters, sparging of pure O_2 and efficient heating-/cooling system, frequent shutdowns due to contamination, cost intensive machinery
18. Flash-ferm process	Rapid recycle of medium through a flash vessel for ethanol removal	Overcomes the problems of vacuum fermentation, air sparging possible, operation at atmospheric pressure, direct venting of CO_2, reduced contamination	Complexity, needs additional flash vessel, pump and other infrastructure
19. Biostill process	Continuous recycle of yeasts and ethanol depleted medium	Overcoming ethanol toxicity, high fermentation efficiency, saving on cost of land, building, water, steam and manpower, less downtime for distillation unit	Buildup of toxic nonfermentable feed component and fermentation byproducts
20. Simultaneous saccharification and fermentation	No prior hydrolysis of starch/cellulose	Process economy, reduced batch time, lower capital and operating expenses	Limitation of fermenation by the slower rate of saccharification, higher residual sugars
21. Solid phase fermentation	Use of feedstock in moist solid phase as such or with continuous stripping of ethanol	Greater fermentor productivity, reduction in fermentor capacity, lower production cost, less stillage output, reduced energy requirement, lowered ethanol toxicity	Lower recovery of ethanol from fermented solids, gradients in fermenting mass, low ethanol concentration in mashes for some feedstock
22. Immobilized cell fermentation	Replacing free yeast cells with immobilized cells	Simplicity, ease of continuous operation, easier control of system, reduced risk of contamination, higher resistance to accidental changes in feedstocks composition or operational parameters, elimination of cell recycle or cell separation, higher efficiency in product recovery	Critical importance of cell immobilization technique, limitation of productivity due to high ethanol concentration

4. Fermentation according to standard parameters.
5. Separation of the cells at the end of the fermentation.
6. Recovery of ethanol from the mash.

In practice, about 20% of the medium is placed in the fermentor, the inoculum is added, and the rest of the medium is charged into the fermentor. The yeast grows during the rest of the charging phase. After fermentation is complete, the fermentor is cleaned and the steps are repeated. The initial sugar concentration is 12–20%, and about 90–95% of this is utilized during fermentation. The cell density is 3–4 g cells/l, and the final ethanol concentration is 6–10% (w/v). Overall productivity is 1.8–2.5 kg ethanol/m^3/h. The batch time is 48–72 h. Most of the world's ethanol is still produced by batch fermentation.

In batch fermentation with cell recycle up to 80% of the cells from a batch are recycled as inoculum for the next. The initial cell density is about 80 billion, compared with 5–10 billion cells/l in batch fermentation. Fermentation time is reduced by 60–70%, and the efficiency of conversion is not altered. Initial sugar concentrations of 22% are possible, and the volumetric productivity is about 6 g of ethanol/l/h.

9.2 Semicontinuous Fermentation

This method uses a battery of fermentors, and a cyclic operation strategy between batch and continuous fermentation is used. The first fermentor is partly filled with medium, inoculated, and filling completed. Before the first fermentor is filled, a continuous flow of fresh medium into it is started, and partly fermented medium overflows into the second fermentor. The flow of medium to the first fermentor is stopped, the fermentation is let go to completion, and the product is recovered. The second fermentor is filled and the medium is allowed to overflow to the third. The fermentation in the second vessel is completed and recovery is started. The process is continued until all of the fermentors have been charged.

9.3 Continuous Fermentation

The initial stages up to the beginning of the exponential growth phase is the same as in batch culture. The cells are maintained in this stage for from 1–12 months by pumping fresh medium into the vessel continuously and removing the surplus medium simultaneously, the volume in the fermentor remaining constant. Two types of continuous processes are used. In the multistage continuous stirred tank process, molasses at 7% sugar concentration and low pH is fed to one fermentor for production of inoculum. Another molasses stream, containing 22–24% sugar, is fed continuously to production vessels, with the inoculum, and yields of 6 l of ethanol/kg sugar are obtained.

In the multiple continuous stirred tank arrangement (series system), fed by a single nutrient stream, molasses is diluted to 21% sugar and passed through a series of two to three vessels operated in the aerobic phase and consuming about 30% of the sugar. The effluent from these fermentors is collected in a battery of eight to ten vessels, where the remaining 70% of the sugar is fermented to ethanol.

In a simple continuous process, the fermentation is completed in a single vessel and the effluent is sent directly to the recovery plant. The growth rate is regulated by the dilution rate. All the sugar is fermented in the tank and enough new cells are produced to replace those lost in the effluent. The residence time is therefore 7–9.5 h, and the volumetric productivity 8.3 g ethanol/l/h.

The overall productivity of continuous fermentation is about 6 g/l/h, about three times greater than that in batch processes. The total volume of medium required to produce a unit of ethanol is about a third of that needed for batch processes.

9.4 Improved Continuous Processes

The methodology of continuous fermentation with yeast recycle is similar to the conventional continuous process except for recovery of the yeast by centrifugation or other means, and recycling up to 150 g of yeast cream/l. The residence time is reduced to 1.5 h, productivity 30–40 g ethanol/l/h and 83 g/l cell density. The process used by Danish Distilleries Ltd. in the Granna plant uses two to three fermentors. The wort is centrifuged before distillation to recover the yeast. It produces 8.4% ethanol (v/v), 10 g dry cell mass, and 0.1% residual sugars. Yield is 28–29 l of ethanol/100 kg molasses or 69 l ethanol/100 kg fermentable sugars.

In the Alcon rapid continuous process, a single stirred tank with yeast recycle reached high cell densities, about 45 g/l. The yeast is separated by gravity settling. Good mixing of the medium is achieved by conventional agitation or recirculation of the liquid (cell densities of more than 100 g dry weight/l can be obtained without cell recycle, under aerobic conditions, by adjusting the sugar concentration in the feed). Production of extracellular metabolities (ethanol) can also be increased in this way.

The continuous tower process resembles the simple continuous fermentation except for the fermentation tower. The process uses a flocculent strain of yeast. The rate of upflow of the medium is adjusted to match the settling rate of the yeast. This makes mechanical separation of the cells from the broth unnecessary. Cell densities of 50–80 g/l are achieved. About 90% of the sugars are utilized in a residence time of less than 1 h. Productivity is 80 times higher than in a simple batch process, with operating times of more than 12 months.

9.5 Integrated Fermentation and Product Recovery Process

Continuous vacuum fermentation (Vacuferm) involves fermentation under vacuum for continuous removal of ethanol, and replacement by continuous feed of concentrated sugar solution (Ramalingham and Finn 1977). The vacuum in the fermentor is maintained at 6.8 kPa and is sufficient to remove ethanol at 35 °C. With feed containing 12% glucose, the cell density increases to 120 g/l, and productivity is about 40 and 80 g ethanol/l/h in batch and continuous operation.

The flashferm process uses fermentation at atmospheric pressure, passage of the fermented medium through an auxiliary flash vessel for ethanol recovery and return of the medium to the fermentor. Air is sparged to supply oxygen, and CO_2 is vented directly. Productivity is about 80 g ethanol/l/h, in laboratory-scale operation.

The Biostill process involves continuous fermentation with cell recycle, and ethanol-depleted medium for closely coupled fermentation and distilling operations. Molasses, 40–50%, is fed continuously to the fermentor, constant aeration of the medium, medium is withdrawn continuously to maintain a constant volume, yeast slurry is separated from the effluent, yeast cream is returned to the fermentor, yeast-free medium is preheated in a heat exchanger, and about 90% of the ethanol as ethanol-water vapor for rectification. The exhausted liquor is returned to the fermentor through a regenerative heat exchanger. The yeast cell density in the fermentor is about 500 billion cells/l, and the ethanol concentration at any time is not more than 5%. An additional 60 l of ethanol/t of molasses is obtained, as compared to batch fermentation.

9.6 Simple Continuous Dialysis Fermentation

This process uses a dialysis membrane allowing diffusion of sugar and ethanol, but not yeast cells. The fermentor is divided into two compartments by the membrane. In part A, the sugar solution is fed continuously through the compartment without yeast cells. The sugars are fermented to ethanol by the yeast in part B. The sugar diffuses from part A to part B, while the ethanol formed diffuses from part B to part A. Compartment B is sparged with air and CO_2, and surplus air is vented from the top.

A pressure dialysis system in which the sugar is fed at high pressure to part B improves the performance. Part A becomes a low-pressure product recovery zone.

The continuous rotofermentor has a rapidly rotating membrane cylinder which prevents fouling of the membrane. Centrifugal force throws large molecules back into the annular zone. Cell densities up to 50.9 g/l and productivity of 36.5 g/l/h have been obtained on a laboratory scale.

9.7 Hollow Fiber Fermentation

This uses tubes of fine hollow fibers of membrane material, and the shell, resembling a shell and tube heat exchanger. The medium is fed into the bottom of the fibers and the diffused sugars are fermented in the shell portion. The ethanol formed diffuses back into the fiber tubes and leaves the fermentor at the top.

A layer of Kieselguhr filter aid on a porous frit support has been used in a plug fermentor to retain the yeast cells in the reactor at high density. The medium is pumped through the reactor under high pressure, from the bottom, and the fermented liquor and CO_2 pass out from the top of the fermentor. Kieselguhr is mixed with the yeast to eliminate dense packing of the cells. The high cell densities and rapid reaction rates give 72 times greater productivities than in a conventional fermentor.

Extractive fermentation: ethanol is extracted from the medium with solvent. The medium is continuously withdrawn from the fermentor, the cells recovered for recycle, and the ethanol extracted from the broth. The phases are separated and the broth returned to the fermentor. Ethanol is distilled off and the solvent is recycled (Barros et al. 1987). The solvent must be immiscible with water, nontoxic to yeasts, nonemulsifying, selective for ethanol, of low volatility, and have a high distribution coefficient for ethanol. The membrane-extractive fermentation uses a selective diffusion membrane in the extraction vessel for the separation of ethanol from the medium. A selective ultrafiltration membrane is used in the selective product recovery membrane fermentation.

9.8 Ex-Ferm Process

Plant material is slurried in water and inoculated with yeasts for extraction of sugars from the feedstock and fermentation to ethanol. The fermented liquor is separated from the chips or segments of the plant material, added to fresh feedstock and the fermentation is continued until the ethanol concentration is high enough. The process is a mixed-phase solid-liquid system and can be operated in batch or continuous modes (Rolz et al. 1979). Substrate are sugar cane, sugar beet, sweet sorghum, and fruits. The ethanol yield is about 10% more than obtained from fermentation of the extracted sugars.

9.9 Solid-Phase Fermentation

Like the Ex-Ferm process, except that no water is added. The feedstock is pulped and fermented as a moist pulp. Conventional fermentors, solid drum fermentors (Er-El et al. 1981), sugar fermentors, piling of the material on platforms (Aidoo et al. 1982), and tanks equipped with progressive cavity pumps or turbine agitators (Ziobro and Williams 1983) have been used. Fodder beets, sugar beets, sugar cane segments, apple pomace, Jerusalem artichokes, sweet sorghum, and tapioca fibrous residues have been used as feedstocks (Lonsane et al. 1992).

9.10 Simultaneous Saccharification and Fermentation

Starch or cellulose is saccharified and fermented in the same vessel. The polysaccharides are pretreated according to the nature of the substrate. Starchy substrates are liquefied by acid or enzymes and partially saccharified by amyloglucosidase. Cellulosic substrates are pretreated with acid or alkali and partially saccharified by cellulolytic enzymes.

9.11 Use of Immobilized Yeast Cells

Yeast cells are immobilized on solid supports and used for continuous ethanolic fermentation. The immobilized cells are packed in columns, the medium fed continuously through the reactor. The fermented beer is collected from the other end of the system and ethanol is recovered (Barros et al. 1987). The production of 114 g ethanol/l with a conversion efficiency of 98%, from a feed containing 25% has been obtained. Productivity is 28.6 g/l/h, as compared with 2.0 and 3.35 in batch and continuous fermentations with free cells. Ethanol production of 50 g/l/h over 6 months has been obtained using immobilized cells.

10 Downstream Processing

10.1 Product Recovery Methods

The ethanol concentration in the effluent broth is 5–12%, depending on operational conditions. It contains yeast cells, insoluble and soluble matter in the substrate, unfermented sugars, and cometabolites of the ethanol. Yeast cells and insolubles are settled out by gravity in the fermentor, or by centrifugation. The product, ethanol, is then recovered from the cell-free broth (wash), usually by distillation (Aldridge et al. 1984).

Distillation of ethanol is expensive, because of the low concentration in the medium. The energy consumed in ethanol recovery is approximately 50% of that required for the whole process. Theoretical steam consumption is 1.8–1.9 kg/l of ethanol produced for wash containing 6–8% ethanol. The energy required for distillation may equal the total combustion energy of the ethanol, especially in small-scale plants. The process is extremely wasteful of energy. Batch distillation in two-stage stills for production of absolute ethanol also consumes more energy than the ethanol contains. The most advanced processes require a total energy of about 10% of the heat of combustion of the ethanol produced.

Recovery of ethanol by distillation is easier in the initial stages, but later becomes more difficult. Ethanol and water form an azeotrope at 95% or 89 mol % ethanol, and the percentage ethanol cannot be increased by distillation. The energy demand rises sharply if the product desired contains more than 80 mol % ethanol. The wash also contains about 100 volatile compounds which appear in the distillate and must be recovered from it if a high degree of purity is required.

Low consumption of energy is essential. The steam requirement for distillation depends on wash feed temperature, reflux ratio, ethanol concentration in the wash, efficiency of the trays in the distillation column, the spent wash temperature, the

Table 16.5. Conventional methods for dehydration of azeotropic ethanol

Entrainer/ solvent method	Number of trays in		Specificity	Energy required kJ/l product X 1000	Reference
	dehydrating column	entrainer recovery column			
Benzene	50	30	With or without vapor reuse/vapor recompression, Additional rectifier column, dehydration cost: 4.24 c/l	2.0	Busche (1985)
Pentane	23	18	Requires smaller column, Additional rectifier column, Excessive foaming problem, dehydration cost: 3.78 c/l	1.8–2.6	Black (1980)
Ether	60	20	With or without vapor reuse, No rectifier column, no recycling of ethanol/water mixture, dehydration cost: 4.72 c/l	0.84	Maiorella et al. (1981)
Ethylene glycol	50	10	No rectifier column, with/ without vapor reuse, dehydration cost: 7.08 c/l	2.71	Black (1980)

overall efficiency of the distillation column, and the loss of ethanol due to faulty operation of the column. Steam consumption in the older distilleries in India is nearly double the theoretical value.

Distillation systems for producing a near-azeotropic mixture of low-quality ethanol may consist of a stripping and rectifying section, combined or separate. There are alternative methods for direct recovery of anhydrous (absolute) ethanol, but they are not in large-scale use. Liquid-liquid separation has considerable potential for energy-saving (Maiorella et al. 1981).

10.2 Ethanol Dehydration

First-stage distillation of ethanol from the fermented wash by simple fractionation gives a near-azeotropic mixture (concentration 95%). This water content is too high for gasohol production. The ethanol and gasoline separate and the water content must be reduced below 1%. Impurities in the gasoline can react with very smaller quantities of water and corrode fuel systems made of two or more metals. Impurities in denaturing agents added to the alcohol accelerate corrosion. Ethanol can be substituted for gasoline, but this causes cold starting difficulties at 0–50 °C. Azeotropic ethanol from the first distillation stage must be dehydrated for making gasohol.

Table 16.6. Alternative nonconventional methods for dehydration of azeotropic ethanol

Method	Specificity/remarks	Reference
Vacuum distillation	Product from two-column conventional distillation is subjected to second vacuum distillation, total energy required is 10 300 kJ/l product	Busche (1985)
Extractive distillation with salt (KAc)	Azeotropic ethanol from IHOSR distillation is starting product, 1700 kJ energy/l product	Lynd and Grethlein (1984)
Gasoline as solvent	Yields gasohol, 2060 kJ energy consumed/l product	Black (1980)
Water adsorption on molecular sieve	Liquid-phase water adsorption in zeolites, 1400 kJ energy/l product	Douglas and Fienberg (1983)
Water adsorption on adsorbent agents	Vapor-phase water adsorption in CaO or cellulose, 340–710 kJ energy/l product. Use of cornmeal with liquid or vaporized feed. 420–1390 kJ energy/l product	Ladisch and Dick (1979) Ladisch et al. (1984)
Water adsorption-ethanol extraction process	Liquid phase ethanol adsorption by saponified starch-g-polyacrylonitrile, ethanol extraction with gasoline	Fanta et al. (1980)
GKSS process	Membrane-pervaporation principle, involves simultaneous application of low pressure and use of membrane	Douglas and Fienberg (1983); Choudhury et al. (1985)
Catalytic conversion process	Catalytic conversion to gasoline, 550 kJ energy usage/l product	Aldridge et al. (1984)

Dehydration requires addition of other materials to the azeotropic mixture to alter the distillation equilibrium. The conventional method is azeotropic distillation with benzene. Water is immiscible with benzene and enhances the vapor pressure of water in the mixture. This allows dehydration of the ethanol in the ternary distillation process. The second-stage distillation consumes 1 kg steam/l of product and is expensive. Methods for dehydration of azeotropic ethanol are given (Tables 16.5, 16.6).

10.3 By-Products

Fusel oil, acetaldehyde, CO_2, and yeast biomass are by-products of ethanol production. Specific byproducts are formed when certain feedstocks are used, the amounts depending on the feedstock, the process, and the recovery technique. They can be separated at different stages of production without affecting the efficiency of ethanol recovery. Some must be separated to maintain the desired quality of ethanol. The by-products must be utilized profitably if fuel alcohol is to compete with other fuels.

10.3.1 Carbon Dioxide

One mol of CO_2 is formed per mol of ethanol produced in yeast fermentation, so 760 g of CO_2 is obtained per liter of ethanol formed. CO_2 is not recovered in most Indian distilleries.

It is purified to remove traces of aldehydes, ethanol, organic acids, fusel oil, water, and compressor lubricating oil, and stored under pressure in cylinders or as solid "dry ice". It has many uses, for carbonation of soft drinks, enhancement of the productivity of greenhouse plants, a feedstock in the chemical industries, as a refrigerant, and in fire extinguishers. Profitable use reduces the cost of ethanol production.

10.3.2 Dry Cell Mass

Dry cell mass, 3–6 g/l, is obtained in batch fermentations. The cells are allowed to settle by gravity and added to the stillage stream in most Indian distilleries. Some ethanol may be lost with the yeast slurry, since the cost of mechanical separation is too high to be economic. The slurry can be dewatered, the liquid from this can be passed to the wash, and the ethanol recovered. Mechanical separators can be used to recycle about 30–40% of the yeast. The remaining yeast and other dewatered yeast is dried for animal feed, the cell mass being about 50% protein.

10.3.3 Fusel Oils

Fusel oils, produced by degradation of amino acids in the feedstocks, are a mixture of isomers of amyl and propyl alcohols. They are formed at rates of 4–20 l/l of ethanol, depending on the pH of the fermentation. They plug the distillation column, and should be removed. Their limited solubility and their low volatility (lower than water) make separation difficult. During distillation, they rise to the center of the ethanol purifying column and can be recovered by bleeding from some of the lower plates of the rectifier column. They can be fed into a fusel oil separator and recovered from the decanter section.

Fusel oils can be used as solvents or as tanning compounds. They can also be used as fuel extenders and denaturants in fuel ethanol, since their compositions are similar to gasoline.

10.3.4 Aldehydes

Aldehydes, principally acetaldehyde, are more volatile than ethanol, and are produced at about 1 l/1000 l of ethanol. Acetaldehyde is easily recovered as a distillation head product. Complete separation is difficult.

Glycerol, another by-product, is formed at a rate of 4l/1000l ethanol. It has a high boiling point and cannot be recovered economically.

10.3.5 Heat

Heat is a by-product in certain stages of ethanol production and can be recovered for reuse in the process itself. Heat in the stillage can be profitably used to preheat fermented wash for reducing the steam requirement.

10.3.6 Pentoses, Furfural, and Lignin

Lignocellulosic feedstocks yield pentoses, furfural, and lignin from the fermentations (Wilke et al. 1983). About 0.37 kg xylose and 0.43 kg of lignin are obtained per kg of degradable sugar from corn stover. Lignin can be used as fuel or ligninformaldehyde binder, while both lignin and furfural are used in the manufacture of plastics. Xylose in the stillage can be used as cattle feed. Fruits and fruit wastes used for ethanol production in solid phase fermentation yield pomace as a by-product, and can be recovered with the yeast cells, dried, and used as animal feed.

11 Effluent Management

11.1 Stillage

Slop or vinasse is the spent liquid remaining after distillation of ethanol and by-products from the wash, and is the major liquid effluent from ethanol production. It is collected from the base of the still, and contains the soluble and nonsoluble fractions present in the fermented wash. It amounts to 10–15l/l of ethanol produced. (Lonsane and Ahmed 1989). It is rich in organic matter. Its composition is related to the nature of the feedstock, the fermentor operation strategy, the efficiency of the process, and the type of downstream processing. It contains about 10% solids, 80% of which arise from the feedstock and spent yeast cells. The remainder of the solids are waxes, fats, fibers, mineral salts, and residual sugars. Unrecovered ethanol is also present.

BOD values of stillage range from 15000 to 100000ppm, depending on the feedstock used. High salt contents interfere with stillage management. Its worst characteristics are its high BOD, large volume, and caramel content (3%), which cause pollution if untreated. Discharge of untreated stillage causes environmental pollution and damage to groundwater. The hazards of such pollution are a matter of great concern (Lonsane and Ahmed 1989). Distilleries are notorious in underdeveloped and developing countries for indiscriminate stillage disposal. The regulations has set limits of 30–100ppm of BOD for effluent to be discharged into the environment (Lonsane and Ahmed 1989). There is no such thing as cost-free

Table 16.7. Methods for utilization of stillage

Ferti-irrigation of agricultural land and soil conditioning
As substrate in fermentation
Feed supplement and enrichment of crop residue
Binder for fuel
Recovery of glycerol and organic acids
As extractant for recovery of sugar from feedstocks
Concentration and incineration

disposal of stillage. Disposal is now a limiting factor in the operation of a distillery, as well as the economics of production. Profitable production of fuel ethanol depends upon it.

Stillage may be treated by aerobic or open anaerobic lagooning, anaerobic digestion, trickling filters, and electro-oxidation, plus pretreatment, primary treatment, and finishing treatment (Lonsane and Ahmed 1989). Two or more of these methods may be combined to increase the efficiency of the process. Sometimes, the treated waste may not meet the requirements for safe discharge, and must be diluted with water. Waste treatment does not bring any financial returns, but does require heavy capital and working costs. Stillage treatment thus violates the vital need for efficient utilization of resources, since valuable organic matter is not utilized, but broken down to simpler, nonpolluting forms.

Research on stillage management has been minuscule. Segregation of research on stillage disposal from fermentation is probably the reason for the unrestricted growth of distilleries. Proper waste management can, of course, bring immense benefits from both the economic and social points of view. Factors in waste management include volume reduction, strength reduction, recycle, and waste exchange (Lonsane and Ahmed 1989), but little practical investigation has been done.

More investigation is essential in profitable utilization of stillage, which may be a way of reducing the cost of production of fuel ethanol (Table 16.7).

12 Overview

Fermentative production of fuel ethanol by yeasts is one promising form of renewable resource for maintaining an energy supply growth. In spite of much research on production of fuel alcohol from many feedstocks, production costs are too high to be economic for industrial use. Possible ways of reducing the cost of fuel alcohol production are: selection of improved yeast strains, genetic improvement of yeast strains for higher ethanol and temperature tolerance, selection of the best and cheapest feedstocks, simplification of preprocessing of feedstocks for lower cost, improvement of fermentor design and control systems, development of efficient fermentor operating techniques, improving the osmotolerance of the strains to allow use of higher substrate concentrations, better methods of downstream processing for ethanol recovery and dehydration, and efficient management of stillage. Some of these factors have been investigated in detail in the last three decades, but more work is needed to make fuel ethanol competitive with fossil fuels. Feedstocks, fermentor operation strategies, and recovery methods which lead to formation and separation of valuable by-products must be improved for maximum efficiency. Of necessity, the use of yeasts for production of fuel ethanol and replacement of fossil fuels will probably occur very soon, especially in countries dependent on others for petroleum fuels.

References

Adams GA, Stanier RY (1945) Production and properties of 2,3-butanediol. III. Studies on the biochemistry of carbohydrate fermentation by *Aerobacillus polymyxa*. Can J Res (B) 23:1–9

Aidoo KE, Hendry R, Wood BJB (1982) Solid substrate fermentations. Adv Appl Microbiol 28:201

Aldridge GA, Verykios XE, Mutharasan R (1984) Recovery of ethanol from fermentation broths by catalytic conversion to gasoline. 2. Energy analysis. Ind Eng Chem Process Des Dev 23(4):733–737

Altermatt HA, Simpson FJ, Neish AC (1955) The anaerobic dissimilation of D-ribose-1-C^{14}, D-xylose-1-C^{14}, D-xylose-2-C^{14} and D-xylose-5-C^{14} by *Aerobacter aerogenes*. Can J Biochem Physiol 33:615–621

Barros MRA, Cabral JMS, Novias JM (1987) Production of ethanol by immobilized *Saccharomyces bayanus* in an extractive fermentation system. Biotechnol Bioeng 29:1097–1104

Batter TR, Wilke CR (1977) A study on the fermentation of xylose to ethanol by *Fusarium oxysporum*. Lawrence Berkeley Lab Production No 6351, University of California

Bauchop T, Dawes EA (1959) Metabolism of pyruvic acid and formic acids by *Zymosarcina ventriculi*. Biochem Biophys Acta 36:294–296

Blackwood AC, Neish AC, Ledingham GA (1956) Dissimilation of glucose at controlled pH values by pigmented and nonpigmented strains of *Escherichia coli*. J Bacteriol 72:497–499

Busche RM (1985) Biotechnol prog 1:165 as cited by Serra et al. 1987

Bryant MP (1976) The microbiology of anaerobic degradation and methanogenesis with special reference to sewage. In: Schlegel HG, Barnea J (eds) Microbial energy conversion. Goltze, Göttingen p 107

Bryant MP, Small N, Bouma C, Robinson IM (1958) Characteristics of ruminal anaerobic cellulolytic cocci and *Cellobacterium cellulosolvens* sp. J Bacteriol 76:529–537

Calleja GB, Levy-Rick S, Lusena CV, Nasim A, Moranelli F (1982) Direct and quantitative conversion of starch to ethanol by the yeast *Schwanniomyces alluvius*. Biotechnol Lett 4:543–546

Castillo FJ (1990) Lactose metabolism by yeasts. In: Verachtert H, De Mot R (eds) Yeasts – biotechnology and biocatalysis. Marcel Dekker, New York, 297pp

Choudhury JP, Gosh P, Guha BK (1985) Separation of ethanol from ethanol water mixture by reverse osmosis. Biotechnol Bioeng 27:1081–1084

Cooney CL, Wang DIC, Wang SD, Gordon J, Juinenez M (1978) Biotechnol Bioeng Symp 8:103–114

Dawes EA (1963) Comparative aspects of alcohol formation. J Gen Microbiol 32:151–155

Douglas I, Fienberg D (1983) Evaluation of non-distillation ethanol separation processes. Rep No De 83011994, Solar Energy Research Institute, US Dept of Commerce

Enebo L (1951) Physiol Plant 4:652–666

Er-El Z, Battat E, Schechter U, Goldberg I (1981) Ethanol production from sugarcane segments in a high-solids drum fermentor. Biotechnol Lett 3:385–390

Fanta GF, Burr RC, Orton WL, Doane WM (1980) Liquid phase dehydration of aqueous ethanol-gasoline mixtures. Science 210:646–647

Ghose TK (1977) Cellulose biosynthesis and hydrolysis of cellulosic substances. Adv Biochem Eng 6:39–76

Gibbs M, Cochrane UW, Paege LM, Wolin H (1954) Fermentation of xylose-1-C^{14} by *Fusarium lini*. Arch Biochem Biophys 50:237–242

Gleason FH (1971) Alcohol dehydrogenase in Mucorales. Mycolagia 63:906–910

Guirad JP, Galzy (1990) Inulin conversion by yeasts. In: Verachtert H, De Mot R (eds) Yeast technology and biocatalysis, Marcel Dekker, New York, p 255

Hespell RB, Canale-Parola E (1970) Carbohydrate metabolism in *Spirochaeta stenostrepta*. J Bacteriol 103:206–226

Hyun HH (1986) Thermo anaerobic bacterial fermentation for production of ethanol and enzymes. Misaengmul Kwa Sanop 12(1):15–22

Jain VK, Nath P (1987) Ethanol production by *Zymomonas mobilis*: effect of high substrate concentration on the kinetic parameter. J Food Sci Technol 24:88–89

Kanno M, Toriyama K (1986) Production of ethanol by a thermophilic anaerobic bacterium and its ethanol tolerant mutant. Agric Biol Chem 50:217–218

Katsuragi T, Kawabata N, Sakai T (1994) Selection of hybrids from protoplast fusion of yeasts by double fluorescence labelling and automatic cell sorting. Lett Appl Microbiol 19:92–94

Ladisch MR, Dick K (1979) Dehydration of ethanol: new approach gives positive energy balance. Science 205:898–900

Ladisch MR, Voloch M, Hong J, Blenkowski P, Tsao GT (1984) Corn meal adsorber for dehydrating ethanol vapors. Ind Eng Chem Proc Des Dev 23:437–443

Laluce C, Mattoon JR (1984) Development of rapidly fermenting strains of *Saccharomyces diastaticus* for direct conversion of starch and dextrins to ethanol. Appl Environ Microbiol 48:17–25

Lamed RJ, Zeikus JG (1980) Glucose fermentation pathway of *Thermoanaerobium brockii*. J Bacteriol 141:1251–1257

Lonsane BK, Ahmed SY (1989) Some neglected aspects of waste management: reduction, recycle, utilization and exchange. In: Souv. National Symp. Impacts of pollution in and from food industries and its management. Assoc Food Scientists and Technologists (India), Mysore, 33 pp

Lonsane BK, Durand A, Almanza S, Maratray J, Desgranges C, Cooke PS, Hong K, Malaney GW, Tanner RD (1992) General principles of reactor design and operation for SSC. In: Doelle HW, Mitchell DA, Rolz CE (eds) Solid substrate cultivation. Elseevier Amsterdam

Lynd LR, Grethlein HE (1984) IHOSR/Extractive distillation for ethanol separation. Chem Eng Prog 60(11):59–62

Maiorella B, Wilke CR, Blanch HW (1981) Alcohol production and recovery. In: Fiechter A (ed) Advances in biochemical engineering, vol 20. Springer, Berlin Heidelberg New York, p 23

Margulies M, Vishiac W (1961) Dissimilation of glucose by the Mx Strain of *Rhizopus*. J Bacteriol 81:1–9

Northrop JH, Ashe LH, Senior JK (1919) Biochemistry of β-acetothylicum with reference to the formation of acetone. J Biol Chem 39:1

Paturau JM (1982) By-products of the cane sugar industry, 2nd edn. Elsevier, New York

Peppler HJ, Reed G (1987) Enzymes in food and feed processing. In: Kennedy JF (ed) Biotechnology, vol 7a. Enzyme engineering, VCH, Weinheim, 547pp

Ramalingham A, Finn RK (1977) The vacuferm process: a new approach to fermentation alcohol. Biotechnol Bioeng 19:583–589

Reynolds H, Werkman CH (1937) The fermentation of xylose by the colon-aerogenes group of bacteria. Iowa State Coll J Sci 11:373–378

Rogers PL, Lee KJ, Tribe DE (1979) Kinetics of alcohol production by *Zymomonas mobilis* at high concentrations. Biotechnol Lett 1(4):165–170

Rolz C, de Cabrera S, Morales E (1979) The ex-ferm process. Adv Biotechnol 2:113–117

Russell I, Crumplen CM, Jones RM, Stewart GG (1986) Efficiency of genetically engineered yeast in production of ethanol from dextrinized cassava starch. Biotechnol Lett 8:169–174

Saucedo-Castaneda G, Lonsane BK, Navarro JM, Roussos S, Raimbault M (1992) Control of carbon dioxide in exhaust air as a method for equal biomass yields at different bed heights in column fermentor. Appl Microbiol Biotechnol 37:580–582

Schneider H, Wang PY, Chan YK, Maleszka R (1981) Conversion of D-xylose into ethanol by the yeast *Pachysolen tannophilus*. Biotechnol Lett 3:89–92

Shirai K, Yorifuji T (1986) Japan Tokyo Koho JP 61(209):544

Slininger PJ, Bothast RJ, Van Cauwenberge JE, Kurtzman CP (1982) Conversion of D-xylose to ethanol by the yeast *Pachysolen tannophilus*. Biotechnol Bioeng 24:371–384

Spivey MJ (1978) The acetone/butanol/ethanol fermentation. Proc Biochem 13(11):2–4, 25

Stanier RY, Adelberg EA, Ingraham JL (1976) The microbial world. Prentice-Hall, Englewood Cliffs

Stephenson MP, Dawes EA (1971) Pyruvic acid and formic acid metabolism in *Sarcina ventriculi* and the role of ferradoxin. J Gen Microbiol 69:331–343

Wilke CR, Maiorella B, Sciamanna A, Tangnu K, Wiley D, Wong H (1983) Enzymic hydrolysis of cellulose: theory and applications. Haigh and Hochland, Manchester

Wood WA (1961) Fermentation of carbohydrates and related compounds. In: Gunsalus IC, Stainer RY (eds) The bacteria, vol 2. Academic Press, New York

Zeikus JG (1979) Thermophilic bacteria: ecology, physiology and technology. Enzyme Microbial Technol 1:243–253

Ziobro GC, Williams LA (1983) Pilot scale fermentation of Jerusalem artichoke tuber pulp mashes. Dev Ind Microbiol 24:313–319

Notes Added in Proof

Chapter 5

Yeasts as Living Objects: Yeast Nutrition

Note 1
Recently a gene has been cloned from *Torulaspora delbrueckii* which complements the deficiencies in a temperature- and salt-sensitive mutant of the same strain, which does not assmilate or accumulate trehalose. The authors suggest that their results show that trehalose plays an important role in stress tolerance (resistance to elevated temperatures and osmotic tension) in this species.

References

Nakata K, Okamura K (1996) Cloning of stress tolerance gene in *Torulopsis delbrueckii* No. 3110. Biosci Biotech Biochem 60(10):1686–1689
Venkov PV, Hadjiolov AA, Battaner E, Schlessinger D (1996) *Saccharomyces ceerevisiae* sorbitol fragile mutants. Biochem Biophys Res Commun 56:599–604

Note 2
Xylose, a major constituent of hemicelluloses, is fermented to ethanol by several yeast and filamentous fungal species, including *Pachysolen tannophilus*, *Pichia stipitis* and *Candida shehatae*, as well as by several species of *Fusarium* (filamentous). Yields are low and fermentations are slow. Therefore, many attempts have been made to construct yeast strains which would hydrolyze xylans, at least partially, and ferment the xylose liberated, to ethanol. Protoplasts of the yeasts *Pachysolen tannophilus* and *Pichia stipitis* have been fused with nuclei of *Fusarium spp* and yeast-like hybrids which would hydrolyze xylans have been obtained. Yeast hybrids which will ferment xylose directly have also been constructed. Recently two groups, one using protoplast fusion (in Thailand) and one using transformation (in Sweden) into the yeast *Saccharomyces cerevisiae*, of bacterial genes, conferring on the yeast the ability to ferment xylose directly (Walfridsson et al., 1996).

References

Walfridsson M, Xiaoming Bao, Anderlund M, Lilius G, Bulow L, and Hahn-Hagerdahl B (1996) Ethanolic fermentation of xylose with *Saccharomyces cerevisiae* harboring the Thermus thermophilus xyl4 gene, which expresses an active xylose (glucose) isomerase. Appl Environ Microbiol 62(12): 4648–4651

Chapter 7

Membranes

The nature of the protein pores or channels through the membranes and the method of introduction of them into the lipid bilayers of the membrane has been determined for a number of these channels. Assuming that the pores (channels) for materials such as hemolysins are similar for these materials and ion channels and transport channels for sugars and other nutrient materials, the structure is as follows: the pore consists of 14 proteins, and is guided into the membrane site by the nature of the proteins at the ends of the preassembled pore. Investigation of the detailed nature of the pores (channels) is being continued.

References

Gorlich D, Mattaj IW (1996) Nucleocytoplasmic transport. Science 271:1513–1518
Schatz G, Dobberstein B (1996) Common principles of protein translocation across membranes. Science 271:1519–1526
Song L, Hobaugh MR, Shustak C, Cheley S, Bayley H, Gouaux JE (1996) Structure of staphylococcal α-hemolysin, a heptameric transmembrane pore. Science 274:1859–1866 – See also other papers in this issue.

Chapter 8

The Yeast: Sex and Nonsex. Life Cycles, Sporulation and Genetics

The details of the control of the cell division cycle are now known to a much greater extent. In particular, the effects and limitations of proteins such as the p53 inhibitor of replication of damaged DNA have been elucidated.

Details of telomere behavior and function, and their role in mitotic and meiotic chromosome segregation are also being elucidated.

References

Nasmyth K (1996) Viewpoint: putting the cell cycle in order. Science 274:1643–1645. – See also other papers in this issue.
Hawleyn RS (1997) Unresolvable endings: Defective telomeres and failed separation. Science 275:1441–1443

Chapter 13

Yeasts and the life of Man: Part II
(Refers to section on protoplast fusion)

Kavanagh and Whittaker (1996) have used treatment with hydroxylamine and cupferron to induce chromosome breaks in yeasts, which are then protoplasted and fused to allow recombination between the broken chromosomes. However, it is possible that the same effect may be achieved by treatment of the yeasts with NaCl or KCl, 1–2 M.

Hydroxylamine and cupferron are only sparingly soluble in water and must be dissolved in a salt solution. Parker and von Borstel (1987) obtained as many mutants from cells mutagenized with salt solutions alone, as from those mutagenized with hydroxylamine or cupferron. The nature of the mutations obtained using salt solutions has not been determined, to the best of our knowledge, but unless there are differences, the use of cupferron or hydroxylamine as "mutagens" may be unnecessary. Salt is much safer.

References

Kavanagh K, and Whittaker, PA (1996) Application of protoplast fusion to the nonconventional yeast. Enzyme Microb Technol 18:45–51

Parker K, and von Borstel RC (1987) Based substitution and frameshift mutagenesis by sodium chloride and potassium chloride in Saccharomyces cerevisiae. Mutat Res 189:11–14

Subject Index

Species Index

Printing: Saladruck, Berlin
Binding: Buchbinderei Lüderitz & Bauer, Berlin